AF821833

Materials in Cultural Heritage: Analysis, Testing, and Preservation

Materials in Cultural Heritage: Analysis, Testing, and Preservation

Guest Editors

Žiga Šmit
Eva Menart

MDPI

Basel • Beijing • Wuhan • Barcelona • Belgrade • Novi Sad • Cluj • Manchester

Guest Editors
Žiga Šmit
Faculty of Mathematics and Physics
University of Ljubljana Jožef Stefan Institute
Ljubljana
Slovenia

Eva Menart
Department of Conservation and Restoration
National Museum of Slovenia
Jožef Stefan Institute
Ljubljana
Slovenia

Editorial Office
MDPI AG
Grosspeteranlage 5
4052 Basel, Switzerland

This is a reprint of the Special Issue, published open access by the journal *Materials* (ISSN 1996-1944), freely accessible at: https://www.mdpi.com/journal/materials/special_issues/H8D4ESWBD7.

For citation purposes, cite each article independently as indicated on the article page online and as indicated below:

Lastname, A.A.; Lastname, B.B. Article Title. *Journal Name* **Year**, *Volume Number*, Page Range.

ISBN 978-3-7258-6236-8 (Hbk)
ISBN 978-3-7258-6237-5 (PDF)
https://doi.org/10.3390/books978-3-7258-6237-5

Cover image courtesy of Eva Menart

© 2026 by the authors. Articles in this book are Open Access and distributed under the Creative Commons Attribution (CC BY) license. The book as a whole is distributed by MDPI under the terms and conditions of the Creative Commons Attribution-NonCommercial-NoDerivs (CC BY-NC-ND) license (https://creativecommons.org/licenses/by-nc-nd/4.0/).

Contents

About the Editors

Žiga Šmit

Žiga Šmit is a retired professor at the Faculty of Mathematics and Physics at the University of Ljubljana. His research was and is still conducted at the Jožef Stefan Institute in Ljubljana. His general field of interest is atomic collisions with light ions. He studied ionization mechanisms of inner shells by proton impact in semiclassical approximation and proposed corrections beyond the first step models. At the same time, he developed numerical models for practical applications of induced X-rays for chemical analysis and produced several methods for the application of PIXE on historic materials, like the analysis of archaeological metals, differential PIXE for the analysis of layered samples, and a combined PIXE-PIGE algorithm for the analysis of historic glasses. Important historical cases include studies of use-wear on flint tools, Roman military equipment metals, glass from almost all historical periods from the Bronze Age to the early 20th century, (semi-) precious stones and paint pigments. Currently, he is interested in the development of procedures for spectral fitting, the modelling of collisions in thick targets, and in the further analysis of historic materials.

Eva Menart

Eva Menart is the head of the Department of Conservation and Restoration at the National Museum of Slovenia and is additionally employed as a researcher at the Jožef Stefan Institute. She began her research, which combines science and cultural heritage, during her studies, after which she obtained a doctoral scholarship from University College London and in 2013 a PhD in Heritage Science. She has researched paper degradation and historical inks and worked on electrochemical gas sensors for detecting heritage material degradation. Her research now mainly focuses on the analysis of heritage objects with an emphasis on non-destructive methods (XRF, PIXE) and the interpretation of results. She is also engaged in research in the field of preventive conservation, where she participates in international and national projects. She was the vice-chair of COST Innovators Grant ENDLESS Metal, focusing on low-cost analytical tools and their dissemination in the metal conservation community.

Preface

This reprint is devoted to the application of analytical techniques in the research of historical and archaeological materials, both with the aim of object characterization and the detection of degradation processes. This is particularly important for avoiding or slowing down degradation, which is a crucial conservation issue. This reprint is a continuation of our previous Special Issue, published by *Materials* in 2023. In the present volume we deal with the characterization of ceramic materials and glasses, archaeological bones, plasters and dating medieval frescos; we also address the conservation of steel in humid environments and the protection of materials prone to degradation with aerogels and ethanol spray. This volume is intended for researchers interested in the material investigation of cultural heritage but may also be interesting for historians, art historians, archaeologists, and conservators. The papers collected here are the result of collective work, but due to space limitations we only mention the first authors (in alphabetical order): M. Abrudeanu, M. Gajić Kvaščev, G. Gorgolis, Hye Ri Yang, T. Leskovar, Liang Zheng, Linyu Xia, T. Palomar, C. Reis Santos, Ž. Šmit, and A. Wawrzyk. The editors would like to thank the authors for their collaboration and to MDPI for their excellent support.

Žiga Šmit and Eva Menart
Guest Editors

materials

MDPI

torial

laterials in Cultural Heritage: Analysis, Testing, and reservation

ga Šmit [1,2,*] and Eva Menart [2,3]

1 Faculty of Mathematics and Physics, University of Ljubljana, Jadranska 19, SI-1000 Ljubljana, Slovenia
2 Department of Low and Medium Energy Physics, Jožef Stefan Institute, Jamova 39, SI-1000 Ljubljana, Slovenia; eva.menart@nms.si
3 National Museum of Slovenia, Prešernova 20, SI-1000 Ljubljana, Slovenia
* Correspondence: ziga.smit@fmf.uni-lj.si

Many things that people achieved in the past are worth admiration, so we are prone to keep, nurse, and preserve them. We can also learn from examples in the past, though many experiences deliberately remain forgotten. The development of analytical techniques has enabled further possibilities for studying how people created, produced, and manipulated certain objects, which then facilitated studies in social sciences and humanities including history, cultural anthropology, or art history. Analytical studies then aim in two directions: the first is to discover properties of the objects that enable their more precise characterization according to their composition, age, or other properties, while the second direction aims toward the preservation of objects, so the analysis tries to uncover the formation of harmful compounds or the degree of degradation. This type of division is also reflected in our volume, so we will first introduce the papers that deal with analytical characterization, followed by papers concerned with conservation; we also distinguish by the historical period.

Clay was the first material than man chemically modified with thermal treatment. The Longshan Culture in Eastern Henan was already studied in the 1990s [1], but knowledge was further expanded in a study by Linyu Xia et al., where the authors employed XRF, XRD, infrared spectroscopy, and SEM-EDS to study Neolithic pottery of the Longshan period from three sites in the Yongcheng area, pointing to the specific differences and cultural relations among them. Slightly younger is the Eneolithic pottery from Velika Humska Čuka near Niš in present-day Serbia [2], which was also investigated by the authors Maja Gajić-Kvaščev et al., who employed EDXRF in combination with statistical techniques, which are indispensable tools in many heritage science studies [3,4]. The aim was to distinguish ceramics using local clay from deposits near the site from imported items, testifying to a long-range connection and exchanges. In the ongoing work, the same authors examined provenance studies of prehistoric ceramics using portable XRF and advanced statistical techniques. Besides artifacts, bone remains can reveal details of people's life such as diet, possible migrations, and even the type of burial (e.g., cremation) [5]. Tamara Leskovar et al. studied the changes in bone exposed to high temperature (caused, for example, during cremation of the deceased) using ATR-FTIR and computed tomography.

Glass and glassy materials are further artificial materials that have accompanied man since prehistory; during the Roman period, glass was a widely used consumable good, subject to a wide-range- trade [6]. Žiga Šmit and Tina Milavec studied the glass from Koper, a port at the northeastern Adriatic, pointing to the transition in the glass industry from the Late Antiquity until the Middle Ages, a theme that has been intensively studied in the Mediterranean. By adding small amounts of chromophores, glass can attain different colors.

 https://doi.org/10.3390/ma18194540

Amber glass has been known since prehistory, and was intensively used for the storage of food and medicaments until the end of the 19th century [7]. Catarin Reis Santos et al. tried to reproduce amber glass following historic recipes from the late 18th century. Enamels are glassy materials characterized by a lower melting point and are usually applied as thin layers on a metal or glassy substrate, which could be studied according to their refractive index [8], and was attempted by Teresa Palomar et al. The authors reviewed different mathematical models for refractive index in lead glasses; they further discuss the influence of glass corrosion and the appearance of alteration layers, with possible application in glass preservation.

Buildings are evident remains of past human activity. Raw walls are usually covered with plasters, and also with frescos in monumental buildings. Such structures in the temple city of Bagan in Ancient Burma have already been investigated [9], and further exploration was carried out by Hye Ri Yang et al. in their study of the calcareous materials at the Phaya Thon Zu Temple in Myanmar. The analysis of mortars showed that they were produced by mixing clay and sandy soil, and the location of the sandy soil for the newly produced plasters was proposed. Corbii de Piatră is a monastery in Romania, renowned for being carved into a huge sandstone rock and decorated with frescos [10]. Marioara Abrudeanu et al. studied the plaster support of the frescos by microscopy and compositional analysis. Carbon dating performed by AMS confirmed two production phases dated to the 14th century.

Analytical studies are particularly important for studies of materials affected by degradation, especially in harsh conditions such as a marine environment [11]. Liang Zheng et al. studied a metal nameplate on the Vila D. Bosco, a modern building from the former Portuguese colony in Macau. The authors applied several analytical techniques, such as XRF, SEM-EDS, XRD, and ATR-FTIR, to analyze the basic material and corrosion products. The plate was made of steel with a high sulfur content, which induced corrosion. The study proposes a restoration plan and protection methodology for objects exposed in a subtropical marine environment. For protection against volatile organic compounds and other environmental factors such as high humidity, hybrid aerogels have previously been investigated [12], and are now being further developed by George Gorgolis et al. as a three-dimensional aerogel based on graphene and transition-metal dichalcogenides, which can absorb harmful substances. The last set of investigated objects is from the darkest period of human history: belongings of the victims from the Nazi concentration camp Auschwitz-Birkenau. As textiles are prone to microbial deterioration [13], the authors Anna Wawrzyk et al. proposed the disinfection of cotton with ethanol mist for their protection and preservation. Subsequent investigation with FTIR-ATR and XPS showed no significant changes in the fabric.

Author Contributions: Draft, Ž.Š.; Final version, Ž.Š. and E.M. Conceptualization, Ž.Š. and E.M.; writing—original draft preparation, Ž.Š.; writing—review and editing, Ž.Š. and E.M. All authors have read and agreed to the published version of the manuscript.

Funding: This work was partly funded by the Slovenian Research and Innovation Agency (ARIS), program P6-0283 "Archaeological Heritage Research" (E.M.).

Conflicts of Interest: The authors declare no conflicts of interest.

References

1. Zheng, Q.S. A Preliminary Study on the Longshan Culture in Eastern Henan and Its Origins. *Cult. Relics Cent. China* **199**5 46–52.

Bulatović, A.; Milanović, D. The cultural and chronological context of sites of Bubanj and Velika Humska Čuka near Niš (southeastern Serbia) and their significance for understanding the emergence and development of the Central Balkans Eneolithic. *Bull. Mosc. Reg. State Univ.* **2021**, 36–64. [CrossRef]
Varmuza, K.; Filzmoser, P. *Introduction to Multivariate Statistical Analysis in Chemometrics*; CRC Press Taylor & Francis Group: Boca Raton, FL, USA, 2009; p. 63, ISBN 978-1420059472.
Ruschioni, G.; Malchiodi, D.; Zanaboni, A.M.; Bonizzoni, L. Supervised learning algorithms as a tool for archaeology: Classification of ceramic samples described by chemical element concentrations. *J. Archaeol. Sci. Rep.* **2023**, *49*, 103995. [CrossRef]
Symes, S.A.; Rainwater, C.W.; Chapman, E.N.; Gipson, D.R.; Piper, A.L. Patterned Thermal Destruction of Human Remains in a Forensic Setting. In *The Analysis of Burned Human Remains*; Schmidt, C.W., Symes, S.A., Eds.; Elsevier: London, UK, 2008; pp. 15–54.
Phelps, M.; Freestone, I.C.; Gorin-Rosen, Y.; Gratuze, B. Natron glass production and supply in the late antique and early medieval Near East: The effect of the Byzantine-Islamic transition. *J. Archaeol. Sci.* **2016**, *75*, 57–71. [CrossRef]
Paynter, S.; Jackson, C.M. Mellow yellow: An experiment in amber. *J. Archaeol. Sci. Rep.* **2017**, *22*, 568–576.
Todorov, R.; Tasseva, J.; Lozanova, V.; Lalova, A.; Iliev, T.; Paneva, A. Ellipsometric Characterization of Thin Films from Multicomponent Chalcogenide Glasses for Application in Modern Optical Devices. *Adv. Condens. Matter Phys.* **2013**, *2013*, 308258. [CrossRef]
Barry, M.B. *Bagan: Temples and Monuments of Ancient Burma*; Paths International: Bedford, UK, 2003; pp. 1–76.
Ilinescu, I.; Gorunescu, I.; Şerban, I.-A. *Monastery of Corbii de Piatră; History Files*; S.C. Zodia Gemenilor S.R.L.: Arges, Romania, 2009.
Chandler, K.A. *Marine and Offshore Corrosion: Marine Engineering Series*; Elsevier: Amsterdam, The Netherlands, 2014.
Androulidakis, C.; Kotsidi, M.; Gorgolis, G.; Pavlou, C.; Sygellou, L.; Paterakis, G.; Koutroumanis, N.; Galiotis, C. Multi-functional 2D hybrid aerogels for gas absorption applications. *Sci. Rep.* **2021**, *11*, 13548. [CrossRef]
Gutarowska, B.; Pietrzak, K.; Machnowski, W.; Milczarek, J.M. Historical textiles–a review of microbial deterioration analysis and disinfection methods. *Text. Res. J.* **2016**, *87*, 2388–2406. [CrossRef]

sclaimer/Publisher's Note: The statements, opinions and data contained in all publications are solely those of the individual hor(s) and contributor(s) and not of MDPI and/or the editor(s). MDPI and/or the editor(s) disclaim responsibility for any injury to ›ple or property resulting from any ideas, methods, instructions or products referred to in the content.

materials

Article

Compositional Analysis of Longshan Period Pottery and Ceramic Raw Materials in the Yongcheng Region, Henan Province

Linyu Xia [1,2], Yinhong Li [1], Ge Zhang [3,4,5,*], Jialing Li [6] and Li Jaang [1,2]

1 School of Archaeology and Cultural Heritage, Zhengzhou University, Zhengzhou 450001, China; xialinyuzk@163.com (L.X.); li06082024@163.com (Y.L.); lijaangchina@gmail.com (L.J.)
2 Historical and Cultural Heritage Protection Research Center, Zhengzhou University, Zhengzhou 450001, China
3 Yellow River Institute of Hydraulic Research, Yellow River Water Conservancy Commission, Zhengzhou 450003, China
4 Key Laboratory of Lower Yellow River Channel and Estuary Regulation, Ministry of Water Resources, Zhengzhou 450003, China
5 Yellow River Laboratory, Zhengzhou 450003, China
6 School of Water Conservancy and Transportation, Zhengzhou University, Zhengzhou 450001, China; jialingli1120@163.com
* Correspondence: gezhangyrihr@163.com

Abstract: This study systematically analyzes the composition and microstructure of Neolithic pottery unearthed from the Dazhuzhuang, Likou, and Biting Sites in the Yongcheng District using techniques such as X-ray fluorescence spectroscopy (XRF), X-ray diffraction (XRD), infrared spectroscopy (IR), and scanning electron microscopy with energy-dispersive spectroscopy (SEM-EDS). The results show that although the raw materials for pottery at the three sites were likely sourced from nearby ancient soil layers, significant differences in chemical composition and manufacturing techniques are evident. Pottery from the Dazhuzhuang Site is mainly composed of argillaceous gray pottery, with relatively loose raw material selection and a wide fluctuation in SiO_2 content (64.98–71.07%), reflecting diversity in raw material sources. At the Likou Site, argillaceous black pottery predominates, characterized by higher Al_2O_3 content (17.78%) and significant fluctuations in CaO content (1.46–2.22%), suggesting the addition of calcareous fluxes and the adoption of standardized manufacturing techniques. Pottery from the Biting Site mainly consists of argillaceous gray pottery, showing higher Al_2O_3 content (17.36%), stable SiO_2 content (65.19–69.01%), and the lowest CaO content (0.84–1.81%). The microstructural analysis further reveals that the black pottery (from the Likou Site) displays dense vitrified regions and localized iron enrichment. In contrast, the gray pottery (from the Dazhuzhuang and Biting Sites) shows clay platelet structures and vessel-type-specific differences in porosity. This research provides important scientific evidence for understanding raw material selection, manufacturing techniques, and regional cultural interactions in the Yongcheng area during the Longshan Culture period.

Keywords: Yongcheng region; Longshan culture; pottery composition analysis; ceramic manufacturing techniques; raw material sources

1. Introduction

The pottery manufacturing techniques during the Neolithic period in China had already reached a high level of maturity, with diverse vessel forms and strong functional designs. As a representative culture of the Late Neolithic period in China, the Longshan Culture played a key role in developing both cultural evolution and pottery technology. The Longshan Culture in the Yongcheng District, as an important and distinctive regional type within the Longshan cultural system, has attracted significant attention from scholars both in China and abroad. In 1936, Li Jingdan conducted archeological surveys in the Shangqiu and Yongcheng areas in search of the origins of the Shang civilization. He discovered several Longshan Culture remains along both sides of the Huai River in Yongcheng and carried out small-scale excavations at the HeiguDui Site, the Zaolütai Site, and the Caoqiao Site, uncovering a collection of pottery, stone tools, and bone artifacts [1]. In the late 1970s, to explore the social conditions in eastern Henan at the end of the primitive society, the Institute of Archaeology of the Chinese Academy of Social Sciences and the Shangqiu Cultural Relics Administration conducted three rounds of surveys in the counties of Shangqiu, discovering 17 Longshan Culture sites [2] and carrying out excavations at the Wangyoufang Site [3] and the Heigudui Site [4] in Yongcheng. The Wangyoufang Site is considered the most representative, with thick cultural deposits and rich artifact types, and the Longshan Culture remains from this site are referred to as the Wangyoufang type. Subsequently, further discoveries of Longshan Culture remains have been made in eastern Henan, including at the Duanzhai Site in Dancheng, the Pingliangtai Site in Huaiyang, and the Luantai Site in Luyi, as well as the Qingliangshan Site in Xiayi and the Lutai Gang Site in Qixian, Kaifeng. These sites have all yielded Longshan Culture remains and are generally classified under the Wangyoufang type. In 1995, a Sino-American joint archeological team excavated the Shantaisi Site in Zhecheng, discovering a large number of Longshan cultural features and artifacts, such as large rammed-earth platforms and sacrificial cattle pits, confirming its central position in the Longshan Culture of eastern Henan [5]. In 2002, the Department of Archaeology at Zhengzhou University surveyed 24 ancient cultural sites in Shangqiu, including four Longshan Culture sites in Yongcheng: the Hongfu Site, the Zhaozhuang Site, the Mingyangsi Site, and the Zaolütai Site [6]. In 2022, to further clarify the distribution of cultural relics and settlement hierarchy of the Longshan Culture in the Huai River basin of Yongcheng, the Department of Archaeology at Zhengzhou University carried out a comprehensive systematic survey and exploration of the cultural remains in the region.

With the continuous publication of archeological excavation data from Longshan Culture sites, the study of Longshan Culture has produced abundant results. Traditional archeological research has primarily focused on the cultural chronology and origins, especially in classifying and forming regional types, such as the Wangyoufang type [7] and the Wangwan type [8] of the Longshan Culture. Secondly, studies have addressed the geographical distribution of the culture. Scholars such as Liang Sicheng [9], An Zhimin [10], and Yang Zifan [11] have proposed different views on the extent of the Longshan Culture, with three prevailing opinions regarding the western boundary of the Haidai Longshan Culture distribution area [12]. Thirdly, many specialized studies have been carried out on the nature and function of sites and have unearthed artifacts and features—especially tombs and typical pottery vessels. By conducting typological research on representative pottery, scholars have explored issues related to their distribution, manufacturing techniques, morphological evolution, functions, and social significance.

With the adoption of experimental archeological methods, Chinese researchers have increasingly employed techniques such as X-ray fluorescence spectroscopy (XRF), wavelength-dispersive X-ray fluorescence (WDXRF), and scanning electron microscopy–energy dispersive spectroscopy (SEM-EDS) to analyze the chemical composition and firing techniques of Longshan Culture pottery. These methods have facilitated deeper investigations into raw material sources, production processes, and technical strategies. For example,

Li Minsheng and Huang Suying conducted chemical and physical analyses on pottery sherds from different periods at the Taosi Site, shedding light on ceramic manufacturing techniques during the Miaodigou Phase II and Longshan periods [13]. Chen Qianqian and Yang Yuzhang used WDXRF to study Neolithic pottery from the Jiahu Site in Henan and the Xiaohuangshan Site in Zhejiang [14]; Gao Shoulei applied XRF to analyze the composition of pottery from the Hongshan Culture at the Niuheliang Site [15]. As pottery is one of prehistoric cultures' most important material remains, its manufacturing technology and social function have become core issues in international archeological research [16–20]. Internationally, scholars widely use techniques such as XRF, X-ray diffraction (XRD), SEM-EDS, and Raman spectroscopy to analyze the composition, mineral structure, and microstructure of pottery, revealing the technological development of ceramic production [21–24]. For example, Maja Gajić-Kvaščev used EDXRF to study the composition of pottery from the Velika Humska Čuka Site in Serbia, providing evidence of cultural exchange influences [25]. Laura Teodorescu and colleagues combined XRD and SEM-EDS to analyze the raw materials and production techniques of Dacian pottery [26]. Philippe Colomban and others applied XRF and Raman spectroscopy to detect different chemical compositions in ceramics, proving that as early as the 12th century CE, Islamic potters had mastered highly advanced ceramic techniques [27]. These international research paradigms have provided important references for the scientific and systematic development of ceramic studies in China [28–32].

However, current scientific archeological research still faces challenges, such as limited analytical methods and lacking a comprehensive technical system that integrates multiple techniques and interdisciplinary approaches. This has hindered a full understanding of the complexity of pottery manufacturing techniques and the diversity of raw material sources. Based on this research context, the present study adopts an integrated approach combining traditional archeology and materials science. By applying both typological and scientific methods and using a combination of analytical techniques, including XRF [33–36], XRD, infrared spectroscopy (IR), and SEM-EDS, we aim to carry out systematic analysis at the levels of chemical composition, mineral structure, and microscopic morphology [37–41]. This multi-technical approach overcomes the limitations of relying on single methods, providing more reliable data for identifying pottery composition and raw material sources.

In 2022, the Department of Archaeology at Zhengzhou University conducted a systematic survey and exploration in the Huai River basin of the Yongcheng District, identifying 17 archeological sites containing cultural remains from the Longshan Culture, as shown in Figure 1. Among them, the Dazhuzhuang Site (16,800 m^2), the Biting Site (40,000 m^2), and the Likou Site (90,000 m^2) are located at the junction of Henan, Anhui, and Shandong Provinces. These sites serve as a crucial area for interaction between the Wangyoufang type of the Longshan Culture in eastern Henan and adjacent cultural regions. With well-defined hierarchical settlement structures, these sites yielded abundant pottery sherds from the Longshan period, featuring diverse types and representative vessel forms. For this reason, pottery from these three sites was selected as the focus of the present study. This research integrates domestic and international advances in pottery studies through the systematic sampling and analysis of typical pottery sherds unearthed from the Dazhuzhuang, Biting, and Likou Sites [42–46]. It applies a multi-technical and interdisciplinary approach to investigate Longshan Culture pottery in the Yongcheng District comprehensively. The objective is to clarify the raw material sources and technological characteristics of these ceramics during the Longshan period, thereby uncovering the intrinsic relationship between ceramic manufacturing and regional cultural interaction. This study aims to provide new empirical evidence for understanding regional variation within the Longshan Culture and the pathways of ceramic technological transmission.

Figure 1. Schematic map of site distribution locations.

The specimens analyzed from the three sites are introduced as follows:

1. Dazhuzhuang Site

The Dazhuzhuang Site is on the northwest side of Dazhuzhuang Village, Peiqiao Town, Yongcheng District. It is approximately 120 m wide from north to south and 140 m long from east to west, covering an area of about 16,800 square meters. The cultural deposits are 0.8–2.1 m thick. Collected artifacts are mainly pottery sherds, with a few shells and animal bone remains. The pottery primarily dates back to the Longshan culture and Han periods. Pottery from the Longshan Culture is mainly argillaceous gray pottery, followed by argillaceous brown pottery and argillaceous black pottery. Decorations are primarily cord marks and plain surfaces, followed by basket patterns, along with a few string patterns and grid patterns. Vessel types include storage jars (ceramic containers for liquids or food), the Yan (a cooking vessel functionally similar to a modern steamer, comprising an upper zeng for holding food and a lower li for water, separated by a perforated grate to allow for steam circulation), the Ding (tripod or quadripod cooking vessels with handles, typically round with three legs and two ears, though rectangular versions exist), the Dou (food-serving vessels with tall ring feet), urns (storage vessels with a narrow mouth and bulging belly), bowls (round concave dishes for food), and cups.

A total of 15 specimens were selected for testing, including vessel types such as storage jars, basins (wide-mouthed, narrow-based containers for liquids, food, or washing), cups (liquid-holding vessels), Ding legs, Yan legs, and handles (lateral protrusions for grasping). The majority are argillaceous gray pottery, with a few argillaceous brown pottery pieces. Decorations are mainly grid patterns and plain surfaces.

Seven storage jars were examined, including six argillaceous gray pottery specimens and one argillaceous brown pottery specimen. The six argillaceous gray pottery pieces exhibit flared rims, folded rims, square lips, contracted necks, and sloping shoulders. Specimen 2023YPDH1:1 features a globular body and is preserved below the belly. The surface is plain (Figure 2(1)). Specimen 2023YPDH1:4 is preserved below the shoulder, which is decorated with grid patterns (Figure 2(6)). Specimen 2023YPDH1:5 has a round lip and is preserved below the shoulder. The shoulder is decorated with diamond-shaped grid patterns (Figure 2(7)). Specimen 2023YPDH1:16 has a square lip with a circumferential groove on the lip surface and is preserved below the shoulder. The shoulder is decorated with grid patterns (Figure 3(1)). Specimen 2023YPDH2:1 has a grid pattern decoration on the shoulder. Specimen 2023YPDH2:5 features a round lip and a curved body, with the lower belly preserved. The exterior is decorated with medium-sized cord marks (Figure 3(3)).

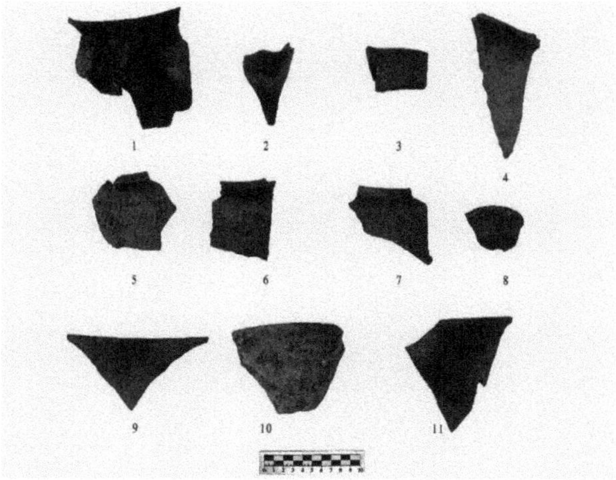

Figure 2. Pottery specimens from the Longshan Culture period at the Dazhuzhuang Site. (**1**) Storage jar (2023YPDH1:1), (**2**) Yan leg (2023YPDH1:23), (**3**) handle (2023YPDH1:17), (**4**) Ding leg (2023YPDH1:21), (**5**) storage jar (2023YPDH1:9), (**6**) storage jar (2023YPDH1:4), (**7**) storage jar (2023YPDH1:5), (**8**) cup (2023YPDH1:19), (**9**) basin (2023YPDH1:13), (**10**) basin (2023YPDH1:12), and (**11**) Basin (2023YPDH1:11).

Figure 3. Pottery specimens from the Longshan Culture period at the Dazhuzhuang Site. (**1**) Storage jar (2023YPDH1:16), (**2**) basin (2023YPDH2:12), and (**3**) storage jar (2023YPDH1:5).

Specimen 2023YPDH1:9 is an argillaceous brown pottery storage jar. It features a flared rim, a folded rim, a square lip with a circumferential groove on the inner edge, a contracted neck, sloping shoulders, and a curved body. The portion below the belly is preserved. The exterior has diamond-shaped grid patterns (Figure 2(5)).

Four basins were examined, including three made of argillaceous gray pottery. These basins have open mouths, round lips, and obliquely straight bodies, with parts preserved below the belly. The surfaces are plain. Specimens include 2023YPDH1:13 (Figure 2(9)), 2023YPDH1:11 (Figure 2(11)), and 2023YPDH2:12.

One specimen is made of argillaceous brown pottery, featuring an open mouth, a round lip, an obliquely straight body, and preservation below the belly. The surface is plain (Specimen 2023YPDH1:12) (Figure 2(10)).

One cup was identified. Specimen 2023YPDH1:19 is an argillaceous gray pottery piece with an open mouth, a square lip, a circumferential groove on the inner edge of the rim, a shallow body, and a flat base. The surface is plain (Figure 2(8)).

One Ding leg was found. Specimen 2023YPDH1:21 is an argillaceous brown pottery piece. It has a side-triangular flat leg with a solid root. There are six depressions on the outer side of the leg root. The surface is plain (Figure 2(4)).

One Yan leg was recovered. Specimen 2023YPDH1:23 is made of argillaceous gray pottery and features a pouch-shaped leg with a tall, tapered, solid root. The surface is plain (Figure 2(2)).

One handle was documented. Specimen 2023YPDH1:17 is an argillaceous gray pottery piece with a bridge-shaped handle. The surface is plain (Figure 2(3)).

2. Biting Site

The Biting Site is located southeast of Bianzhuang Village, Xinqiao Town, Yongcheng District. The central, western, and northern parts of the site are overlain by the town's grain storage facility. The site is irregular in shape and covers an area of approximately 40,000 square meters. Collected artifacts mainly date back to the Longshan and Shang culture periods. Pottery sherds from the Longshan Culture period are primarily composed of argillaceous gray and argillaceous gray-black pottery, along with some argillaceous polished black pottery and argillaceous brown pottery. There are also small quantities of sand-tempered gray/brown pottery and shell-tempered brown/gray pottery. Decorative patterns are dominated by basket patterns, square patterns, and plain surfaces, followed by cord-marked patterns and a few string patterns. Common vessel types include storage jars, basins, bowls, and Ding.

A total of eight specimens were selected for testing, including storage jars, basins, bowls, and Ding legs. Most are made of argillaceous gray pottery, with decorative patterns mainly featuring square patterns, cord marks, and plain surfaces.

Five storage jars, all made of argillaceous gray pottery, have flaring mouths, folded rims, and square lips (Specimen 2023YXBP⑤:7: argillaceous gray pottery). A groove is present inside the lip, and it has a contracted neck and a sloping shoulder but is missing below the shoulder. The shoulder is decorated with a square pattern (Figure 4(3)) (Specimen 2023YXBP⑤:21: argillaceous gray pottery). It has a contracted neck and a sloping shoulder but is missing below the shoulder. The shoulder is decorated with a cord-marked pattern (Figure 4(4)) (Specimen 2023YXBP⑤:13: argillaceous gray pottery). A groove is present inside the lip, and it has a contracted neck, a sloping shoulder, and a plain surface but is missing below the shoulder (Figure 4(5)) (Specimen 2023YXBP⑤:8: argillaceous gray pottery). A groove is present inside the lip, and it has a contracted neck and a sloping shoulder but is missing below the shoulder. The shoulder is decorated with a square pattern (Figure 4(6)) (Specimen 2023YXBP⑤:14: argillaceous gray pottery). A groove is present on the surface of the square lip, and it has a contracted neck, a round shoulder, and a plain surface; it is missing below the shoulder (Figure 4(7)).

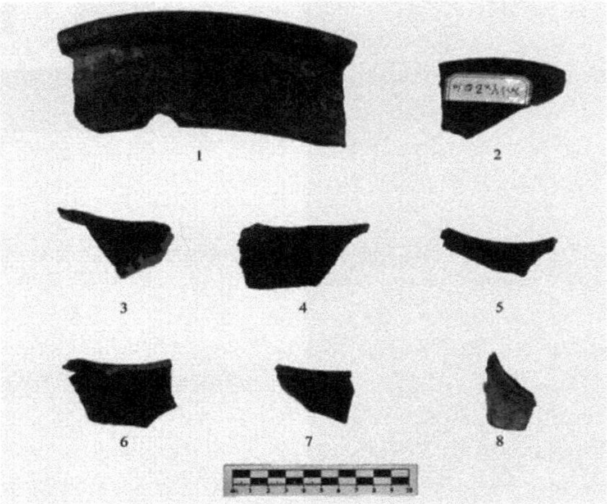

Figure 4. Pottery samples from the Longshan Culture period. (**1**) Basin (2023YXBP⑤:1), (**2**) bowl (2023YXBP⑤:10), (**3**) storage jar (2023YXBP⑤:7), (**4**) storage jar (2023YXBP⑤:21), (**5**) storage jar (2023YXBP⑤:13), (**6**) storage jar (2023YXBP⑤:8), (**7**) storage jar (2023YXBP⑤:14), and (**8**) Ding leg (2023YXBP⑤:19).

One basin: Specimen 2023YXBP⑤:1: argillaceous gray pottery: It has a flaring mouth, a folded rim, a round lip with an inner groove, an oblique straight belly, and a plain surface but is missing below the belly (Figure 4(1)). One bowl: Specimen 2023YXBP⑤:10: argillaceous gray pottery: It has an open mouth, a square lip with a groove on the lip surface, an oblique belly, and a plain surface but is missing below the belly (Figure 4(2)). One ding leg: Argillaceous gray pottery: It has a side-triangular flat leg with a solid foot root. The tip of the foot is broken. A depression is visible on the upper part of the outer side of the leg (Figure 4(8)).

3. Likou Site

The Likou Site is located in Likou Village, Houling Community. The Huai River lies to the south, a village to the west, and the northern and eastern areas border the Huaibei City of Anhui Province. It is situated on a high terrace surrounded by water on two sides. The site measures about 300 m in length and width, covering an area of approximately 90,000 square meters, with cultural deposits ranging from 0.3 to 3.8 m thick. Collected artifacts include pottery sherds, stone tools, shells, deer antlers, and animal bones. Pottery sherds mainly date back to the Yangshao, Longshan, and Shang cultures. Pottery from the Longshan Culture period is primarily argillaceous gray pottery, along with polished black pottery, argillaceous brown pottery, and small quantities of sand-tempered gray/brown pottery and shell-tempered gray/brown pottery. The decorations include basket, square, and cord-marked patterns, with minor plain surfaces and scattered string patterns. Vessel types include storage jars, urns, basins, bowls, pottery lids (a cover for containers), and Ding.

A total of five specimens were selected for testing, including urns, storage jars, and a pottery lid.

Two urns, both made of argillaceous black pottery, have straight mouths and sharp lips. Specimen 2023YHL⑥:1: The outer side of the rim has two raised ridges. It has a short neck and a sloping shoulder but is missing below the shoulder. The surface is plain and polished (Figure 5(1)). Specimen 2023YHL⑥:2: The inner side of the rim has a groove; it has a tall neck and a sloping shoulder but is missing below the shoulder (Figure 5(2)).

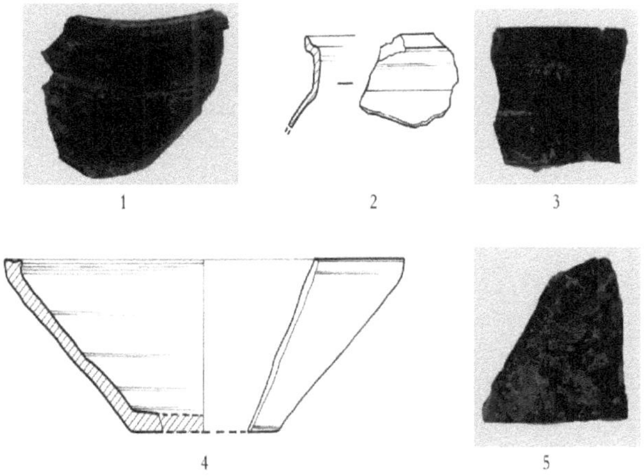

Figure 5. Pottery samples from the Longshan Culture period at the Likou Site. (**1**) Urn (2023YHL⑥:1), (**2**) urn (2023YHL⑥:2), (**3**) rim of a storage jar (B6:4), (**4**) pottery lid (2023YHLH2:2), and (**5**) body sherd of a storage jar (B6:3).

One pottery lid: Specimen 2023YHLH2:2: argillaceous gray pottery: It has an open mouth with a slightly constricted rim, a square lip with a groove on the lip surface, an oblique straight belly, and a flat base, along with wheel marks on the inside and a plain surface (Figure 5(4)).

Two storage jars are made of argillaceous black pottery with plain polished surfaces: Specimen B6:3: a body sherd (Figure 5(5)). Specimen B6:4: a flaring mouth, a rounded sharp lip, a contracted neck, and missing below the neck (Figure 5(3)).

2. Experimental

2.1. Sample Preparation

Firstly, the ceramic sample is cut into small pieces of approximately 2 cm^2 using a cutting machine, and the sections are ground flat. Secondly, the test samples are cleaned by ultrasonic treatment in purified water twice, followed by cleaning with alcohol. For SEM-EDS analysis, the samples are broken into smaller pieces using pliers. For XRD, XRF, and IR tests, the smaller pieces were sequentially pulverized, ground, and sieved through a 0.075 mm sieve. Table 1 shows the grouping of performance index tests, including the size and number of each test and specimen.

Table 1. Grouping of the performance index tests.

Performance Index	Specimen Size	Quantity
Energy-dispersive X-ray fluorescence (XRF)	20 mm × 20 mm × 20 mm	84
Infrared spectroscopy (IR)	20 mm × 20 mm × 20 mm	57
X-ray diffraction analysis (XRD)	20 mm × 20 mm × 20 mm	57
Scanning electron microscopy (SEM)	40 mm × 40 mm × 40 mm	49

2.2. Energy-Dispersive X-Ray Fluorescence Analysis

Energy-dispersive X-ray fluorescence (XRF) (Shimadzu EDX-8100, Kyoto, Japan) was used to test the chemical compositions of the bodies and surface coatings, with an X-ray spot diameter of 1.2 mm, an X-ray tube voltage of 30 kV, an X-ray tube current of 0.029 mA, and a data acquisition time of 100 s. The standard sample is Corning Glass D, which is used to calibrate the content of the samples.

2.3. Infrared Spectroscopy Analysis

An infrared spectrometer (XploRA PLUS, Horiba, Longjumeau, France) was used to analyze the phase of surface coatings. Infrared spectra were collected at room temperature under the 785 nm excitation line in the 100–4000 cm^{-1} spectral range. The laser beam was focused by a 50× objective lens with a laser spot diameter of 1 μm and a laser power of 1 mW.

2.4. X-Ray Diffraction Analysis

To use the Japanese physical X-ray diffractometer (Shimadzu Corporation, Kyoto, Japan), we took the sample after drying and grinding treatment and placed it into the glass groove for testing. The sampling interval was 0.04° (2θ); the sampling speed was 2 °/min, and the scanning angle range was 5–70° (2θ).

2.5. Scanning Electron Microscopy Test

The microstructure of the samples was observed using a Sigma 300 field emission environmental scanning electron microscope (Carl Zeiss AG, Oberkochen, Germany). The signal encompassed secondary electrons and backscattered electrons; the acceleration voltage was 15 kV; the vacuum level was maintained at 1 Pa; the amplification range was 18–30,000 times, and the maximum resolution was 3 nm. A thin layer of gold was deposited on the samples to enhance their conductivity for scanning electron microscopy. Backscattered electron images were obtained at a voltage of 15 kV. EDS was performed at 20 kV. The samples analyzed by SEM-EDS were prepared as polished resin blocks. The results were obtained from a single bulk chemical analysis.

3. Results and Discussion

3.1. Analysis of Chemical Composition Result

A total of 28 pottery samples were analyzed in this study, including 15 pieces from the Dazhuzhuang Site in Yongcheng, 5 pieces from the Likou Site in Yongcheng, and 8 pieces from the Biting Site in Yongcheng, all dating to the Longshan Culture period of the Neolithic Age. The 15 samples from the Dazhuzhuang Site include vessel types such as storage jars, Yan legs (tripod leg of "Yan," a ritual cooking vessel with hollow legs), handles ("pan", horizontal non-perforated clay attachments for gripping), Ding legs, cups, and basins, with the majority made of argillaceous gray pottery and a small number made of argillaceous brown pottery. The five samples from the Likou Site consist of pottery lids, urns, and body sherds and are mainly composed of argillaceous black pottery. The eight samples from the Biting Site include basins, bowls, storage jars, and Ding vessels, all made of argillaceous gray pottery.

From a typological perspective, these samples exhibit distinct regional characteristics. Artifacts from the Dazhuzhuang Site display a rich diversity of vessel types, covering these major functional categories, including cooking (Ding legs and Yan legs), storage (storage jars and basins), and drinking (cups), possibly reflecting the multifunctionality of the settlement. Artifacts from the Likou Site are predominantly storage vessels, such as pottery lids, urns, and body sherds, made from thin-walled, polished black pottery. In contrast, artifacts from the Biting Site are mainly utilitarian vessels for daily use, such as basins, bowls, storage jars, and Ding vessels, characterized by regular forms and consistent body structures.

3.1.1. Dazhuzhuang Site

XRF was used to measure the chemical composition of 16 elements in the pottery samples unearthed from the Dazhuzhuang site. The contents of 10 elements, including SiO_2, Al_2O_3, Fe_2O_3, K_2O, MgO, CaO, Na_2O, TiO_2, and P_2O_5, are shown in Table 2. The detailed statistical results of chemical composition are shown in Table 3.

Table 2. XRF experimental data results of the Dazhuzhuang Site in the Yongcheng area (content: wt.%).

Sample Number	Pottery Type	Part	SiO_2	Al_2O_3	Fe_2O_3	K_2O	MgO	CaO	Na_2O	TiO_2	P_2O_5
2023YPDH1:1-1	argillaceous gray pottery	Storage jar	68.41	16.28	5.44	3.52	1.63	1.75	1.30	0.74	0.63
2023YPDH1:1-2	argillaceous gray pottery	Storage jar	68.48	16.25	5.45	3.50	1.57	1.81	1.24	0.75	0.66
2023YPDH1:1-3	argillaceous gray pottery	Storage jar	68.37	16.31	5.40	3.54	1.59	1.77	1.31	0.75	0.68
2023YPDH1:23-1	argillaceous gray pottery	Yan leg	68.01	16.67	5.49	3.70	1.67	1.49	1.15	0.76	0.77
2023YPDH1:23-2	argillaceous gray pottery	Yan leg	68.04	16.69	5.47	3.66	1.68	1.53	1.13	0.75	0.79
2023YPDH1:23-3	argillaceous gray pottery	Yan leg	68.17	16.59	5.37	3.72	1.69	1.48	1.20	0.75	0.77
2023YPDH1:17-1	argillaceous gray pottery	Handle	68.25	16.54	5.33	3.66	1.63	1.44	1.16	0.74	0.98
2023YPDH1:17-2	argillaceous gray pottery	Handle	68.50	16.56	5.19	3.58	1.71	1.41	1.11	0.73	0.98
2023YPDH1:17-3	argillaceous gray pottery	Handle	68.29	16.58	5.33	3.66	1.64	1.44	1.10	0.74	0.98
2023YPDH1:21-1	argillaceous brown pottery	Ding leg	68.24	16.81	5.52	3.57	1.60	1.45	1.16	0.74	0.65
2023YPDH1:21-2	argillaceous brown pottery	Ding leg	68.30	16.87	5.40	3.55	1.61	1.42	1.16	0.73	0.72
2023YPDH1:21-3	argillaceous brown pottery	Ding leg	68.39	16.86	5.29	3.51	1.63	1.41	1.12	0.74	0.83
2023YPDH1:9-1	argillaceous brown pottery	Storage jar	68.84	16.54	5.40	3.60	1.63	1.42	1.13	0.73	0.55
2023YPDH1:9-2	argillaceous brown pottery	Storage jar	68.88	16.47	5.39	3.62	1.65	1.39	1.14	0.73	0.55
2023YPDH1:9-3	argillaceous brown pottery	Storage jar	69.00	16.44	5.33	3.62	1.63	1.40	1.14	0.74	0.53
2023YPDH1:4-1	argillaceous gray pottery	Storage jar	67.29	17.09	5.92	3.62	1.65	1.50	1.29	0.78	0.60
2023YPDH1:4-2	argillaceous gray pottery	Storage jar	67.76	16.95	5.82	3.60	1.62	1.49	1.22	0.77	0.54
2023YPDH1:4-3	argillaceous gray pottery	Storage jar	67.55	17.15	5.79	3.64	1.63	1.48	1.23	0.76	0.54
2023YPDH1:5-1	argillaceous gray pottery	Storage jar	67.93	16.81	5.11	3.68	1.62	1.64	1.12	0.74	1.10
2023YPDH1:5-2	argillaceous gray pottery	Storage jar	68.18	16.74	5.20	3.50	1.61	1.63	1.09	0.74	1.06
2023YPDH1:5-3	argillaceous gray pottery	Storage jar	68.05	16.70	5.19	3.65	1.59	1.63	1.13	0.74	1.08
2023YPDH1:19-1	argillaceous gray pottery	Cub	68.64	16.63	5.64	3.42	1.80	1.19	1.10	0.71	0.65
2023YPDH1:19-2	argillaceous gray pottery	Cub	68.56	16.62	5.71	3.51	1.74	1.21	1.11	0.74	0.61
2023YPDH1:19-3	argillaceous gray pottery	Cub	68.43	16.57	5.75	3.51	1.77	1.21	1.16	0.73	0.62
2023YPDH1:13-1	argillaceous gray pottery	Basin	67.64	16.99	5.61	3.74	1.53	1.55	1.12	0.75	0.76
2023YPDH1:13-2	argillaceous gray pottery	Basin	67.54	17.09	5.64	3.74	1.53	1.53	1.17	0.74	0.76
2023YPDH1:13-3	argillaceous gray pottery	Basin	67.90	17.01	5.53	3.67	1.49	1.53	1.08	0.73	0.84
2023YPDH1:12-1	argillaceous brown pottery	Basin	67.71	17.01	5.52	3.75	1.53	1.51	1.17	0.74	0.78
2023YPDH1:12-2	argillaceous brown pottery	Basin	67.90	16.94	5.54	3.76	1.50	1.51	1.10	0.75	0.75
2023YPDH1:12-3	argillaceous brown pottery	Basin	67.73	17.02	5.55	3.77	1.54	1.51	1.14	0.75	0.75
2023YPDH1:11-1	argillaceous gray pottery	Basin	67.47	16.93	5.72	3.71	1.59	1.54	1.07	0.73	0.96
2023YPDH1:11-2	argillaceous gray pottery	Basin	67.75	16.81	5.66	3.73	1.60	1.54	1.05	0.72	0.94
2023YPDH1:11-3	argillaceous gray pottery	Basin	67.64	16.78	5.68	3.72	1.60	1.55	1.09	0.73	0.97
2023YPDH1:16-1	argillaceous gray pottery	Storage jar	67.08	17.90	7.03	2.94	1.43	1.32	1.11	0.76	0.22
2023YPDH1:16-2	argillaceous gray pottery	Storage jar	67.01	17.85	7.07	3.00	1.43	1.30	1.10	0.77	0.27
2023YPDH1:16-3	argillaceous gray pottery	Storage jar	67.22	17.87	7.02	2.96	1.41	1.30	1.08	0.76	0.17
2023YPDH2:12-1	argillaceous brown pottery	Basin	65.08	18.59	6.80	3.35	1.68	1.77	1.00	0.79	0.69
2023YPDH2:12-2	argillaceous brown pottery	Basin	65.08	18.83	6.71	3.27	1.68	1.77	0.95	0.80	0.71
2023YPDH2:12-3	argillaceous brown pottery	Basin	64.98	18.68	6.81	3.36	1.72	1.74	1.00	0.79	0.68
2023YPDH2:1-1	argillaceous gray pottery	Storage jar	66.94	18.17	6.52	3.18	1.67	1.28	1.08	0.79	0.17
2023YPDH2:1-2	argillaceous gray pottery	Storage jar	66.74	18.22	6.68	3.20	1.68	1.25	1.04	0.80	0.18
2023YPDH2:1-3	argillaceous gray pottery	Storage jar	66.67	18.24	6.71	3.18	1.70	1.25	1.06	0.79	0.18
2023YPDH2:5-1	argillaceous gray pottery	Storage jar	70.19	15.79	5.23	2.95	1.40	1.55	1.24	0.78	0.65
2023YPDH2:5-2	argillaceous gray pottery	Storage jar	70.97	15.37	5.08	2.90	1.37	1.50	1.20	0.77	0.63
2023YPDH2:53	argillaceous gray pottery	Storage jar	71.07	15.18	5.10	2.91	1.33	1.50	1.24	0.77	0.66

Table 3. Statistical results of the chemical composition of the pottery shards from the Dazhuzhuang Site in the Yongcheng area (wt.%).

Statistical Measures	SiO_2	Al_2O_3	Fe_2O_3	K_2O	MgO	CaO	Na_2O	TiO_2
Number of Samples	45	45	45	45	45	45	45	45
Mean	67.93	16.96	5.74	3.49	1.60	1.49	1.14	0.75
Maximum	71.07	18.83	7.07	3.77	1.80	1.81	1.31	0.80
Minimum	64.98	15.18	5.08	2.90	1.33	1.19	0.95	0.71
Standard Deviation	1.19	0.78	0.58	0.26	0.11	0.16	0.08	0.02

According to Table 3, the SiO_2 content at the Dazhuzhuang Site ranges from 64.98% to 71.07%, with an average of 64.98% and a standard deviation of 1.19%. The Al_2O_3 content ranges from 15.18% to 18.83%, with an average of 16.96% and a standard deviation of 0.78%. The Fe_2O_3 content varies between 5.08% and 7.07%, with an average of 5.74% and a standard deviation of 0.58%. The K_2O content ranges from 2.90% to 3.77%, with an average of 3.49% and a standard deviation of 0.26%. The MgO content ranges from 1.33% to 1.80%, with an average of 1.60% and a standard deviation of 0.11%. The CaO content varies between 1.19% and 1.81%, with an average of 1.49% and a standard deviation of 0.16%. The Na_2O content ranges from 0.95% to 1.31%, with an average of 1.14% and a standard deviation of 0.08%.

In terms of oxide composition, the SiO_2/Al_2O_3 ratio of 3.96 at the Dazhuzhuang Site indicates the use of clay with moderate weathering. The relatively high Fe_2O_3 content (5.74%) suggests that the raw materials may contain a significant amount of hematite, while the stable alkali metal content indicates that illite is the dominant clay mineral. Notably, SiO_2 exhibits the largest standard deviation (1.19%) among the oxides, reflecting the most significant variability in silicon content. This variability may be attributed to the following technical characteristics in raw material processing: (1) an uneven distribution of quartz particles within the raw materials, (2) relatively lenient selection criteria for siliceous materials by local potters, and (3) the potential mixing of raw materials from different batches or sources. In contrast, the relatively stable Al_2O_3 content (standard deviation of 0.78%) suggests that potters in this region likely prioritized stricter quality control for clay components, while showing greater flexibility in the selection of siliceous temper materials.

3.1.2. Likou Site

The chemical compositions of 16 elements in pottery samples excavated from the Likou Site were measured using X-ray fluorescence (XRF). Among these, the contents of ten elements, including SiO_2, Al_2O_3, Fe_2O_3, K_2O, MgO, CaO, Na_2O, TiO_2, and P_2O_3, are presented in Table 4, with detailed statistical results of their chemical compositions provided in Table 5.

Table 4. XRF experimental data results of the Likou Site in the Yongcheng area (content: wt.%).

Sample Number	Pottery Type	Part	SiO_2	Al_2O_3	Fe_2O_3	K_2O	MgO	CaO	Na_2O	TiC
2023YHLH2:2-1	argillaceous gray pottery	Pottery lid	67.06	18.11	6.17	3.01	1.70	1.50	1.22	0.8
2023YHLH2:2-2	argillaceous gray pottery	Pottery lid	67.04	18.04	6.24	3.01	1.70	1.51	1.22	0.8
2023YHLH2:2-3	argillaceous gray pottery	Pottery lid	66.91	18.09	6.26	3.03	1.70	1.51	1.27	0.8
2023YHL⑥:1-1	argillaceous black pottery	Urn	67.89	17.37	5.14	3.47	1.70	1.76	1.15	0.8
2023YHL⑥:12	argillaceous black pottery	Urn	67.90	17.43	5.16	3.43	1.69	1.79	1.10	0.8
2023YHL⑥:1-3	argillaceous black pottery	Urn	67.97	17.46	5.02	3.47	1.67	1.75	1.16	0.7
2023YHL⑥:2-1	argillaceous black pottery	Urn	67.48	17.83	5.69	3.34	1.85	1.47	1.08	0.7
2023YHL⑥:2-2	argillaceous black pottery	Urn	67.00	17.91	5.93	3.43	1.91	1.46	1.06	0.7
2023YHL⑥:2-3	argillaceous black pottery	Urn	66.97	17.90	5.97	3.44	1.87	1.47	1.09	0.8
2023YHLB6:3-1	argillaceous black pottery	Body sherd	65.93	17.75	6.00	3.60	1.71	2.22	1.12	0.8
2023YHLB6:3-2	argillaceous black pottery	Body sherd	65.99	17.70	6.01	3.59	1.70	2.19	1.14	0.8
2023YHLB6:3-3	argillaceous black pottery	Body sherd	65.97	17.76	5.97	3.60	1.70	2.21	1.12	0.8
2023YHLB6:4-1	argillaceous black pottery	Body sherd	67.44	17.75	5.30	3.46	1.58	1.82	1.20	0.8
2023YHLB6:4-2	argillaceous black pottery	Body sherd	67.37	17.85	5.29	3.45	1.58	1.82	1.19	0.8
2023YHLB6:4-3	argillaceous black pottery	Body sherd	67.53	17.81	5.23	3.49	1.59	1.82	1.09	0.8

Table 5. Statistical results of the chemical composition of the pottery shards from the Likou Site in the Yongcheng area (wt.%).

Statistical Measures	SiO_2	Al_2O_3	Fe_2O_3	K_2O	MgO	CaO	Na_2O	TiO_2
Number of Samples	15	15	15	15	15	15	15	15
Mean	67.10	17.78	5.69	3.39	1.71	1.75	1.15	0.81
Maximum	67.97	18.11	6.26	3.60	1.91	2.22	1.27	0.83
Minimum	65.93	17.37	5.02	3.01	1.58	1.46	1.06	0.78

As shown in Table 5, the SiO_2 content at the Likou Site in the Yongcheng region ranges from 65.93% to 67.97%, with an average of 67.10% and a standard deviation of 0.68%. The Al_2O_3 content varies between 17.37% and 18.11%, averaging 17.78% (standard deviation:

0.23%). Fe_2O_3 levels span 5.02–6.26%, with a mean value of 5.69% and a standard deviation of 0.45%. The K_2O content ranges from 3.01% to 3.60% (average: 3.39%; SD: 0.20%), while MgO concentrations fall within 1.58–1.91% (average: 1.71%; SD: 0.10%). CaO values are recorded between 1.46% and 2.22% (mean: 1.75%; SD: 0.28%), and the Na_2O content ranges from 1.06% to 1.27% (average: 1.15%; SD: 0.06%). Regarding major elemental characteristics, the high Al_2O_3 content (17.78%) at the Likou Site suggests using high-quality kaolin or well-weathered sedimentary clay in pottery production. The SiO_2/Al_2O_3 ratio of 3.77 aligns with typical clay ranges, effectively excluding the possibility of primary loess utilization.

Compared to the Dazhuzhuang site, the pottery from the Likou site exhibits significant technological differences. Firstly, in terms of elemental composition, although the levels of SiO_2 and Fe_2O_3 are similar, the contents of Al_2O_3 and CaO are notably higher. Secondly, regarding process control, the standard deviations of Fe_2O_3, SiO_2, and Al_2O_3 are significantly lower than those at the Dazhuzhuang site, whereas the standard deviation of CaO is relatively higher. This phenomenon may reflect the following technological characteristics during the firing process of the pottery: (1) The Likou site used high-quality clay materials with higher Al_2O_3 content. (2) Potters in the region demonstrated more standardized selection and processing of silico-aluminous raw materials. (3) The greater fluctuation in CaO content may indicate the intentional addition of calcareous tempering agents (such as shell powder or lime) during firing or could result from natural variations in raw material batches.

3.1.3. Biting Site

The chemical composition of 16 elements in the pottery samples unearthed from the Biting site was measured using X-ray fluorescence (XRF). Among them, the contents of 10 major elements—SiO_2, Al_2O_3, Fe_2O_3, K_2O, MgO, CaO, Na_2O, TiO_2, and P_2O_5—are presented in Table 6. Detailed statistical results of the chemical composition are shown in Table 7.

Table 6. XRF experimental data results from the Yiting Site in the Yongcheng area (content: wt.%).

ample Number	Pottery Type	Part	SiO_2	Al_2O_3	Fe_2O_3	K_2O	MgO	CaO	Na_2O	TiO_2	P_2O_5
2023YXBP⑤:1-1	argillaceous gray pottery	Basin	67.12	17.60	5.51	3.37	1.73	1.64	1.31	0.82	0.67
2023YXBP⑤:1-2	argillaceous gray pottery	Basin	67.06	17.49	5.62	3.39	1.74	1.62	1.36	0.82	0.67
2023YXBP⑤:1-3	argillaceous gray pottery	Basin	67.20	17.55	5.65	3.28	1.73	1.63	1.27	0.81	0.65
023YXBP⑤:10-1	argillaceous gray pottery	Bowl	67.95	16.97	5.45	3.35	1.76	1.53	1.33	0.82	0.62
023YXBP⑤:10-2	argillaceous gray pottery	Bowl	67.86	16.91	5.55	3.34	1.81	1.51	1.34	0.81	0.62
023YXBP⑤:10-3	argillaceous gray pottery	Bowl	67.59	16.96	5.68	3.39	1.81	1.57	1.29	0.83	0.64
2023YXBP⑤:7-1	argillaceous gray pottery	Storage jar	67.97	17.67	5.38	3.43	1.50	1.43	1.28	0.82	0.29
2023YXBP⑤:7-2	argillaceous gray pottery	Storage jar	68.49	17.43	5.28	3.27	1.52	1.41	1.34	0.77	0.25
2023YXBP⑤:7-3	argillaceous gray pottery	Storage jar	67.93	17.65	5.30	3.39	1.55	1.44	1.41	0.79	0.31
023YXBP⑤:21-1	argillaceous gray pottery	Storage jar	65.22	17.77	7.08	4.31	1.95	1.13	0.96	0.93	0.45
023YXBP⑤:21-2	argillaceous gray pottery	Storage jar	65.20	17.69	7.10	4.28	1.98	1.13	1.04	0.90	0.45
023YXBP⑤:21-3	argillaceous gray pottery	Storage jar	65.19	17.67	7.11	4.32	1.97	1.13	1.02	0.93	0.46
023YXBP⑤:13-1	argillaceous gray pottery	Storage jar	67.02	17.33	5.98	3.66	1.89	1.38	1.21	0.79	0.51
023YXBP⑤:13-2	argillaceous gray pottery	Storage jar	66.92	17.34	6.08	3.67	1.88	1.39	1.24	0.79	0.49
023YXBP⑤:13-3	argillaceous gray pottery	Storage jar	67.04	17.32	6.04	3.66	1.88	1.37	1.16	0.79	0.51

Table 7. Statistical results of the chemical composition of the pottery shards from the Yiting Site in the Yongcheng area.

Statistical Measures	SiO_2	Al_2O_3	Fe_2O_3	K_2O	MgO	CaO	Na_2O	TiO_2
Number of Samples	24	24	24	24	24	24	24	24
Mean	67.29	17.36	5.98	3.49	1.72	1.41	1.25	0.81
Maximum	69.01	18.08	7.11	4.32	1.98	1.81	1.52	0.93
Minimum	65.19	15.88	5.28	3.11	1.41	0.84	0.96	0.70
Standard Deviation	1.30	0.62	0.59	0.34	0.19	0.28	0.16	0.06

As shown in Table 7, the SiO_2 content in pottery from the Biting site in the Yongcheng area ranges from 65.19% to 69.01%, with an average of 67.29% and a standard deviation of 1.30%. The Al_2O_3 content ranges from 15.88% to 18.08%, with an average of 17.36% and a standard deviation of 0.62%. Fe_2O_3 ranges from 5.28% to 7.11%, averaging 5.98% with a standard deviation of 0.59%. The K_2O content falls between 3.11% and 4.32%, with a mean of 3.49% and a standard deviation of 0.34%. MgO ranges from 1.41% to 1.98%, averaging 1.72% with a standard deviation of 0.19%. CaO ranges from 0.84% to 1.81%, with an average of 1.41% and a standard deviation of 0.28%. The Na_2O content ranges from 0.96% to 1.52%, with a mean of 1.25% and a standard deviation of 0.16%. Regarding major elements, the high Al_2O_3 content (up to 17.78%) in pottery from the Biting site suggests the use of high-quality kaolinite or well-weathered sedimentary clay as raw materials. The SiO_2/Al_2O_3 ratio of 3.87 indicates a high degree of weathering of the raw materials. The notable Na_2O content (1.25%) may reflect the presence of sodium feldspar minerals in the clay.

Compared to the Dazhuzhuang and Likou Sites, the pottery from the Biting Site exhibits the highest Al_2O_3 content and the lowest CaO content, although with greater variability. In contrast, the distribution of SiO_2 content is the most stable, suggesting that calcareous tempering materials were rarely used during the firing process.

3.2. *Two-Dimensional Scatter Analysis of XRF Data*

In archeological science research, two-dimensional scatter analysis of elemental concentrations is important for revealing the characteristics of raw materials used in ancient artifacts. Based on the chemical composition of the pottery, this study selects four major oxides—SiO_2, Al_2O_3, CaO, and Fe_2O_3—for two-dimensional scatter analysis [47].

Figure 6a presents a two-dimensional scatter plot of the SiO_2 and CaO content. As shown in the figure, the pottery unearthed from the Dazhuzhuang, Likou, and Biting Sites in the Yongcheng area generally has SiO_2 concentrations between 65.00% and 69.00% and CaO concentrations mostly ranging from 1.10% to 1.85%. The SiO_2/CaO ratios vary widely, ranging from 35.14 to 62.72. Specifically, while the CaO content at the Dazhuzhuang site is relatively concentrated, the SiO_2 distribution is significantly more scattered than at the other two sites, with a standard deviation of up to 1.19. In contrast, artifacts from the Biting Site generally show lower CaO content, whereas those from the Likou Site display higher CaO levels.

Figure 6b illustrates the bivariate distribution of SiO_2 and Al_2O_3 for pottery from the three sites. The analysis reveals that most samples from Dazhuzhuang, Likou, and Biting fall within the ranges of 65.00–69.00% for SiO_2 and 15.70–18.25% for Al_2O_3. The SiO_2/Al_2O_3 ratios range from 3.56 to 4.40, which is significantly lower than the typical 7–10 range found in the loess of the middle Yellow River region [48]. According to geochemical indicators proposed by Liu Dongsheng [49], such low ratios may reflect the following formation mechanisms: (1) The pottery raw materials were derived from highly weathered sediments. (2) The materials formed under humid climatic conditions and underwent intense chemical weathering. (3) There may have been deliberate human processes of raw material selection or levigation. Notably, the three sites exhibit different distribution patterns. Dazhuzhuang shows the greatest elemental dispersion, possibly due to varied raw material sources or processing methods. In contrast, Likou shows the most concentrated distribution, which, along with its significantly larger site scale, indirectly suggests a more standardized raw material selection and processing system. These differences provide important clues for understanding ceramic technological traditions in the Neolithic settlements of the Yongcheng region and offer new perspectives for exploring the relationship between settlement size, hierarchy, and craft production in the Neolithic period.

Figure 6c presents a two-dimensional scatter plot of the SiO_2 and Fe_2O_3 content. The pottery from all three sites shows SiO_2 concentrations mainly between 65.00% and 69.00%, and Fe_2O_3 concentrations between 5.00% and 6.00%, with SiO_2/Fe_2O_3 ratios mostly in the range of 11.5–13. Among them, Dazhuzhuang shows the greatest dispersion in both SiO_2 and Fe_2O_3 contents, while Likou and Biting show more concentrated SiO_2 distributions, though the Fe_2O_3 content still varies considerably.

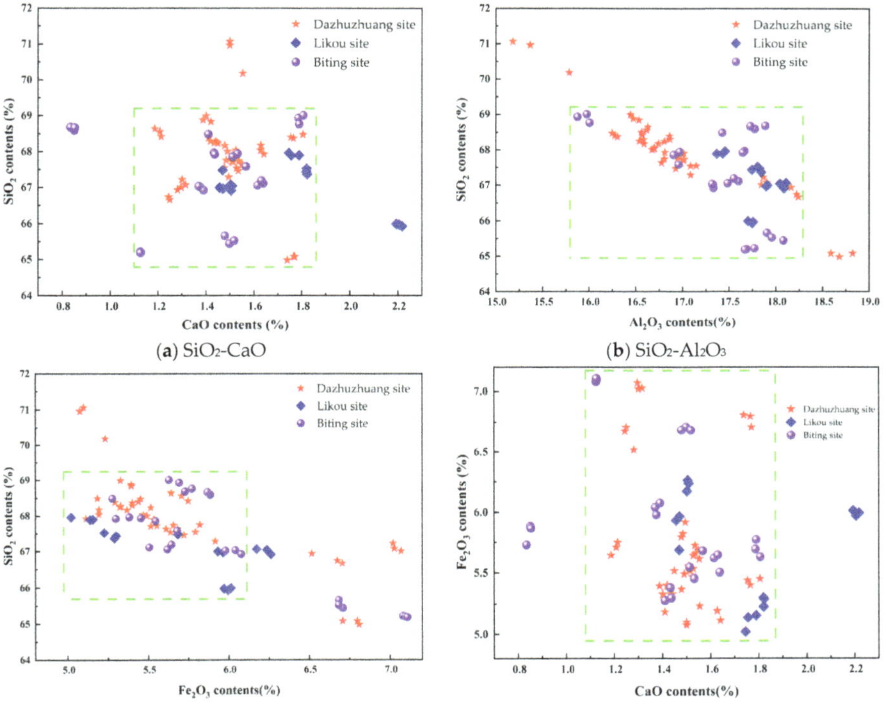

(a) SiO_2-CaO (b) SiO_2-Al_2O_3 (c) SiO_2-Fe_2O_3 (d) Fe_2O_3-CaO

Figure 6. Two-dimensional compositional analysis of oxides in pottery unearthed from the Dazhuzhuang, Likou, and Biting Sites in the Yongcheng District.

Figure 6d shows the two-dimensional scatter plot of the Fe_2O_3 and CaO content. Pottery from all three sites generally has Fe_2O_3 concentrations between 5.00% and 6.00% and CaO concentrations between 1.10% and 1.85%, with Fe_2O_3/CaO ratios ranging from 2.70 to 5.45. From the figure, it can be seen that the Fe_2O_3 content at Dazhuzhuang is more widely dispersed compared to Likou and Biting. In contrast, the CaO content is more tightly clustered, indicating that potters at Dazhuzhuang had stricter control over calcareous components in their raw materials.

In summary, although the Dazhuzhuang, Likou, and Biting Sites are all located in the Yongcheng area of Shangqiu and their pottery dates to the Longshan period, two-dimensional scatter analysis of elemental composition reveals notable differences. These distribution patterns reflect the diversity of ceramic raw materials in the region and suggest differentiated strategies of raw material selection or technological traditions across the sites. Dazhuzhuang, in particular, exhibits the highest degree of elemental variability, indicating a relatively loose standard for raw material selection—likely involving the mixing of various clays or sourcing materials from multiple locations. Notably, no kiln remains have been discovered at Dazhuzhuang despite a surveyed site area of 16,800 square meters. Whether a site of this scale during the Longshan culture possessed the technology and facilities for independent pottery production remains a topic for further investigation.

3.3. *Microcomposition-IR*

Infrared (IR) analysis was conducted on pottery artifacts unearthed from the Dazhuzhuang, Likou, and Biting Sites in the Yongcheng area, with the results shown in Figure 4. Figure 7a,b display the IR spectra of samples from the Dazhuzhuang Site; Figure 7c presents the spectrum from the Likou Site, and Figure 7d shows the spectrum from the Biting Site. The analysis of Figure 7a,b reveals that the IR spectra of artifacts from the Dazhuzhuang Site exhibit notable heterogeneity. Under identical conditions, there are fluctuations in the intensity of major absorption peaks and the widths of absorption bands, indicating a diversity in raw material sources. As shown in Figure 7c, the argillaceous black pottery from the Likou Site exhibits a pronounced absorption trough near 3400 cm^{-1}, while the argillaceous gray pottery also presents a distinct absorption band in the same region. This feature is typically attributed to the stretching vibrations of O–H bonds, which likely originate from the presence of adsorbed or structural water within the materials. Further analysis of the IR data from the Biting Site (Figure 7d) shows that some distinctions are evident, while peak positions across different vessel types largely coincide. Jars and tripods exhibit stronger peaks at around 1000 cm^{-1}, whereas basins and bowls show weaker peaks at the same position, but with significantly broadened absorption bands. These variations reflect differences in raw material composition, forming techniques, and thermal conditions tailored to the functional requirements of different vessel types [50].

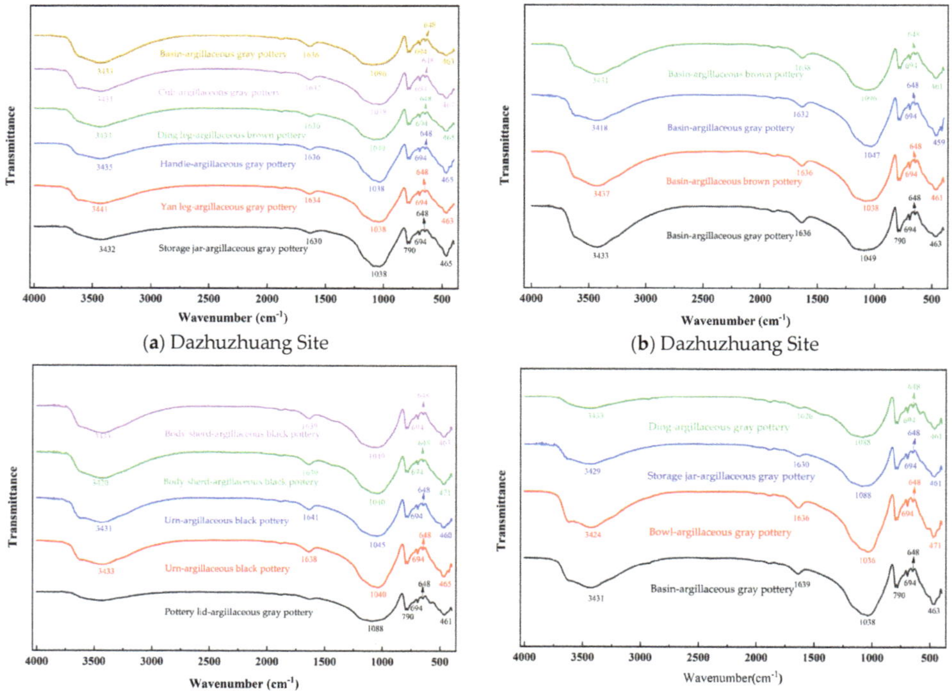

(**a**) Dazhuzhuang Site (**b**) Dazhuzhuang Site

(**c**) Likou Site (**d**) Biting Site

Figure 7. Infrared spectral analysis of pottery artifacts unearthed from the Dazhuzhuang, Likou, and Biting Sites in the Yongcheng area.

3.4. Phase Analysis Using XRD

XRD analysis was conducted on pottery artifacts unearthed from the Dazhuzhuang, Likou, and Biting Sites in the Yongcheng area, with the results shown in Figure 8. Figure 8a,b display the XRD spectra of samples from the Dazhuzhuang Site; Figure 8c presents data from the Likou Site, and Figure 8d shows results from the Biting Site. The interpretation of the XRD spectra indicates a high degree of consistency in the primary mineral composition of pottery from all three sites. Common characteristic phases include mica ($KAl_2(AlSi_3O_{10})(OH)_2$), mullite ($3Al_2O_3{\cdot}2SiO_2$), quartz ($SiO_2$), and hematite ($Fe_2O_3$). Kaolinite, which was a key raw material for ceramic bodies in northern China, undergoes significant phase transformations during firing: At 450–650 °C, it loses structural water ($Al_2Si_2O_5(OH)_4 \rightarrow Al_2Si_2O_7 + 2H_2O$), forming amorphous metakaolinite. Beyond 900 °C, metakaolinite decomposes into free Al_2O_3 and SiO_2 ($Al_2Si_2O_7 \rightarrow Al_2O_3 + 2SiO_2$) and recrystallizes into mullite around 1100 °C, with excess SiO_2 remaining as quartz. The pronounced intensity of quartz peaks in the XRD spectra may stem from crystallized free SiO_2 from the thermal decomposition of kaolinite and the partial transformation of original quartz particles. Although the characteristic peaks of mullite are relatively weak, the combined presence of mullite and quartz forms the ceramic skeleton structure [51]. The residual presence of hematite further confirms that firing temperatures commonly exceeded 900 °C [52]. This aligns with the advanced high-temperature firing techniques of the Late Neolithic period in the Yellow River basin.

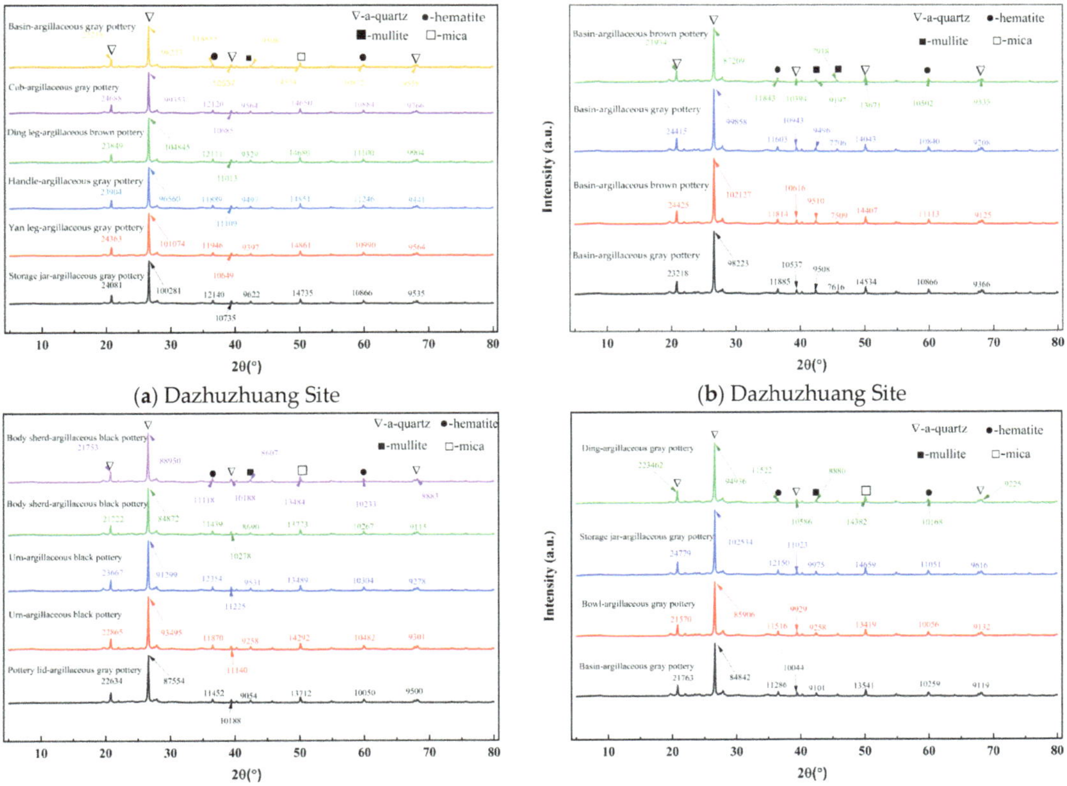

(**a**) Dazhuzhuang Site (**b**) Dazhuzhuang Site (**c**) Likou Site (**d**) Biting Site

Figure 8. XRD pattern analysis of pottery artifacts unearthed from the Dazhuzhuang, Likou, and Biting Sites in the Yongcheng area.

Figure 8a,b show that due to the heterogeneous nature of raw materials used at the Dazhuzhuang Site, no clear patterns emerge in the XRD spectra. In Figure 8c, the clayey black pottery sample from the Likou Site exhibits a significantly stronger quartz peak at 2θ = 26.6° compared to contemporaneous gray pottery. Generally, black pottery shows stronger quartz peaks than gray pottery. Notably, different vessel types also show variability; urns have stronger quartz peaks than body sherds. Further analysis of the XRD data from the Biting Site, as shown in Figure 8d, reveals that under similar clayey gray pottery conditions, jars exhibit distinct phase characteristics: (1) Quartz peaks (2θ = 26.6°) are significantly stronger than those of tripods, bowls, and basins. (2) The intensity of the hematite peak (2θ = 33.2°) also increases. (3) A clear positive correlation is observed between the intensities of quartz and hematite peaks. These findings suggest that as representative storage vessels, jars may have incorporated more quartz temper to enhance their mechanical strength. Meanwhile, the elevated hematite content could be associated with specific surface treatment techniques used for jars, which may improve their density and extend their service life.

In conclusion, XRD phase analysis of pottery from the Dazhuzhuang, Likou, and Biting Sites reveals important characteristics of ceramic production techniques. All three sites had mastered kiln technologies capable of reaching high temperatures (900–1100 °C). The residual hematite peak at 2θ = 33.2° suggests that oxidizing atmospheres were deliberately maintained during firing, implying that temperature control techniques were likely adapted to suit different vessel types. The preservation of the mica phase (2θ = 8.8°) indicates that the firing temperature did not reach full vitrification levels (<1200 °C).

3.5. Microstructure Analysis via SEM

3.5.1. Dazhuzhuang Site

Figure 9 presents SEM images of pottery from the Dazhuzhuang Site. The object types from Figure 9a–e are a storage jar, a Yan leg, a handle, a cup, a basin, and a ding leg, respectively. In terms of pottery fabrics, except for the ding leg, which is made of argillaceous brown pottery, all other samples are made of argillaceous gray pottery. From Figure 9a–f, it can be observed that at the level of the clay matrix, clay aggregates and partially vitrified clay platelets are commonly present within the matrix, displaying microstructural characteristics typical of the Longshan Culture. Notable technological variations are evident among different vessel types. In the argillaceous gray pottery samples (Figure 9a–e), the clay platelets generally exhibit layered accumulation yet show vessel-specific differences: The storage jar (Figure 9a) exhibits a typical heterogeneous structure with numerous partially vitrified clay platelets (5–15 μm) and aggregates (20–50 μm in diameter), suggesting a relatively low firing temperature [53]. The Yan leg (Figure 9b) shows unevenly distributed and loosely packed inclusions, which may be related to functional adjustments in manufacturing. The handle (Figure 9c) exhibits not only the layered platelet structure but also a small number of rod-shaped features, likely mullite whiskers formed at high temperatures (>1000 °C), indicating targeted high-temperature treatment for this part. The cup (Figure 9d) displays well-developed clay mineral growth with denser layered structures. The basin (Figure 9e) and the ding leg (Figure 9f) possess orderly layered structures and uniformly distributed inclusions. Although the argillaceous brown pottery ding leg differs chemically from the argillaceous gray pottery basin, their microstructures are highly similar. These findings suggest the following: (1) Potters at the Dazhuzhuang Site had mastered techniques to adjust manufacturing processes according to the functional requirements of different vessel types. (2) The appearance of mullite whiskers indicates that localized high-temperature treatments were employed to enhance the mechanical properties of critical components. (3) Despite differences in chemical composition, argillaceous brown pottery and gray pottery likely shared similar forming techniques, with variations mainly in firing atmosphere control.

(**a**) Storage jar-argillaceous gray pottery

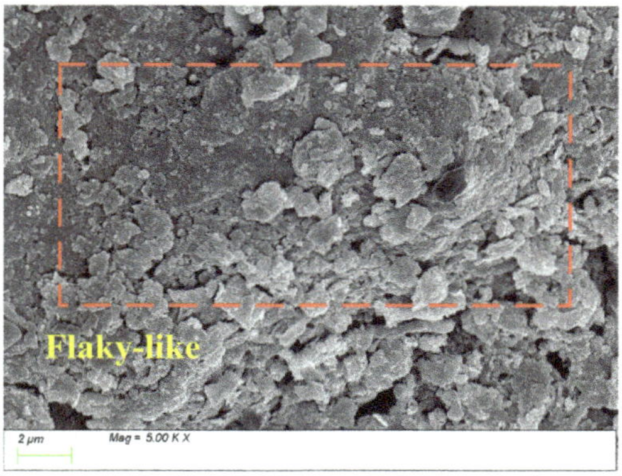

(**b**) Yan leg-argillaceous gray pottery

(**c**) Handle-argillaceous gray pottery

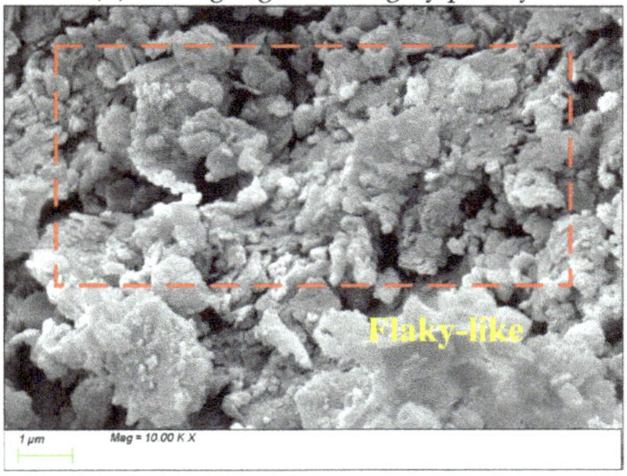

(**d**) Cup-argillaceous gray pottery

(**e**) Basin-argillaceous gray pottery

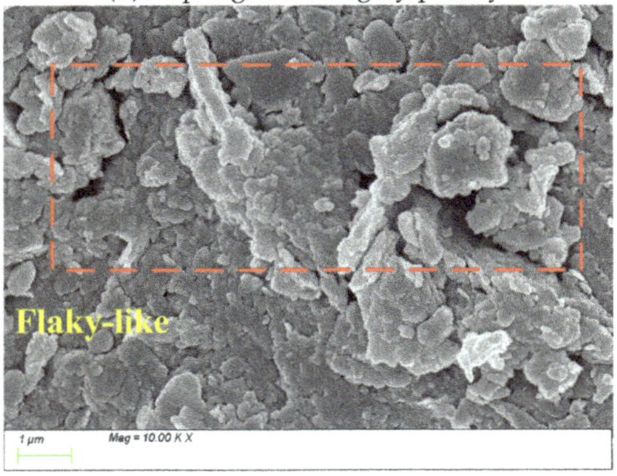

(**f**) Ding leg-argillaceous brown pottery

Figure 9. SEM images of pottery sherds unearthed from the Dazhuzhuang Site.

3.5.2. Likou Site

Figures 10–12 present SEM images of pottery sherds excavated from the Likou Site. Figure 7 shows a sample identified as a pottery lid that is made of argillaceous gray pottery. A comparative analysis reveals significant differences in the microstructural characteristics of the artifacts from the Likou Site compared to those from Dazhuzhuang Site. As shown in Figure 10a, although similar to Dazhuzhuang in exhibiting flaky and platy structures, the pottery lid from Likou displays a notable decrease in quantity but an increase in the size of the platy phases. The appearance of these large platy structures suggests that the clay minerals used as raw materials had a higher degree of crystallinity and that the firing process was controlled more precisely [54], potentially involving specialized forming techniques. In Figure 10b, the matrix predominantly exhibits a porous structure, with significant variability in pore size distribution. Combined with the elevated CaO content identified in the compositional analysis, it is inferred that a calcium-based flux was intentionally added during the firing process to promote foaming and enhance porosity [55].

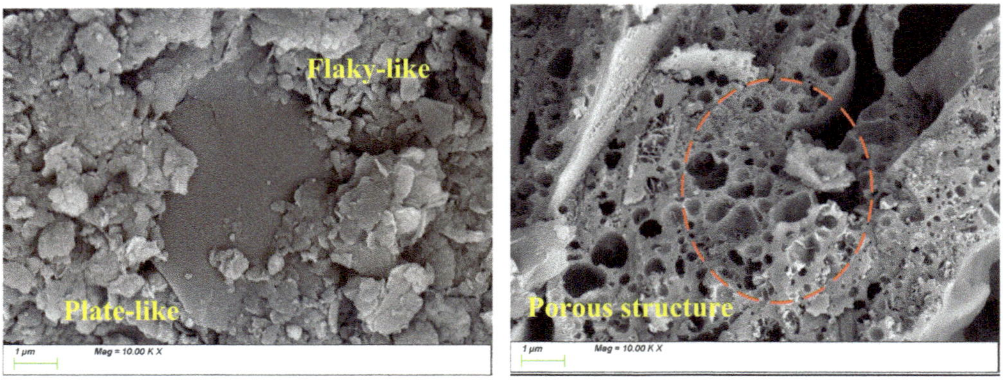

(**a**) Pottery lid-argillaceous gray pottery (**b**) Pottery lid-argillaceous gray pottery

Figure 10. An SEM image of a pottery lid unearthed from the Likou Site.

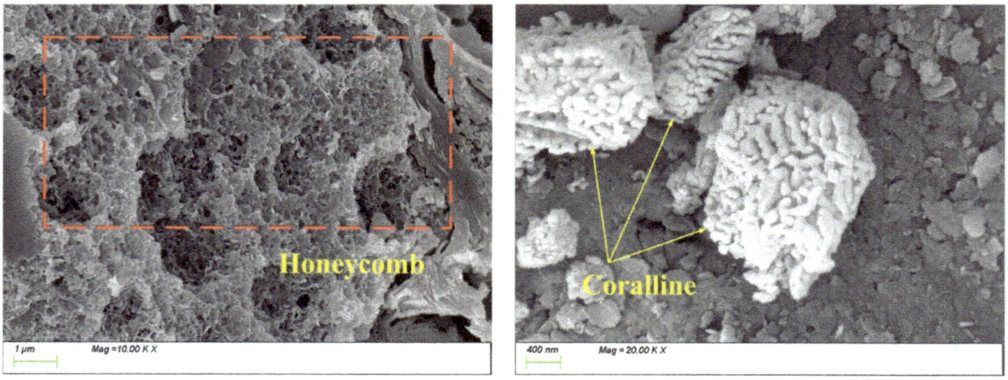

(**a**) Urn-argillaceous black pottery (**b**) Urn-argillaceous black pottery

Figure 11. An SEM image of an urn unearthed from the Likou Site.

The vessel type in Figure 11 is an urn that is made of argillaceous black pottery. Microstructural and characteristic phase analysis of Figure 11 reveals that while the urn shares a porous structure similar to that of the gray pottery storage jar, it exhibits more densely vitrified regions with fewer pores (Figure 11a). As shown in Figure 11b, coral-like accumulations are observed within the matrix. Energy-dispersive spectroscopy (EDS) analysis indicates that the Fe content in the coral-like regions (8–10%) is significantly higher than that of the surrounding matrix (5–6%), confirming a segregation effect of iron. The formation mechanisms of these special coral-like structures may involve (1) phase separation and recrystallization of iron-bearing minerals at high temperatures, (2) compositional segregation within the melt, and (3) non-equilibrium solidification induced by specific cooling rates.

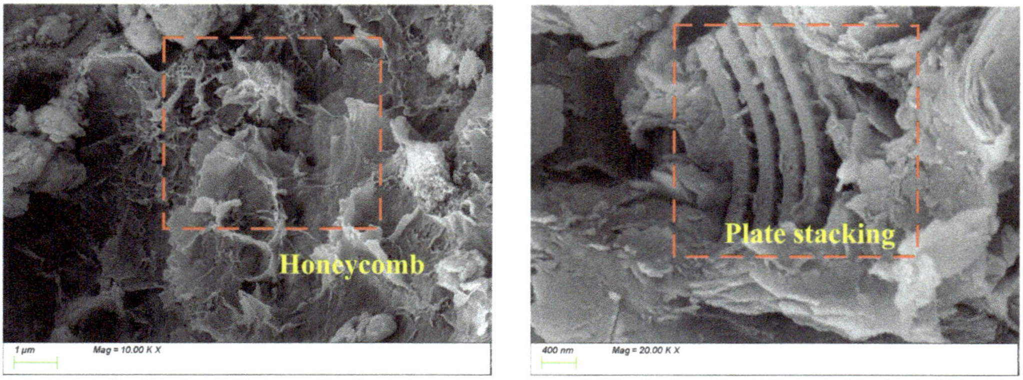

(**a**) Body sherd-argillaceous black pottery (**b**) Body sherd-argillaceous black pottery

Figure 12. SEM image of a body sherd unearthed from the Likou Site.

The vessel type in Figure 12 is a body sherd from a pottery storage jar that is made of argillaceous black pottery. As shown in Figure 12a, the matrix structure exhibits a honeycomb-like growth pattern. Figure 12b reveals the presence of numerous plate-like products with a preferred orientation within the matrix. Energy-dispersive spectroscopy (EDS) analysis indicates that the plate-like phases are rich in Al_2O_3 (22 $\pm$ 2%) and exhibit a dense structure [56], while the surrounding matrix is primarily composed of SiO_2 (68 $\pm$ 3%). This compositional differentiation at the microscale may result from (1) preferentially oriented growth of kaolinite-derived phases at high temperatures, (2) phase separation induced by compositional gradients in the melt, and (3) non-equilibrium crystallization triggered by controlled cooling processes.

In summary, comparative analysis indicates that compared to argillaceous gray pottery, argillaceous black pottery exhibits more densely vitrified regions, fewer pores, and a reduced pore size distribution. These structural features confer improved mechanical strength and thermal stability to the black pottery. Collectively, the observed microstructural characteristics suggest that the production of black pottery at the Likou Site involved sophisticated firing control techniques, reflecting a highly advanced material design achieved during the late Longshan Culture period.

3.5.3. Biting Site

Figure 13 presents SEM images of pottery from the Biting Site. The vessel types from Figure 13a–d are a basin, a bowl, a storage jar, and a Ding, all made of argillaceous gray pottery. Although all samples are classified as argillaceous gray pottery, significant technological differences are observed among different vessel types at the microstructural level [57]. The basin (Figure 13a) exhibits a distinctive "block-plate" synergistic accumulation structure, where blocky phases and thin plate-like phases interlock to form a network. The bowl (Figure 13b) displays a typical clay platelet stacking pattern with moderate porosity, balancing requirements for lightweight construction and thermal insulation. The storage jar (Figure 13c) also shows a platelet stacking structure but with a denser arrangement and a high degree of platelet orientation, which is favorable for enhancing mechanical strength and thermal stability. In contrast, the Ding (Figure 13d) presents the densest matrix structure, where well-developed plate-like products form strong interfacial bonding with the matrix. The extensive growth and tight integration of the plate-like phases contribute significantly to improved mechanical properties and thermal performance.

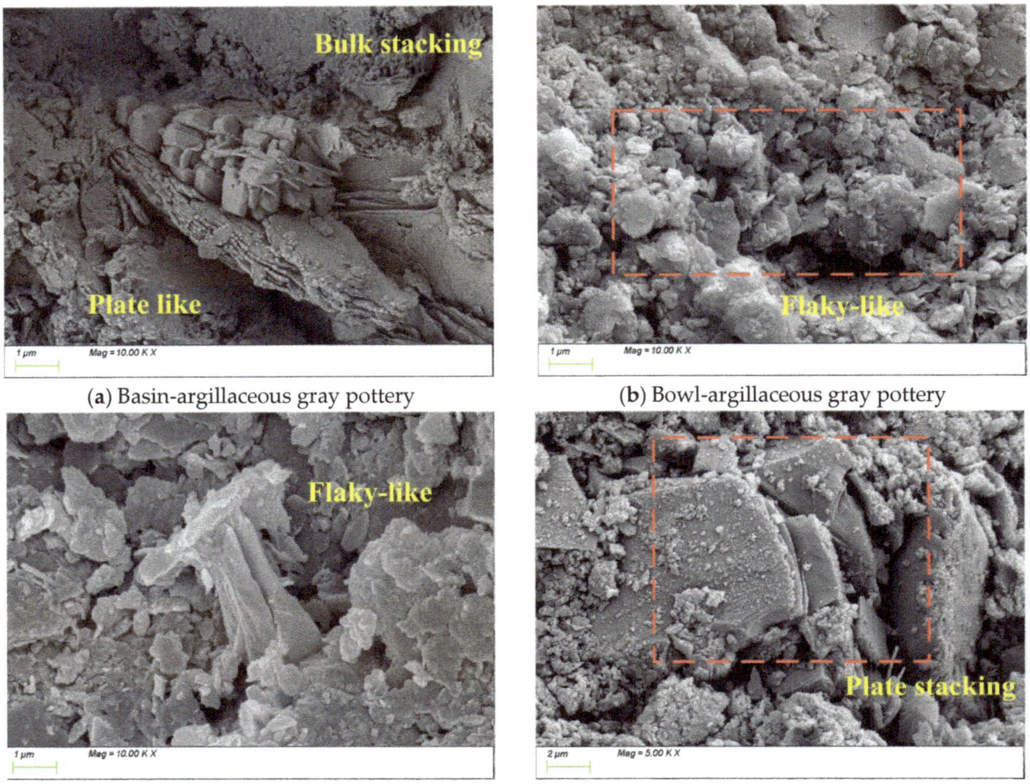

(**a**) Basin-argillaceous gray pottery (**b**) Bowl-argillaceous gray pottery

(**c**) Storage jar-argillaceous gray pottery (**d**) Ding-argillaceous gray pottery

Figure 13. SEM image of pottery unearthed from the Biting Site.

In summary, these microstructural differences indirectly reflect the potters' profound understanding of the functional requirements of different vessel types and their mastery of targeted raw material preparation and forming techniques. Notably, the densification observed in the storage jar and ding indicates specialized technological adaptations for cooking and storage vessels. In contrast, the relatively more porous structures of the bowl and basin reflect different functional priorities for daily use items. This precise matching of material properties, functional demands, and technological choices demonstrates an advanced understanding of performance optimization and reveals a highly developed capability to systematically integrate material characteristics with vessel functions. From an archeological perspective, such a highly specialized technical system marks the transition of Neolithic pottery production from empirical practices to a more scientific approach, providing a significant example for understanding the technological evolution of prehistoric craft industries.

4. Conclusions

This study conducted a comprehensive multidisciplinary analysis of pottery unearthed from three Longshan Culture sites—Dazhuzhuang, Likou, and Biting—in the Yongcheng District of Henan Province. By integrating modern analytical techniques such as X-ray fluorescence (XRF), X-ray diffraction (XRD), infrared spectroscopy (IR), and scanning electron microscopy with energy-dispersive spectroscopy (SEM-EDS), together with archeological typology and materials science approaches, the following major conclusions were drawn:

(1) Raw Material Sources:

Chemical composition analysis reveals variations in raw material selection among the sites. The pottery from the Dazhuzhuang Site exhibits significant fluctuations in SiO_2 content, indicating flexible raw material choices. The Likou Site pottery shows high Al_2O_3 and CaO content, suggesting the use of high-quality kaolinite clay and the addition of calcium-based fluxes. In contrast, the Biting Site pottery displays more weathered raw materials and the lowest CaO content among the three sites.

(2) Technological Aspects:

Pottery from the Dazhuzhuang Site, predominantly argillaceous gray pottery, exhibits an uneven distribution of clay platelets and aggregates at the microstructural level, indicating relatively low firing temperatures (around 900 °C) and less controlled manufacturing processes. The Likou Site pottery, primarily argillaceous black pottery, features dense vitrified structures and iron segregation phenomena, suggesting the application of high-temperature reducing atmospheres and a high degree of technological standardization. At the Biting Site, the vessel-specific microstructural variations (e.g., dense laminar arrangements in storage jars) reflect precise adjustments in raw material preparation and firing techniques tailored to the functional demands of different pottery forms.

(3) Cultural Significance:

The high Al_2O_3 content in the black pottery from the Likou Site is comparable to that of the Shandong Longshan Culture, implying possible technological interactions between regions. In contrast, the gray pottery from the Dazhuzhuang and Biting Sites reflects localized production traditions.

(4) Research Implications:

This study highlights the technological diversity and complexity of pottery production in the Yongcheng area during the Longshan period. It provides critical empirical evidence for exploring the technological evolution, raw material strategies, and functional adaptations of late Neolithic craft industries. Future research could benefit from comparative studies across broader site distributions and incorporating advanced techniques, such as stable isotope tracing, to identify clay sources accurately. Ultimately, reconstructing ancient pottery production chains will enable a more comprehensive understanding of the technological evolution of prehistoric crafts and their role in early social complexity and civilization development. Advancing these research directions will help uncover the intrinsic mechanisms of technological innovation and its significance in the origins of early civilizations.

Author Contributions: Conceptualization, L.X., G.Z. and L.J.; methodology, G.Z. and Y.L.; validation, L.X. and G.Z.; formal analysis, G.Z., Y.L. and L.X.; investigation, Y.L., G.Z., J.L. and L.J.; Data curation, L.X., Y.L. and J.L.; writing—original draft preparation, L.X.; writing—review and editing, L.X., Y.L. and G.Z.; funding acquisition, L.X. All authors have read and agreed to the published version of the manuscript.

Funding: This research was funded by the Natural Science Foundation of Henan (grant number 242300421462), the Science and Technology Development Fund of the Yellow River Institute of Hydraulic Research (202112), and the Henan Provincial Special Fund for Cultural Heritage Conservation (Yu Wen Wu Han [2021] No. 114).

Institutional Review Board Statement: Not applicable.

Informed Consent Statement: Not applicable.

Data Availability Statement: The original contributions presented in the study are included in the article; further inquiries can be directed to the corresponding author.

Conflicts of Interest: The authors declare no conflicts of interest.

References

1. Li, J.D. Investigation of Yongcheng in Shangqiu, Eastern Henan and Small-Scale Excavations at Zaolütai, Hegu Dui and Caoq *J. Chin. Archaeol.* **1947**, *2*, 83–120+7–8.
2. Zheng, Q.S. A Preliminary Study on the Longshan Culture in Eastern Henan and Its Origins. *Cult. Relics Cent. China* **199** 46–52.
3. Henan Team of the Institute of Archaeology, Chinese Academy of Social Sciences; Cultural Relics Administration of Shang Region. *Excavation Report of Wangyoufang Site in Yongcheng, Henan. Collection of Archaeological Studies*; China Social Sciences Pr Beijing, China, 1981; Volume 5.
4. Henan Team of the Institute of Archaeology, Chinese Academy of Social Sciences; Cultural Relics Administration of Shang Region. Summary of Archaeological Work in Eastern Henan. *Archaeology* **1981**, *5*, 385–397+481.
5. Institute of Archaeology, Chinese Academy of Social Sciences; Peabody Museum, Harvard University. Study on Longsl Culture at Shantaishi. *Archaeology* **2010**, *10*, 52–60.
6. Department of Archaeology, School of History, Zhengzhou University. Brief Report on Archaeological Survey in the Shang Area, Eastern Henan. *Huaxia Archaeol.* **2005**, *2*, 13–27. [CrossRef]
7. Zhao, J.Y. Re-discussion on the Formation of the Wangyoufang Type in Eastern Henan. *Huaxia Archaeol.* **2024**, *1*, 77–86. [Cross
8. Gao, T.L.; Meng, F.R. A Preliminary Study on the "Wangwan Type" of Longshan Culture in Henan. *Cult. Relics Cent. China* **1** 2, 52–60.
9. Liang, S. Longshan Culture—The Prehistoric Stage of Chinese Civilization. *Proc. Sixth Pac. Sci. Congr.* **1939**, *4*.
10. An, Z.M. A Preliminary Discussion on the Neolithic Cultures in the Yellow River Basin. *Archaeology* **1959**, *10*, 559–565.
11. Yang, Z.F.; Wang, S.L. A Tentative Discussion on Longshan Culture. *Archaeology* **1963**, *7*, 377–380.
12. Luan, F.S. A Comprehensive Study on Haidai Longshan Culture. *Dongyue Trib.* **1995**, *3*, 77–82.
13. Li, M.S.; Huang, S.Y. Testing and Analysis of Pottery Shards from the Taosi Site, Xiangfen, Shanxi. *Archaeology* **1993**, *2*, 176–1
14. Chen, Q.Q.; Yang, Y.Z.; Zhang, J.Z.; Cui, W. WDXRF Analysis on the Chemical Composition of Early Neolithic Pottery Uneart from Xiaohuangshan, Zhejiang and Jiahu, Henan. *Spectrosc. Spectr. Anal.* **2011**, *31*, 3140–3144.
15. Gao, S.L.; Xiong, Z.L.; Zhang, T.X. Composition Analysis and Study of Hongshan Culture Pottery from the Niuheliang Site. *N Cult. Relics* **2021**, *6*, 80–91. [CrossRef]
16. Migliori, A.; Bonanni, P.; Carraresi, L.; Grassi, N.; Mandò, P.A. A novel portable XRF spectrometer with range of detect extended to low-Z elements. *X-Ray Spectrom.* **2011**, *40*, 107–112. [CrossRef]
17. Wang, W.; Fu, L.; Huang, Y.; Xu, W. Compositional analysis of black-slipped pottery from the Hulushan and Maoershan kiln s Insights into social complexity in northern Fujian. *J. Archaeol. Sci. Rep.* **2023**, *51*, 104149. [CrossRef]
18. Xenogiannopoulou, E.; Andreouli, C.; Stournaras, C. Application of LIBS Technique for the Compositional Analysis of an A Black Pottery. *J. Nano Res.* **2009**, *868*, 61–70. [CrossRef]
19. Pierce, D.E.; Mehta, H.H.; Ferguson, J.R. Local pottery, hybrid identities: Exploring the Zapo-Teotihuacáno community a Tesoro, Hidalgo through ceramic compositional analysis. *J. Archaeol. Sci. Rep.* **2022**, *45*, 103598. [CrossRef]
20. Forte, V.; Tarquini, O.; Botticelli, M.; Medeghini, L. The technology of Copper Age funerary pottery from Central Italy: integrated study of compositional analyses and manufacturing traces. *Archaeometry* **2020**, *62*, 712–730. [CrossRef]
21. Capobianco, G.; Pelosi, C.; Agresti, G.; Bonifazi, G.; Santamaria, U.; Serranti, S. X-ray fluorescence investigation on yel pigments based on lead, tin and antimony through the comparison between laboratory and portable instruments. *J. Cult. H* **2018**, *29*, 19–29. [CrossRef]
22. Marques, R.; Rodrigues, A.L.; Russo, D.; Gméling, K.; Valera, A.C.; Dias, M.I.; Prudêncio, M.I.; Basílio, A.C.; Fernandes, P.G.; R F. Fingerprinting Ceramics from the Chalcolithic Santa Vitória Enclosure (SW Iberia). *Minerals* **2024**, *14*, 399. [CrossRef]
23. Mutin, B.; Minc, L. The formative phase of the Helmand Civilization, Iran and Afghanistan: New data from compositic analysis of ceramics from Shahr-i Sokhta, Iran. *J. Archaeol. Sci. Rep.* **2019**, *23*, 881–899. [CrossRef]
24. Andric, V.; Gajic-Kvascev, M.; Korolija Crkvenjakov, D.; Maric-Stojanovic, M.; Gadzuric, S. Evaluation of pattern recognit techniques for the attribution of cultural heritage objects based on the qualitative XRF data. *Microchem. J.* **2021**, *167*, 106 [CrossRef]
25. Kvaščev, G.M.; Mladenović, O.; Milojević, P.; Bulatović, A. Where Did Vessels Come from? A Study of Pottery Provenance fr the Site of Velika Humska Čuka, Serbia. *Materials* **2025**, *18*, 1083. [CrossRef]

Teodorescu, L.; Ben Amara, A.; Cantin, N.; Chapoulie, R.; Ducu, C.; Ciucă, S.; Tulugea, C.; Terteci, C.; Abrudeanu, M. Characterization of Archaeological Artefacts Using Methods Specific to Materials Science: The Case Study of Dacian Ceramics from 2nd c. BC to 1st c. AD. *Materials* **2021**, *14*, 3908. [CrossRef]
Colomban, P.; Franci, G.S.; Ngo, A.-T.; Gallet, X. Non-Invasive Raman and XRF Study of Mīnā'ī Decoration, the First Sophisticated Painted Enamels. *Materials* **2025**, *18*, 575. [CrossRef]
Olin, J.S.; Blackman, M.J.; Mitchem, J.E.; Waselkov, G.A. Compositional Analysis of Glazed Earthenwares from Eighteenth-Century Sites on the Northern Gulf Coast. *Hist. Archaeol.* **2002**, *36*, 79–96. [CrossRef]
Chiriu, D.; Pisu, F.A.; Ricci, P.C.; Carbonaro, C.M. Application of Raman Spectroscopy to Ancient Materials: Models and Results from Archaeometric Analyses. *Materials* **2020**, *13*, 2456. [CrossRef]
Sarathi, D.P.; Acharya, A.R. Application of Instrumental Neutron Activation Analysis for Chemical Composition Analysis of Ancient Potteries from Buddhist Sites of Andhra Pradesh: Part I. *J. Nucl. Radiochem. Sci.* **2008**, *9*, 7–12.
Galli, A.; Caccia, M.; Alberti, R.; Bonizzoni, L.; Aresi, N.; Frizzi, T.; Bombelli, L.; Gironda, M.; Martini, M. Discovering the material palette of the artist: A p-XRF stratigraphic study of the Giotto panel 'God the Father with Angels'. *X-Ray Spectrom.* **2017**, *46*, 435–441. [CrossRef]
Gajic-Kvascev, M.; Andric, V.; Jancic-Heinemann, R.; Mladenovic, O.; Bulatovic, A. Comparison of Quantitative and Qualitative EDXRF Analysis for Provenance Study of Archaeological Ceramics. *Materials* **2024**, *17*, 3725. [CrossRef] [PubMed]
Merk, V.; Berkh, K.; Rammlmair, D.; Pfeifer, L. Chemical and Mineralogical Analysis of Samples Using Combined LIBS, Raman Spectroscopy and μ-EDXRF. *Minerals* **2023**, *13*, 729. [CrossRef]
Forster, N.; Grave, P.; Vickery, N.; Kealhofer, L. Non-destructive analysis using PXRF: Methodology and application to archaeological ceramics. *X-Ray Spectrom.* **2011**, *40*, 389–398. [CrossRef]
Bonizzoni, L. ED-XRF analysis for Cultural Heritage: Is quantitative evaluation always essential? *J. Phys. Conf. Ser.* **2015**, *630*, 012001. [CrossRef]
Orsilli, J.; Galli, A. Angle-Dependent XRF Analyses: Pros and Cons of a Novel Technique in the Field of Cultural Heritage. *Sustainability* **2024**, *16*, 1460. [CrossRef]
Ruschioni, G.; Malchiodi, D.; Zanaboni, A.M.; Bonizzoni, L. Supervised learning algorithms as a tool for archaeology: Classification of ceramic samples described by chemical element concentrations. *J. Archaeol. Sci. Rep.* **2023**, *49*, 103995. [CrossRef]
Visco, G.; Avino, P. Employ of multivariate analysis and chemometrics in cultural heritage and environment fields. *Environ. Sci. Pollut. Res.* **2017**, *24*, 13863–13865. [CrossRef]
Finley, B.J.; Scheiber, L.L.; Ferguson, R.J. Compositional analysis of Intermountain Ware pottery manufacturing areas in western Wyoming, USA. *J. Archaeol. Sci. Rep.* **2018**, *18*, 587–595. [CrossRef]
Gjesfjeld, E. The compositional analysis of hunter-gatherer pottery from the Kuril Islands. *J. Archaeol. Sci. Rep.* **2018**, *17*, 1025–1034. [CrossRef]
Monette, Y.; Richer-LaFlèche, M.; Moussette, M.; Dufournier, D. Compositional analysis of local redwares: Characterizing the pottery productions of 16 workshops located in Southern Québec dating from late 17th to late 19th-century. *J. Archaeol. Sci.* **2006**, *34*, 123–140. [CrossRef]
Li, T.; Yao, S.; He, L.; Yu, X.; Shan, S. Compositional study of household ceramic assemblages from a Late Neolithic (5300–4500 cal BP) earthen walled-town in the middle Yangtze River valley of China. *J. Archaeol. Sci. Rep.* **2021**, *39*, 103159. [CrossRef]
Zvietcovich, F.; Navarro, L.; Saldana, J.; Castillo, L.J.; Castaneda, B. A novel method for estimating the complete 3D shape of pottery with axial symmetry from single potsherds based on principal component analysis. *Digit. Appl. Archaeol. Cult. Herit.* **2016**, *3*, 42–54. [CrossRef]
Moradi, H.; Dadian, H.S.; Ramli, Z.; Rahman, N.H.S.N.A. Compositional Analysis of the Pottery Shards of Shahr-I Sokhta, South Eastern Iran. *Res. J. Appl. Sci. Eng. Technol.* **2013**, *6*, 654–659. [CrossRef]
Ma, Q.; Yan, A.; Hu, Z.; Li, Z.; Fan, B. Principal component analysis and artificial neural networks applied to the classification of Chinese pottery of neolithic age. *Anal. Chim. Acta* **2000**, *406*, 247–256. [CrossRef]
Bray, T.L.; Minc, L.D.; Ceruti, M.C.; Chávez, J.A.; Perea, R.; Reinhard, J. A compositional analysis of pottery vessels associated with the Inca ritual of capacocha. *J. Anthropol. Archaeol.* **2004**, *24*, 82–100. [CrossRef]
Zhao, W.J.; Lu, X.K.; Li, G.X.; Guo, M.; Xie, J.Z.; Gao, Z.Y.; Sun, X.M.; Guo, M.S.; Chneg, H.S.; Zhang, B. Major chemical composition of celadon glaze from Qingliangsi Kiln and Zhanggongxiang Kiln. China Science. *Ser. G* **2005**, *35*, 167–175.
Liu, D.S. *Loess Deposits in China (Explanatory Notes for the Map of Loess Distribution in China)*; Science Press: Beijing, China, 1965; p. 219.
Liu, D.S. *Loess and Environment*; Science Press: Beijing, China, 1985; p. 237.
Vlase, D.; Vlase, T.; Bunoiu, M.; Sfîrloagă, P.; Ursuţ, A.-G.; Vlase, G. Characteristics of late Neolithic pottery of the Zau culture: Analysis of pottery samples from Iernut-Site II (Mures County, Romania). *J. Therm. Anal. Calorim.* **2022**, *147*, 5351–5363. [CrossRef]

51. Zvereva, I.A.; Kuznetsov, V.M.; Zhukov, Y.M.; Mazurkevich, A.N.; Dolbunova, E.V. Thermal analysis and multi-analyt comparison of samples of Neolithic ceramics from Dnepr–Dvina and Low Don regions. *J. Therm. Anal. Calorim.* **2019**, 1879–1886. [CrossRef]
52. Lu, X.K.; Li, W.D.; Luo, H.J.; Xu, H.; Zhao, H.T.; Yuan, J. Study on White Pottery, Stamped Hard Pottery and Proto-Porce Unearthed from the Erlitou Site. *Archaeology* **2012**, *10*, 89–96.
53. Druc, I.; Underhill, A.; Wang, F.; Luan, F.; Lu, Q.; Hu, Q.; Guo, M.; Liu, Y. Late Neolithic white wares from southeastern Shand China: The tricks to produce a white looking pot with not much kaolin. Results from petrography, XRD and SEM-EDS analy *J. Archaeol. Sci. Rep.* **2021**, *35*, 102673. [CrossRef]
54. Guo, M.J.; Wen, D.A.; Luan, F.S.; Zhu, T.Q.; Druc, I.C.; Wan, J.L.; Lu, Q.Y.; Sun, B.; Liang, Z.H. Preliminary Study on Provenance of Longshan Culture White Pottery: A Case Study Based on White Pottery Samples from the Liangchengzhen and Others. *Archaeology* **2024**, *2*, 89–106+2.
55. Wu, H.; Han, J.; Zhang, X.; Cui, J.; Chen, G. Scientific research on white potteries from the Nanzuo site in Qingyang, Ga province. *Archaeometry*, 2024; *Early View*. [CrossRef]
56. Lu, X.K.; Fang, Y.M.; Li, W.D. Scientific and Technological Study of Longshan Culture White Pottery from the Wadian Site. *Hu Archaeol.* **2015**, *4*, 118–124.
57. Chaves, R.C.; Veiga, J.P.; Monge, S.A. Characterization of Chalcolithic Ceramics from the Lisbon Region, Portugal: An Arch metric Study. *Heritage* **2022**, *5*, 2422–2443. [CrossRef]

Disclaimer/Publisher's Note: The statements, opinions and data contained in all publications are solely those of the individ author(s) and contributor(s) and not of MDPI and/or the editor(s). MDPI and/or the editor(s) disclaim responsibility for any injur people or property resulting from any ideas, methods, instructions or products referred to in the content.

materials

MDPI

icle

/here Did Vessels Come from? A Study of Pottery Provenance ·om the Site of Velika Humska Čuka, Serbia

ja Gajić-Kvaščev [1,*], Ognjen Mladenović [2], Petar Milojević [2] and Aleksandar Bulatović [2]

[1] Vinča Institute of Nuclear Sciences, National Institute of the Republic of Serbia, University of Belgrade, 11001 Belgrade, Serbia
[2] Institute of Archaeology, National Institute of the Republic of Serbia, 11000 Belgrade, Serbia; mladenovic40@gmail.com (O.M.); pertinax1983@gmail.com (P.M.); abulatovic3@gmail.com (A.B.)
* Correspondence: gajicm@vin.bg.ac.rs

Abstract: The archaeological materials from the Velika Humska Čuka site on the northern fringe of the Niš Basin in southeastern Serbia were analyzed to reveal the provenance of ceramics and other artifacts. This study focused on the elemental analysis of 61 samples, including local clay pits, potsherds, and whole vessels. Samples were chosen based on stylistic and typological characteristics to distinguish local and "foreign" pottery. Elemental analysis was conducted using energy-dispersive X-ray fluorescence (EDXRF) spectrometry, complemented by principal component analysis (PCA) for data interpretation. Results indicated that the majority of pottery samples, over 80%, were produced using local clay from deposits near the site. However, approximately 20% of the analyzed vessels were made using clay from deposits near the Bubanj site, 8 km south of Velika Humska Čuka. A vessel on a hollow high foot combining stylistic elements of the Bubanj-Hum I group and Early Eneolithic Pannonian groups was made of clay not sourced from any identified local deposits, suggesting its non-local origin. While the predominance of local materials suggests self-sufficient production, the use of non-local clays and stylistic influences highlights long-distance connections and exchanges. The study emphasizes the importance of Velika Humska Čuka in understanding the development of ceramic traditions and the cultural dynamics of the Early Eneolithic in the Central Balkans.

Keywords: Velika Humska Čuka; elemental composition; EDXRF spectrometry; Early Eneolithic; pottery provenance; Central Balkans

1. Introduction

The multilayered site of Velika Humska Čuka is located approximately 10 km east of the South Morava River and 7 km north of the present-day city of Niš, on the northern fringe of the Niš Basin in southeastern Serbia (Figure 1). The site lies on a dominant and hardly accessible elevation that comprises four terraces on different altitudes. In total, the plateau covers an area of 200 × 160 m^2, with the highest altitude of around 445 m above sea level. The site is surrounded by the Hum River on the northern and western sides and is relatively easily accessible only from the northeastern side. In the north, the site is connected with a smaller elevation, the Mala Humska Čuka [1]. The dominant position that the site holds enables good visual connection with the surroundings, particularly the sites of Bubanj and Kremenac (Figure 1). The site of Kremenac is known as a source of quality raw materials for the production of lithic tools [2].

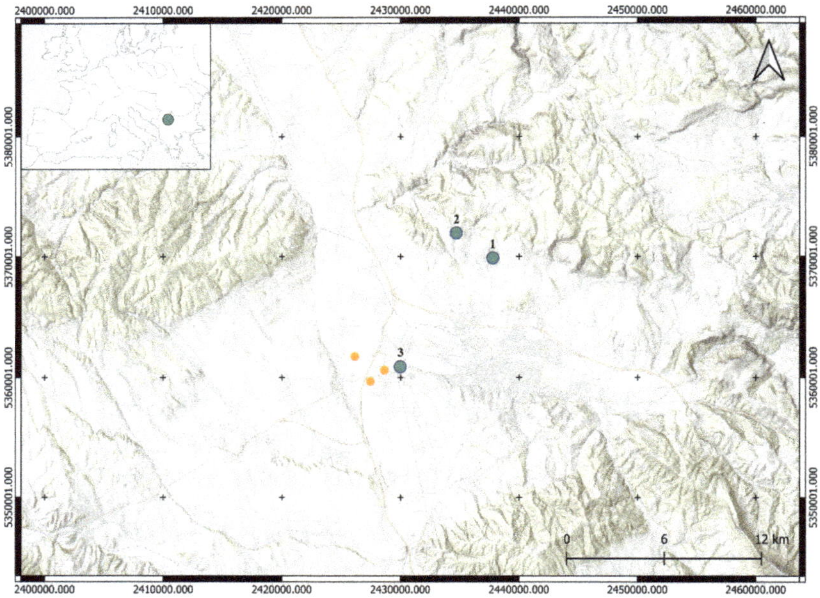

Figure 1. The sites of Velika Humska Čuka (1), Kremenac (2), and Bubanj (3); clay pits near the site of Bubanj—orange dots (Esri Topo Map/Open Topo Map; EPSG: 3857—WGS 84/Pseudo-Mercator; QGIS 3.28. Firenze).

The first data about the site originates from the beginning of the 20th century and the survey conducted by V. Fewkes. Soon after, the first archaeological excavations were conducted in 1932 and 1933 by V. Grbić and the National Museum in Niš. In 1938, the excavations were continued by B. Gojković. Of particular importance are the excavations in 1956 by M. Garašanin, which served as a basis for the definition of the Bubanj-Hum group that marks the Copper and Bronze Age in this part of the Central Balkans [1,3,4]. Finally, the ongoing excavations were renewed in 2009 by the Institute of Archaeology in Belgrade and the National Museum in Niš [1]. The excavations resulted in the discovery of numerous archaeological features from different prehistoric and historic periods, as the site was settled throughout the Copper Age, Bronze Age, Iron Age, Antique Period, and Medieval Period [1,5]. However, the importance of the site and the neighbouring site of Bubanj is particularly reflected in their potential for the study of the development and chronology of the Early Eneolithic in the region [6].

Due to the multilayered nature of the site, well-defined stratigraphy and archaeological features, and the variety of the material culture, the site of Velika Humska Čuka acted as one of the pillars of the FLOW project (THE FLOW is the acronym for the multidisciplinary project Interactions-Transmission-Transformation: Long-distance connections in the Copper and Bronze Ages of the Central Balkans, which is financed by the ScienceFund of the Republic of Serbia, IDEAS call, Grant no. 7750074; the project runs from 2022 to 2025 and incorporates the joint work of archaeologists, physicists, and chemists). Namely, different categories of the material culture from the site were incorporated into the project, which included the provenance analyses of copper, bronze, obsidian [7], and now ceramic artifacts.

While examining archaeological ceramics, various scientific analytical techniques can be used to reveal their chemical and/or mineralogical composition [8–10]. Currently, non-invasive techniques offer more advantages and are more used in practice [11–13]. This approach was used in the study to examine the characteristics of materials from which vessels and potsherds were made. The materials were characterized in terms of elemental composition determined using energy-dispersive X-ray fluorescence (EDXRF) spectrometry. By comparing them with the characteristics of clay samples, the usage of local raw materials for production was discussed. It was discovered that a whole assemblage of potsherds of different kinds and periods was locally produced, while certain vessels had a non-similar chemical composition with the local clay. The analytical results confirm the archaeological study of these findings.

2. Materials and Methods

2.1. Samples

The provenance analyses based on the comparison of elemental compositions were conducted on a total of 61 samples, of which 5 samples originate from local and nearby clay pits, 45 samples originate from targeted potsherds from all prehistoric periods registered at the site, 10 samples originate from Early Eneolithic vessels of unusual stylistic or typological characteristics from different archaeological contexts, and 1 control potsherd. The idea was to compare the elemental composition on a diachronic and cross-cultural scale represented at the site, and hence, the selection of samples from the site was based on those parameters. Further, the choice of samples was based on the stylistic and typological characteristics that indicated both the "foreign" and domestic origin of pottery. The "foreign" pottery was defined by certain forms and ornamentation styles that demonstrate the cultural attribution uncommon for the existing cultural stratigraphy of the South Morava Basin.

Two samples from local clay pits near Velika Humska Čuka, marked as Jezero 1 and Jezero 2, were collected on-field and selected based on communication with the local population of the village of Hum, where the site is located. Location Jezero 1 is located approximately 700 m northwest of the site, and location Jezero 2 is located around 330 m to the west (Figure 2). The sample from location Jezero 1 was used to make two clay tiles for analysis, while Jezero 2 represented a control sample that resembled soil rather than clay. An additional three clay samples from the nearby multilayered prehistoric site of Bubanj (Crepana in Novo Selo, Crepana in Lalinac, and Tri Bresta) were included in the study [14].

Figure 2. Position of clay pits Jezero 1 (1) and Jezero 2 (2) in relation to the Velika Humska Čuka (Google Satellite; EPSG: 3857—WGS 84/Pseudo-Mercator; QGIS 3.28. Firenze).

The 45 samples that are taken from potsherds originate from different archaeological features and the cultural layer at the site (Figure 3). The number of samples covered a wide chronological (c. 4500–1000 BCE) and cultural span, including the Early Eneolithic (Bubanj-Hum I/Bubanj-Sălcuţa-Krivodol complex) (9 samples), Middle Eneolithic (Sălcuţa IV group) (3 samples), Late Eneolithic (Coţofeni-Kostolac group) (4 samples), Early Bronze Age (Bubanj-Hum II, Bubanj- Hum III groups) (10 samples), Middle/Late Bronze Age (Verbicioara group) (8 samples), Late Bronze Age (Brnjica group) (7 samples), and the Transitional Period (Gava-Belegiš II group) (4 samples).

Figure 3. Examples of the potsherds used in the study (each scale is 5 cm).

The final set of samples is represented by a group of specific lavishly decorated and painted vessels from House 3, a vessel from House 4, and vessels from the pottery accumulation in the trench 1/18 on the southern edge of the site. Most vessels were discovered in connection with Early Eneolithic House 3 in two pits that were dated to the first half of the 44th century BCE. Those vessels were sampled since most of them bear characteristics of the Bubanj-Hum I group, a local manifestation of the Bubanj-Sălcuţa-Krivodol complex of the Central Balkans, yet certain forms indicate the permeation with the Early Eneolithic groups of the Lower Danube Region, such as the Gumelniţa group, or groups from the Pannonian regions, such as the Tiszapolgár and Lasinje groups. Further, the ornamentation of some of the vessels, represented by asymmetrical geometric motifs painted in several different colors (purple, red, white, blue), at the moment, besides a few vessels from nearby Bubanj site, represents a unique manifestation of the Early Eneolithic in the wider region (Figure 4) [1]. Vessels from House 4 and pottery accumulation in trench 1/18 show exclusively stylistic and typological features characteristic of the local Bubanj-Hum I group.

2.2. Sample Preparation

The ceramic potsherds were prepared for analysis in a standardized manner for elemental characterization using a handy sander equipped with a diamond blade. Suitable parts of the potsherds were polished to achieve a maximally flat surface and then cleaned. Where possible, the potsherds were polished in one to three different parts. Each potsherd was analyzed at several points, and the average value of collected spectra was used for further analysis [15].

Figure 4. A set of Early Eneolithic vessels in situ during 2018 excavations (7573337.0822, 4804286.2416—coordinate of the central vessel with two handles) (EPSG: 31277—WGS 85/Balkans Zone 7).

The whole vessels were not treated before the analysis. The suitable fractured parts that remained visible were chosen for the analysis, and care was taken to select the flattest possible.

The clay samples were prepared in the form of tiles with dimensions around 5 × 4 cm and a thickness of 3 cm (Figure 5a). After the clay is dried and impurities are removed, the material is slowly hydrated to form the paste with a suitable consistency for molding. After tile shaping and air drying for several days, they were fired at 400 °C [16]. The tiles were analyzed under the same conditions as the ceramic material to test the assumption about local raw material sources.

2.3. Analytical Technique and Data Analysis

The ceramic material was characterized in terms of elemental composition. To establish these results, EDXRF spectrometry was employed for qualitative analysis. An in-house-developed EDXRF spectrometer was used (developed at Vinča Institute of Nuclear Sciences, Figure 5b). The spectrometer has a milli-beam spot of an air-cooled X-ray tube (Oxford Instruments, Scotts Valley, CA, USA, Rh anode, maximum 50 kV, maximum current 1 mA) with a pin-hole collimator. The instrument is equipped with a Si-PIN X-ray detector (6 mm^2/500 μm, Be window 12.5 μm thickness, with an energy resolution of 160 eV at 5.89 keV). The detector is associated with a DSP (X123, Amptek Inc., Bedford, MA, USA) for spectra acquisition. Two laser pointers, mounted onto the spectrometer head, serve for proper sample positioning. ADMCA software (Amptek Inc., version 1, 0, 0, 16) was used for spectra acquisition and processing. The following parameters, X-ray tube voltage of 40 kV and 800 μA current, without filter, together with measuring time of 120 s, were kept constant during all measurements.

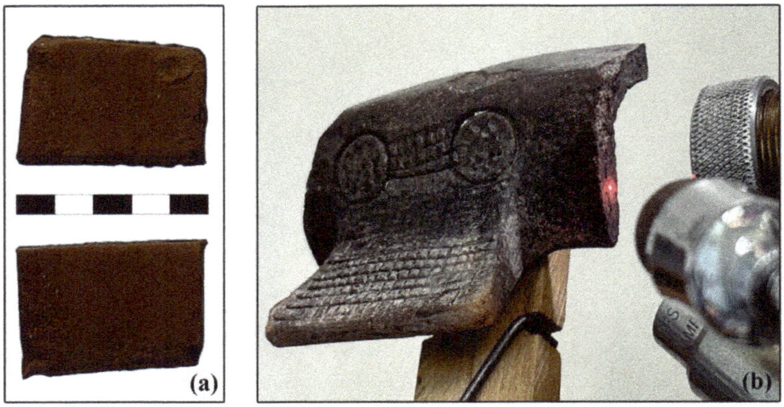

Figure 5. (**a**) Clay tiles from Jezero 1 clay pit and control sample Jezero 2 (scale is 5 cm); (**b**) measurement of an Early Bronze Age potsherd using EDXRF spectrometer.

All collected EDXRF spectra were pretreated by the peak balancing procedure (using the original MATLAB (ver. R2023b) code based on peak alignment) to minimize any contributions of the experimental setup. All 61 aligned measurements were organized in a 61 × 2048-dimensional matrix.

Due to the lack of certified reference materials for the quantification of the chemical composition of archaeological ceramic materials, provenance studies are based on the similarities/differences between the analyzed samples and raw materials. To reveal similarities/differences between examined samples, the analytical results were subjected to a multivariate analysis to transform multidimensional space into lower-dimensional space. Principal component analysis (PCA) was chosen for this purpose. It is the most often used unsupervised multivariate technique for dimension reduction [17–19]. This technique was chosen for the study because dimension reduction was performed without prior knowledge about the objects' membership in a specific group. Therefore, the resulting picture of grouping the lower-dimensional space can be put into an archaeological context. The PCA-based dimension reduction was performed on raw spectra containing overall spectral information [20,21].

3. Results and Discussion

The EDXRF spectra collected on the randomly selected vessels and potsherd body, as well as clay samples, show similar elemental compositions. The Si, K, Ca, Ti, Mn, Fe, Rb, Sr, Y, and Zr are detected in all the measured samples (Figure 6).

As we mentioned in the above text, the provenance study was performed based on the qualitative EDXRF results, as its results are comparable to the quantitative ones [14]. The PCA-based dimension reduction, performed on the dataset formed as described above, is shown in Figure 7. During PCA, a new two-dimensional space was created from the initial 2048-dimensional one, following the variance of the dataset. The first two principal components (PC1 and PC2) account for the maximal variance of the initial data. According to the results of the PCA, the first two principal components account for nearly 61% of the maximal variance, giving an acceptable picture of the initial dataset structure in a newly formed space. However, the projection of 61 dataset elements in the PC1–PC2 space shows that two groups are formed and clearly separated, indicating a difference between members in newly formed groups caused by differences in the elemental composition.

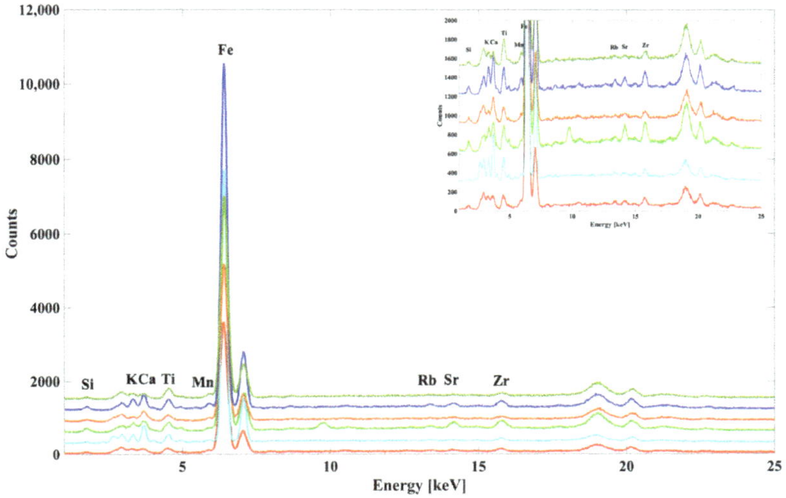

Figure 6. Representative EDXRF spectra collected at vessels, potsherds, and clay samples.

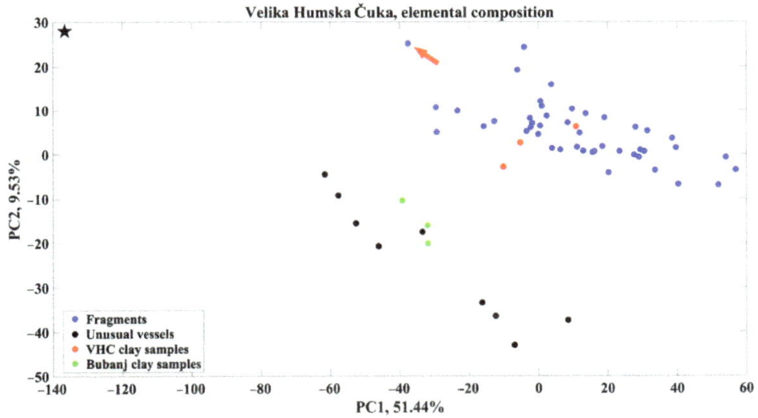

Figure 7. PCA-based dimension reduction.

The results show that the analyzed potsherds from different prehistoric periods (blue dots) have a similar elemental composition as the clay samples (Jezero 1 and Jezero 2, red dots), indicating the use of local raw materials and local production of ceramic pots for various utilities. As can be seen from Figure 7, the group that contains potsherds is quite coherent. This result indicates continuity in the ceramics production during the long period at the site of Velika Humska Čuka. Only one potsherd, denoted as A-502 (marked with an arrow and shown in Figure 3), deviates slightly from this finding, making the group slightly less coherent. A slight deviation in the elemental composition of the A-502 sample might bear connections with their cultural attribution. Namely, analyzed sherds bear certain ornamental characteristics of the Vučedol group, such as polished surface and ornamentation organized in vertical friezes. The Vučedol group is primarily connected with the territory of Western Balkans, from where it spread towards the Adriatic, Carpathian Basin, and parts of Central Europe [22,23]. However, its penetration towards the east has been recorded on multiple sites, although as an exception rather than a common feature. Interestingly, the Bubanj-Hum II group, which marked the earlier phase of the Early Bronze Age in the region, bears certain ornamental characteristics of the Vučedol group and the Coţofeni-Kostolac group, but analyzed samples of the Bubanj-Hum I group were made by local clay [4,6,24]. It is important to highlight that the absolute and relative chronologies of the given groups overlap in the first quarter of the 3rd millennium BCE [6,25–28], and hence, such a permeation and continuation of ornamental elements might indicate certain contacts between different yet contemporary communities.

The group formed by the EDXRF measurements on the Early Eneolithic vessels of uncommon characteristics is less coherent than the one comprising the potsherds. The group is divided into two parts. The differences between the members of these two groups might originate even from measuring conditions since the measuring spots were not prepared prior to analysis.

For the complete vessel marked as A-931 (denoted as a star mark in the upper left corner in Figure 7 and shown in Figure 4—bottom right corner), additional consideration is necessary, as it is far from any other sample in the study. The vessel originates from the large pit within House 3, dated to the first half of the 44th century BCE. Sample A-931 is a vessel whose stylistic and typological characteristics of the upper portion, meaning the recipient can be connected with the local Early Eneolithic group (Bubanj-Hum I group). The high hollow foot of the vessel is characteristic of the Pannonian groups of the Early Eneolithic.

As can be seen from Figure 7 (black dots), the vessels seem to be made of raw materials that differ in the elemental composition of local ones. In addition to vessel A-931, the pit contained vessels made from both local (Jezero 1 and 2) and 'foreign' clay. This suggests that vessels made from different clays, or those of different origins, were treated similarly within the Early Eneolithic household at the site. This was the reason for introducing the EDXRF measurements of the samples prepared from the clay sampled from the nearby simultaneous archaeological site Bubanj. The introduction of the results of the analyses of clays from deposits in the vicinity of the nearby site of Bubanj (Figure 7, green dots) indicate that their composition almost perfectly matches the composition of unusual vessels from the site of Velika Humska Čuka which differentiate in the PC1-PC2 picture (Figure 7, black dots).

The remaining samples from both the earlier and the later phase of the Early Eneolithic, were most likely made of local clay from the aforementioned clay pits (Jezero 1 and 2, red dots), and though in small number, from the clay from clay pits in the vicinity of Bubanj (Crepana in Novo Selo, Crepana in Lalinac and Tri Bresta, green dots). Hence, there is no emerging pattern regarding the utilization of a specific clay for a certain type of vessel or a certain phase of the Early Eneolithic. Although the number of samples is currently too low for some final conclusions, it can be noticed that the graphite-painted pottery, which is one of the main characteristics of the local Bubanj-Hum I cultural group of Early Eneolithic from the site of Velika Humska Čuka, is made exclusively from local clay. On the other hand, pottery painted in red, purple, white, and other colors, decorated with motifs uncommon for the Bubanj-Hum I group, is made exclusively from clay in the vicinity of the site of Bubanj. These data are not surprising, considering that the pottery of similar stylistic characteristics, painted in different colors and decorated with similar geometric motifs, was recorded sporadically at the site of Bubanj during the excavations in the 1930s [29].

Interestingly, the samples of vessels which were based on the stylistic and typological characteristics marked as "foreign", or of northern origin, such as the vessel with the *Scheibenhankel* type of pottery attributed to the late phase of the Early Eneolithic of Oltenia and Banat, vessels with characteristics of the Late Eneolithic Vučedol group, or the Gava-Belegiš II group from the end of the Late Bronze Age [30–33], were in fact made of local clay, from the clay pits of Jezero 1 and 2. This also illustrates the contacts between the communities of the Nišava Valley and the communities of the Morava Valley and the Pannonian Plain (Figure 8), which resulted in the production of imitations of foreign pottery made of local clay.

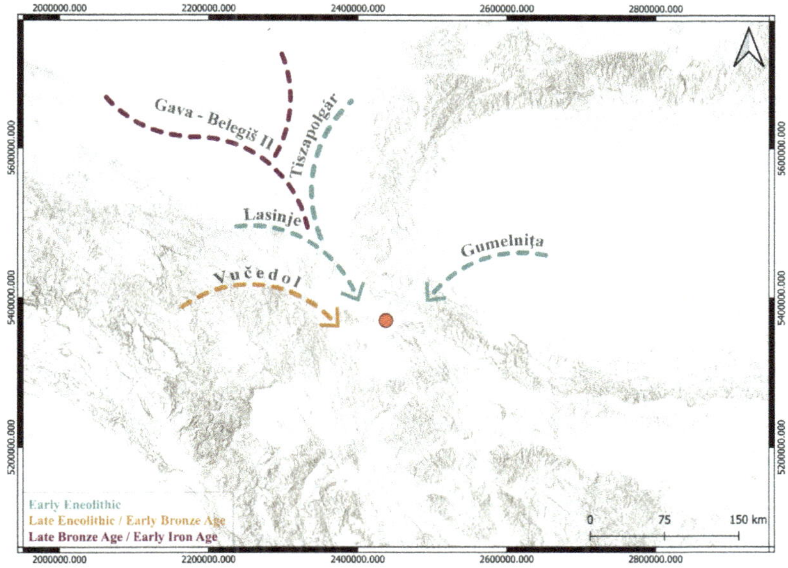

Figure 8. Representation of cultural influences mentioned in the study (Esri Topo Map; EPSG: 3857—WGS 84/Pseudo-Mercator; QGIS 3.28. Firenze).

4. Conclusions

The presented data call for certain conclusions, which are not final due to the upcoming excavations at this site yet provide certain insights into the possible origin of clay for pottery production at the site of Velika Humska Čuka.

Based on the picture of the dataset structure in the PC1-PC2 space following maximal variance among the data (with nearly 61% preserved variance), the following conclusions can be withdrawn. The majority of pottery samples, 45 out of 55 (more than 80%), from different periods of prehistory, and "foreign" or unusual stylistic and typological characteristics were made of local clay from the vicinity of the site (Jezero 1 and 2).

Only 10 samples (less than 20%) were made of clay whose elemental composition matches a different clay deposit, in this case, the deposits in the vicinity of the site of Bubanj, located around 8 km south of Velika Humska Čuka. These vessels originate from three different archaeological contexts, dated to the 44th century BCE. The stylistic common feature of most of these vessels (five examples and all vessels from pits within House 3) is their decoration comprising uncommon geometric motifs and paintings in different colors, similar to the pottery recorded at the site of Bubanj in the 1930s.

This leads to the assumption that those vessels were exchanged as final products with the inhabitants from the site of Bubanj or that the clay was due to the quality brought from those distant deposits.

Such origin of pottery from the site of Velika Humska Čuka indicates intensive contact with the population at the site of Bubanj, whether it was an exchange of goods, seasonal movements between the two sites, or a shared use of clay pits in the vicinity of Bubanj.

On the other hand, the vessel on a hollow high foot (Figure 7, black star—A931), whose recipient is common for the Early Eneolithic Bubanj-Hum I group, and the foot characteristic for the contemporary groups in Pannonia, was made of clay whose elemental composition does not match either of the presented clay deposits and most likely represents a vessel of a non-local origin.

Author Contributions: Conceptualization, M.G.-K. and A.B.; methodology, M.G.-K. and O.M.; formal analysis, M.G.-K. and P.M.; investigation, O.M., P.M. and A.B.; resources, P.M. and A.B.; data curation, M.G.-K.; writing—original draft preparation, M.G.-K. and O.M.; writing—review and editing, P.M.; supervision, A.B. All authors have read and agreed to the published version of the manuscript.

Funding: This research was funded by the Science Fund of the Republic of Serbia (Programme IDEAS, Grant No. 7750074)—the project THE FLOW—Interactions—Transmission—Transformation: Long-distance connections in Copper and Bronze Age of the Central Balkans, and the Ministry of Science, Technological Development, and Innovations, Republic of Serbia (Grant No. 451-03-136/2025-03/200017 dated 4 February 2025, Research Program No. 1, Contract No. 110-10/2019-000, T0602403).

Institutional Review Board Statement: Not applicable.

Informed Consent Statement: Not applicable.

Data Availability Statement: The original contributions presented in this study are included in the article. Further inquiries can be directed to the corresponding author.

Acknowledgments: M.G.K. wishes to express her gratitude to Danica Maksimovic, who helped with the preparation of clay samples.

Conflicts of Interest: The authors declare no conflicts of interest.

References

1. Garašanin, M.; Đurić, N. *Bubanj i Velika Humska Čuka*; Catalog; National Museum Niš: Niš, Serbia, 1983.
2. Šarić, J. *Kremenac, Donjepaleolitsko Nalazište*; Institute of Archaeology: Belgrade, Serbia, 2013; ISBN 978-86-80093-85-7.
3. Garašanin, M.; Garašanin, D. *Arheološka Nalazišta u SRBIJI*; Prosveta: Belgrade, Serbia, 1951.
4. Garašanin, M.; Garašanin, D. Nova iskopavanja na Velikoj Humskoj Čuki kod Niša. *Starinar* **1959**, *IX–X*, 243–254.
5. Stojić, M.; Jocić, M. *Niš Kulturna Stratigrafija Praistorijskih Lokaliteta u Niškoj Regiji*; Institute of Archaeology: Belgrade, Ser National Museum Niš: Niš, Serbia, 2006; ISBN 9788680093253.
6. Bulatović, A.; Milanović, D. *Bubanj. The Eneolithic and the Early Bronze Age Tell in Southeastern Serbia*; Austrian Academy of Scier Press: Vienna, Austria, 2020; ISBN 978-3-7001-8451-5.
7. Bulatović, A.; Gajić-Kvaščev, M.; Kapuran, A.; Ljuština, M.; Filipović, V.; Mladenović, O.; Milojević, P.; Milić, B. The F project—A contribution to the study of the cultural transmission of the Cental Balkan communities and the neighbouring regi in later prehistory. *Archaeol. Sci.* **2022**, *18*, 101–112. [CrossRef]
8. Vlase, D.; Vlase, G.; Ursut, G.; Sfirloaga, P.; Manea, F.; Budiul, M.; Roatru, A.; Vlase, T. The in-depth study of Romanian prehist ceramics: Late Neolithic/Eneolithic pottery and clay materials from the Foeni Tell-Orthodox cemetery in Timis county. *Ceram.* **2023**, *49*, 13941–14956. [CrossRef]
9. Merk, V.; Berkh, K.; Rammlmair, D.; Pfeifer, L. Chemical and Mineralogical Analysis of Samples Using Combined LIBS, Rar Spectroscopy and μ-EDXRF. *Minerals* **2023**, *13*, 729. [CrossRef]
10. Saraiva, A.S.; Coutinho, M.L.; Tavares da Silva, C.; Soares, J.; Duarte, S.; Veiga, J.P. Archaeological Ceramic Fabric Attribu Through Material Characterisation—A Case-Study from Vale Pincel I (Sines, Portugal). *Heritage* **2025**, *8*, 84. [CrossRef]
11. Emmitt, J.J.; McAlister, A.J.; Phillipps, R.S.; Hodaway, S.J. Sourcing without sources: Measuring ceramic variability with pXR *Archaeol. Sci. Rep.* **2018**, *17*, 422–432. [CrossRef]
12. Długosz-Lisiecka, M.; Sikora, J.; Krystek, M.; Płaza, D.; Kittel, P. Novel method of ancient pottery analysis based on radioac isotope ratios: A pilot study. *Herit. Sci.* **2022**, *10*, 113. [CrossRef]
13. Reedy, C.L.; Reedy, C.L. Micro-Computed Tomography with 3D Image Analysis to Reveal Firing Temperature Effects on F Systems in Archaeological and Ethnographic Ceramics. *Appl. Sci.* **2022**, *12*, 11448. [CrossRef]
14. Gajić-Kvaščev, M.; Andrić, V.; Jančić-Heinemann, R.; Mladenović, O.; Bulatović, A. Comparison of Quantitative and Qualita EDXRF Analysis for Provenance Study of Archaeological Ceramics. *Materials* **2024**, *17*, 3725. [CrossRef] [PubMed]
15. Forster, N.; Grave, P.; Vickery, N.; Kealhofer, L. Non-destructive analysis using PXRF: Methodology and application to arch logical ceramics. *X-Ray Spectrom.* **2011**, *40*, 389–398. [CrossRef]
16. Gliozzo, E. Ceramic technology. How to reconstruct the firing process. *Archaeol. Anthropol. Sci.* **2020**, *12*, 260. [CrossRef]

Fukunaga, K. *Introduction to Statistical Pattern Recognition*, 2nd ed.; Academic Press Inc.: Cambridge, MA, USA, 1990; ISBN 0-12-269851-7.
Varmuza, K.; Filzmoser, P. *Introduction to Multivariate Statistical Analysis in Chemometrics*; CRC Press Taylor & Francis Group: Boca Raton, FL, USA, 2009; p. 63. ISBN 978-1420059472.
Brereton, R.G. *Chemometrics: Data Analysis for the Laboratory and Chemical Plant*; John Wiley & Sons, Ltd.: Hoboken, NJ, USA, 2003; p. 229. ISBN 978-0-470-84574-5.
Bonizzoni, L. ED-XRF analysis for Cultural Heritage: Is quantitative evaluation always essential? *J. Phys. Conf. Ser.* **2015**, *630*, 012001. [CrossRef]
Panchuk, V.; Yaroshenko, I.; Legin, A.; Semenov, V.; Kirsanov, D. Application of chemometric methods to XRF-data—A tutorial review. *Anal. Chim. Acta* **2018**, *1040*, 19–32. [CrossRef] [PubMed]
Tasić, N. *Badenski i Vučedolski Kulturni Kompleks u Jugoslaviji*; Arheološko Društvo Jugoslavije: Belgrade, Serbia; Zajednica Muzeja Vojvodine: Novi Sad, Serbia, 1967.
Dimitrijević, S. Vučedolska kultura i vučedolski kulturni kompleks. In *Praistorija Jugoslavenskih Zemalja III. Eneolitsko Doba*; Benac, A., Ed.; Centar za Balkanološka Ispitivanja: Sarajevo, Yugoslavia, 1979; pp. 267–341.
Garašanin, M. Neolithikum und Bronzezeit in Serbien und Makedonien. *Ber. Römisch-Ger. Komm.* **1959**, *39*, 1–130. [CrossRef]
Benkő, L.; Horvath, F.; Horvatnić, N.; Obelić, B. Radiocarbon and thermoluminescence dating of prehistoric sites in Hungary and Yugoslavia. *Radiocarbon* **1989**, *31*, 992–1002. [CrossRef]
Durman, A.; Obelić, B. Radiocarbon dating of the Vučedol culture complex. *Radiocarbon* **1989**, *31*, 1003–1009. [CrossRef]
Ciugudean, H.I. *Eneoliticul Final in Transilvania si Banat: Cultura Cotofeni*; Muzeul Național al Banatului: Timişoara, Romania, 2000; ISBN 973-585-178-4.
Bulatović, A.; Gori, M.; Vander Linden, M. Radiocarbon dating the 3rd millennium BC in the Central Balkans: A re-examination of the Early Bronze Age sequence. *Radiocarbon* **2020**, *62*, 1163–1191. [CrossRef]
Oršić-Slavetić, A. *Bubanj, eine Prähistorische Ansiedlung bei Niš*; Mitteilungen der prähist; Kommission der Akademie Wissenchaften: Vienna, Austria, 1940; pp. 1–42. ISSN 0065-5376.
Bulatović, A.; Molloy, B.; Filipović, V. Serbia the Balkan-Aegean migrations revisited: Changes in material culture and settlement patterns in the Late Bronze Age Central Balkans in light of new data. *Starinar* **2021**, *LXXI*, 61–105. [CrossRef]
Bulatović, A.; Milanović, D. The cultural and chronological context of sites of Bubanj and Velika Humska Čuka near Niš (southeastern Serbia) and their significance for understanding the emergence and development of the Central Balkans eneolithic. *Bull. Mosc. Reg. State Univ.* **2021**, 36–64. [CrossRef]
Raczky, P. The Cultural and Chronological Relations of the Tisza Region During the Middle and Late Neolithic as Reflected by the Excavations at Öcsöd-Kováshalom. 1988. Available online: https://www.researchgate.net/publication/292447339_The_cultural_and_chronological_relations_of_the_Tisza_region_during_the_Middle_and_Late_neolithic_as_reflected_by_the_excavations_at_Ocsod-Kovashalom (accessed on 20 February 2025).
Jovanović, B.; Salkuta, I.V.; Krivodol; Bubanj, J. *James Harvey Gaul in Memoriam*; Stefanovich, M., Todorova, H., Hauptmann, H., Eds.; J. Harvey Foundation: Sofia, Bulgaria, 1988; pp. 197–202. ISBN 9544910263/978-9544910266.

sclaimer/Publisher's Note: The statements, opinions and data contained in all publications are solely those of the individual hor(s) and contributor(s) and not of MDPI and/or the editor(s). MDPI and/or the editor(s) disclaim responsibility for any injury to ple or property resulting from any ideas, methods, instructions or products referred to in the content.

materials

Article

Comparison of Quantitative and Qualitative EDXRF Analysis for Provenance Study of Archaeological Ceramics

Maja Gajic-Kvascev [1,*], Velibor Andric [1], Radmila Jancic-Heinemann [2], Ognjen Mladenovic [3] and Aleksandar Bulatovic [3]

[1] Department of Chemical Dynamics and Permanent Education, Vinča Institute of Nuclear Sciences, National Institute of the Republic of Serbia, University of Belgrade, 11000 Belgrade, Serbia
[2] Faculty of Technology and Metallurgy, University of Belgrade, 11000 Belgrade, Serbia
[3] Institute of Archaeology, National Institute of the Republic of Serbia, 11000 Belgrade, Serbia
* Correspondence: gajicm@vinca.rs

Abstract: The most common scientific analysis of archaeological ceramics aims to determine the raw material source and/or production technology. Scientists and archaeologists widely use XRF-based techniques as a tool in a provenance study. After conducting XRF analysis, the results are often analyzed using multivariate analysis in addition to interpretation and conclusions. Various multivariate techniques have already been applied in archaeological ceramics provenance studies to reveal different raw material sources, identify imported pieces, or determine different production recipes. This study aims to evaluate the results of multivariate analysis in the provenance study of ceramics that belong to three cultures that settled in the same area during various prehistoric periods. Portable energy-dispersive X-ray fluorescence spectrometry (pEDXRF) was used to determine the elemental composition of the ceramic material. The ceramic material was prepared in two different ways. The ceramic body material was ground into powder, homogenized, and then pressed into tablets. After that, the same fragments are polished in suitable places. Quantitative and qualitative analyses were performed on the tablets and polished pieces. The results were subjected to both unsupervised and supervised multivariate analysis. Based on the results, it was concluded that qualitative analysis of the well-prepared shards' surface using EDXRF spectrometry could be utilized in provenance studies, even when the ceramic assemblages were made of similar raw materials.

Keywords: archaeological ceramics; clay; EDXRF spectrometry; pattern recognition; classification

1. Introduction

The production of ceramic in prehistory can be considered state-of-the-art technology. Understanding this process provides valuable insights into the skills of the manufacturers, trading patterns, and overall development of early civilizations [1]. Since ceramics do not change over time, analyzing the materials provides reliable conclusions about whether they were made from local raw materials or resulted from trade with others. Scientific examining archaeological ceramics usually involves various analytical techniques to reveal their chemical and/or mineralogical composition [2–5]. Nowadays, these analyses can be conducted either by sampling the ceramic materials or in a non-invasive manner [6,7]. Moreover, provenance studies can be performed using both quantitative and qualitative analytical results [8–10].

Many documented provenance studies have focused on differentiating between ceramic assemblages or identifying imported ceramics. Different material characteristics, such as elemental, chemical, or mineralogical composition, were analyzed using pattern recognition techniques to draw conclusions about variations in raw materials and/or production technology among ceramic findings. The present study aims to evaluate the effectiveness of various pattern recognition techniques used for the provenance study of archaeological ceramics from different prehistoric cultures that were made of similar raw materials. For that purpose, archaeologists carefully selected and grouped the ceramic fragments according to stratigraphy, typology, and dating. They chose a suitable site where three cultures settled in prehistory for a relatively short period. The Bubanj archaeological site is in the southern Pomoravlje region, in the central part of the Niš Valley, approximately 7 km southeast of the confluence of the Nišava and South Morava rivers (today's Serbia, Figure 1a). Early excavations in this area (which started in the 1930s) revealed multilayered stratigraphy. This site was inhabited in the Middle Neolithic period, with the earliest findings associated with the Starcevo culture (in the first half of the 6th millennium BC). The Eneolithic and early Bronze Age layers follow mentioned cultural layers, dated to the Middle Neolithic [11]. The first cultural layer in this study belongs to the so-called Bubanj Hum I culture, corresponding to the Early Eneolithic and being a regional representative of the widely spread Bubanj–Salcuţa–Krivodol cultural complex. The findings from the upper layer, from the Middle Eneolithic, belong to Cernavoda III, with elements of Baden and Boleraz cultures. The latest group included in this study is from the Late Eneolithic period and belongs to the Coţofeni–Kostolac cultural layer. All three layers were well dated to the corresponding period and cultures [12]. The ceramic assemblages were formed from fragments that belong to each cultural layer. According to archaeologists, Bubanj Hum I culture settled on this site for the longest time, implying that local raw materials were used for ceramic production. The Cernavoda III settled the site long after Bubanj Hum I culture; therefore, they could not take over previous knowledge but rather found local clay sources nearby the site. Only the fragments from the Coţofeni–Kostolac cultural layer might have been brought to the site during the settlement. Those archaeological assumptions were tested in this study to evaluate analytical and multivariate protocols for the provenience study.

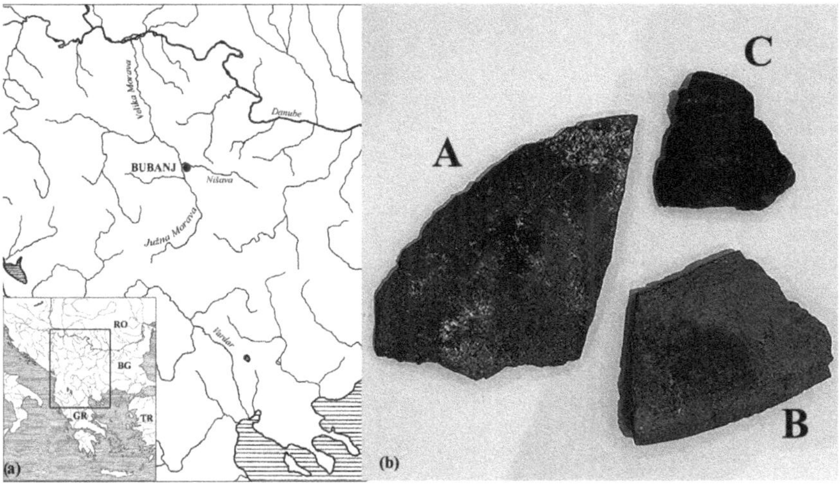

Figure 1. (**a**) Location of the site of Bubanj, Serbia (drawing: A. Bulatović); (**b**) Representative ceramic fragments from the Bubanj Hum I (A), Cernavoda III (B) and Coţofeni–Kostolac (C) culture.

The ceramic material was analyzed using EDXRF spectrometry. The quantitative and qualitative analytical results were subjected to dimension reduction to evaluate suitability for classification. The most-used unsupervised and supervised techniques were chosen for the evaluation [13–15]. These techniques were evaluated based on several parameters to measure their efficiency in reducing dimensions and their ability to separate different classes, which is crucial for designing classifiers. The classification's effectiveness was evaluated by considering its recognition ability and its rate of misclassification.

This study aims to examine how the method of preparing archaeological ceramics for analysis affects classification efficiency. For this purpose, samples were taken from the material of the selected fragments from which tablets were made. Their elemental composition was quantified using standard reference material (CRM). The elemental composition was also determined for the analyses performed on well-polished places on the fragments. The qualitative and quantitative results obtained in this way are classified using different dimension reduction techniques. The selection of ceramic materials from different archaeological stratigraphies, made from similar raw materials, is such that it does not prejudice an effective classification. Based on the achieved results, qualitative analytical results can be used in provenience studies as effectively as quantitative ones.

2. Materials and Methods

2.1. Sample Description

The ceramic fragments are grouped according to the stratigraphy layers in which they were found, and special attention was paid to selecting the fragments from reliable archaeological contexts corresponding to each cultural layer with certainty. The first group, labeled BI, contains 18 fragments belonging to the so-called Bubanj Hum I culture and comes from a structure (pit) marked as 69. The second one, denoted as CV, contains 18 pottery vessel shards (structure 108-pit) from the upper layer, considered Cernavoda III. The last group, labeled as KK, is formed of the 13 shards that belong to the Coţofeni–Kostolac cultural layer. These shards were taken from the oldest (first) phase found on the house floor (structure 15), which can be considered brought in during settlement. Macroscopic archaeological analysis indicates similarities between the findings from the BI and CV ceramic fragments but differences with the KK group (Figure 1b).

The ceramic fragments were prepared for analysis in two different ways. Small amounts of the ceramic material were powdered by grinding the ceramic body from its various parts. A handy sander equipped with a diamond blade was used. The fragments were cleaned before grinding. The obtained ceramic powder was fine for tableting, which was performed without prior sieving. The homogenized material was then pressed into large tablets, each 3 cm in diameter, containing an equal amount of the material. After that, suitable parts of the fragments were polished and cleaned. Each fragment was analyzed at three points, and the average values were used for further analysis [16]. To quantify the chemical composition, the IAEA PT ancient Chinese ceramic certificated reference material (CRM, [17]) was tableted in the same manner as the ceramic samples and analyzed under the same conditions. Three clay samples from nearby clay pits were also analyzed under the same conditions as the ceramic material to test the assumption about local raw material sources. The nearest clay pit, Crepana, is about 800 m west–southwest of Bubanj. This pit was used until the 1940s. Today, it is closed but easily accessible for sampling. The Tri Bresta clay pit is 2 km from the site in the same direction, which also worked until the middle of the last century. The farthest clay deposit, Crepana near Lalinac, is located 2.6 km west of the Bubanj site and remains active. All three clay deposits are located at a 15–35 min walk from the site on flat surroundings, so the probability of their exploitation in the Eneolithic is high.

2.2. Analytical Technique and Datasets Forming

The EDXRF spectrometry was employed for quantitative and qualitative elemental analysis of the ceramic material using an in-house-developed milli-beam spot instrument. The spectrometer consists of an air-cooled X-ray tube (Oxford Instruments, Scotts Valley, CA, USA, Rh anode, maximum voltage 50 kV, maximum current 1 mA) with a pin-hole collimator and a Si-PIN X-ray detector (6 mm2/500 m, Be window 12.5 m thickness, with energy resolution of 160 eV at 5.89 keV), associated with a DSP (X123, Amptek Inc., Bedford, MA, USA) for spectra acquisition. Two laser pointers were used for the accurate sample, and tablet positioning. ADMCA software (Amptek Inc., version 1, 0, 0, 16)) was used for spectra acquisition and processing. The following parameters: X-ray tube voltage of

35 kV and 800 μA current, without filter, were kept constant during all measurements and measuring time of 120 s.

An original MATLAB code was developed to align the EDXRF spectra using the peaks balancing procedure to minimize any experimental setup contributions. All EDXRF spectra were aligned before further analysis. The dimension reduction procedure was performed on raw spectra containing overall spectral information along with the datasets containing characteristic elemental composition. This was done to examine the influence of the feature selection process, which involves potentially losing some information about the dataset structure. The feature selection was performed by quantifying the chemical composition using CRM and using a radial-basis neural network (RBNN) procedure. The EDXRF results were quantified using the Net Peak Area parameter. This parameter was determined by ADMCA software by marking the peak area. For each EDXRF spectrum, the Net Peak Area parameter was calculated for nine peaks, corresponding to the elements determined in the CRM. These nine values can be considered as selected features, and their informativeness for classification purposes will be evaluated in this study. The data selection using RBNN was performed as another suitable feature selection method. The detailed procedure is described in [18]. The RBNN was designed and trained to reach maximal reconstruction of the initial EDXRF spectrum using normal distribution functions. The maximal height parameter of the nine functions was utilized to characterize ceramic fragment material. The procedure is used because it is much faster and more reliable than calculating the Net Peak Area parameter, so it is suitable for testing.

2.3. Pattern Recognition Techniques

Principal component analysis (PCA) was selected among unsupervised dimension reduction techniques as it is widely used in analyzing cultural heritage objects [19,20]. PCA creates a reduced space of maximal variance, where the first few components account for most of the variation in the original datasets. Based on the above definition, applying PCA to the provenience studies of ceramic materials made from different raw materials will result in considerable classification possibilities. In this study, the classification possibilities of the PCA will be evaluated for the dataset of ceramics made of the same or similar raw material.

PCA may not be able to accurately represent the dataset's group membership due to the small variance expected. Therefore, a supervised method called scattering matrices-based dimension reduction was tested, aiming to preserve group coherence as much as possible during dimension reduction. More mathematical details about PCA and scattering matrices-based dimension reduction can be found in [21,22].

To determine which dimension reduction technique is more effective for provenance studies, it is important to understand how much of the initial dataset structure is preserved during the dimension reduction transformation. The effectiveness of PCA transformation to lower dimensional space is quantified by the percentage of the dataset's total variance preserved along the projection axes [23]. The amount of information lost during dimension reduction is measured by the index of informativeness, a parameter defined as:

$$I_{d/n} = \frac{\sum_{i=1}^{d} \lambda_i}{\sum_{j=1}^{n} \lambda_j} \quad (1)$$

where λ_i—eigenvalues of the covariance matrix, set in descending order, d—dimension of reduced space, and n—dimension of initial space. The higher value of this parameter indicates that dimension reduction was performed in a way that minimized information loss. The Bhattacharyya distance $\mu(1/2)$ (a measure of separability between groups in the space of reduced dimensions) is defined as [24]:

$$\mu(1/2) = \frac{1}{8}(M_2 - M_1)^T \left[\frac{\Sigma_1 + \Sigma_2}{2}\right]^{-1} (M_2 - M_1) + \frac{1}{2}\ln\frac{\left|\frac{\Sigma_1+\Sigma_2}{2}\right|}{\sqrt{|\Sigma_1||\Sigma_2|}} \quad (2)$$

where M_1 and M_2 denote expected vectors and Σ_1 and Σ_2 denote covariance matrices. A larger parameter value indicates that two classes are separable, and a space between them is large enough for classifier design. The $\mu(1/2)$ parameter value will be discussed according to the possibility of linear classifier design. The recognition ability parameter (defined as the percentage of correctly classified dataset members and mathematically represented by the ratio of correctly classified and the total number of dataset members) was calculated to evaluate the linear classification effectiveness, together with the percentage of the misclassified results.

3. Results and Discussion

3.1. EDXRF Spectrometry Analysis

According to the collected EDXRF spectra (Figure 2a), Si, K, Ca, Ti, Mn, Fe, Rb, Sr, and Zr were detected in both the tablets and ceramic fragments. The same chemical composition contains the tablet of the CRM, which was used to quantify ceramics' chemical composition. The raw EDXRF spectral data collected for both tablets and ceramic fragments, aligned prior, were reduced to exclude non-informative channels before Si and after Zr. The data were organized in a matrix with the dimensions 36 × 1700, denoted as RT and RF, for tablets and fragments, respectively. Further, RBNN was employed for feature selection from raw spectra. A comparison of the original and RBNN reconstructed spectra is presented in Figure 2b. The reconstruction enabled total superposition of the spectra and high confidence in the data selection. The RBNN was used to extract the maximum value of the most distinctive peak for the same nine chemical elements from the EDXRF spectra. Using the described procedure, two more datasets were organized as a matrix with the dimension of 36 × 9 for both fragments (denoted as NF) and tablets (denoted as NT). The Net Peak Area parameter was used to quantify the tablets' elemental composition; the results are presented in Table 1. The dataset organized in the matrix with dimensions of 36 × 9 was denoted as QT.

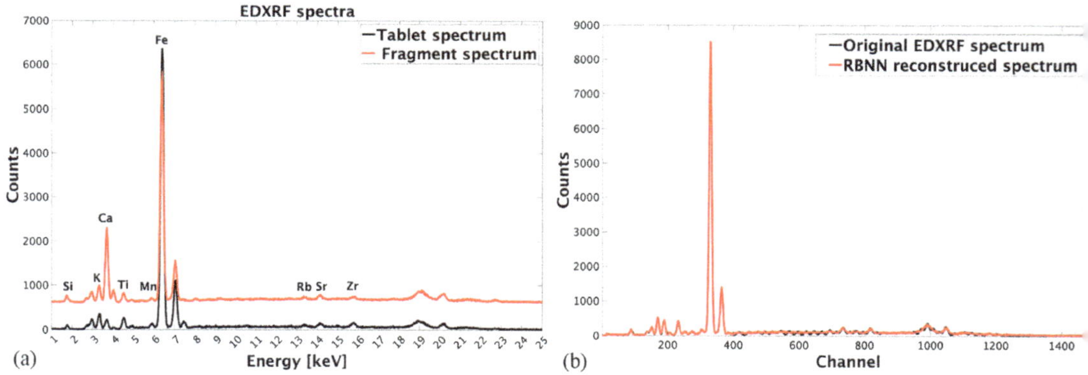

Figure 2. (**a**) EDXRF spectra aligned using original spectra alignment procedure; (**b**) Radial basis neural network spectra reconstruction for feature selection.

Table 1. Chemical composition of the ceramic materials from BI and CV ensemble.

	SiO_2, %	K_2O, %	CaO, %	TiO_2, %	MnO, %	Fe_2O_3, %	Rb, ppm	Sr, ppm	Zr, ppm
BI1	86.82	3.98	1.29	0.86	0.05	5.30	50.22	94.51	153.59
BI2	69.51	3.29	2.56	0.79	0.04	5.48	66.84	95.19	169.15
BI3	78.77	3.42	2.11	0.96	0.05	5.11	65.82	60.53	144.78
BI4	90.31	3.26	2.12	0.80	0.05	4.95	80.41	113.33	152.69
BI5	71.66	3.54	8.73	0.52	0.03	3.63	72.47	111.26	170.44
BI6	79.44	2.65	12.21	0.45	0.03	3.11	63.24	128.71	109.27
BI7	71.39	3.31	2.10	0.76	0.06	5.67	52.53	107.67	128.58
BI8	54.35	2.85	11.11	0.43	0.04	2.96	88.07	97.80	159.04
BI9	67.77	3.61	9.30	0.52	0.03	3.67	50.59	102.62	104.73
BI10	60.12	3.58	2.45	0.83	0.04	5.12	61.12	101.93	111.60
BI11	72.47	4.41	1.91	0.72	0.04	4.64	61.21	116.85	111.86
BI12	75.95	3.35	1.98	0.76	0.04	5.43	66.38	68.34	182.37
BI13	85.62	3.13	1.39	0.72	0.05	4.66	95.92	80.20	142.19
BI14	68.71	3.56	2.21	0.83	0.05	5.33	80.60	98.87	132.60
BI15	77.70	2.84	1.91	1.14	0.05	6.12	72.84	64.51	202.07
BI16	88.30	3.56	2.04	0.70	0.04	4.58	73.21	74.69	167.85
BI17	55.96	1.14	6.57	0.54	0.06	7.12	45.98	68.11	91.77
BI18	84.01	3.78	1.19	0.80	0.13	5.31	90.20	61.22	215.03
CV1	78.91	3.84	2.16	0.83	0.03	5.59	63.89	108.66	152.95
CV2	69.51	4.97	2.44	0.94	0.07	6.50	73.39	76.06	152.17
CV3	88.84	4.14	2.51	0.70	0.10	5.19	78.38	97.41	203.50
CV4	97.56	4.68	2.49	0.96	0.07	6.13	84.38	87.31	149.84
CV5	80.11	4.25	2.27	0.81	0.08	5.56	87.70	100.63	145.69
CV6	83.47	3.84	3.48	0.85	0.05	4.34	81.61	92.90	197.02
CV7	73.67	4.24	2.20	0.86	0.06	6.02	77.73	92.29	121.19
CV8	86.02	4.34	2.21	0.81	0.07	5.10	58.53	109.58	142.84
CV9	85.48	3.90	2.17	0.88	0.07	5.30	71.18	66.42	143.74
CV10	75.95	3.84	1.96	0.71	0.05	4.82	61.39	91.75	120.67
CV11	72.20	4.31	2.61	0.72	0.11	4.88	68.69	89.53	147.89
CV12	70.05	3.59	1.83	0.83	0.03	4.58	64.07	64.05	197.02
CV13	87.09	4.07	7.95	0.71	0.04	5.10	63.52	80.50	146.34
CV14	53.54	4.38	2.21	0.66	0.06	5.64	78.93	92.98	114.19
CV15	75.42	3.02	2.09	0.70	0.05	5.27	82.35	99.86	117.17
CV16	58.91	3.80	3.35	0.65	0.08	5.15	75.43	88.23	150.09
CV17	87.50	4.41	3.37	0.69	0.09	5.88	59.64	97.34	100.58
CV18	69.92	3.72	1.72	1.08	0.05	5.25	98.14	87.62	277.12
RM-CC	67.50	2.30	0.62	0.95	0.03	2.70	113.00	103.00	337.00

All the experimental data subjected to PCA and SMB dimension reduction were autoscaled before analysis.

3.2. Dimension Reduction and Classification Results

The dimension reduction results of a dataset containing elemental composition quantified values (QT) can be considered the reference because the analysis was performed on homogenized material, and the exact composition is determined. The results of the PCA dimension reduction of the QT dataset are presented in Figure 3a. As can be seen, the between-class distance and group coherence are small, with 52.88% of the total variance preserved. The Bhattacharya distance value is 0.31. Nevertheless, the linear classifier is designed with a recognition ability of nearly 36% for the BI dataset and 33% for the CV. The classification enabled the misclassification of 28% of BI and 33% of CV samples. The supervised SMB dimension reduction of the same datasets achieved slightly higher between-class distance, with the Bhattacharya distance value of 0.39. At the same time, group coherence remained small in the new y1–y2 space (Figure 3b). The linear classification parameters (together with errors given in brackets) are shown in Table 2. Supervised SMB dimension reduction enables only 11% of BI and 22% of CV samples to be misclassified

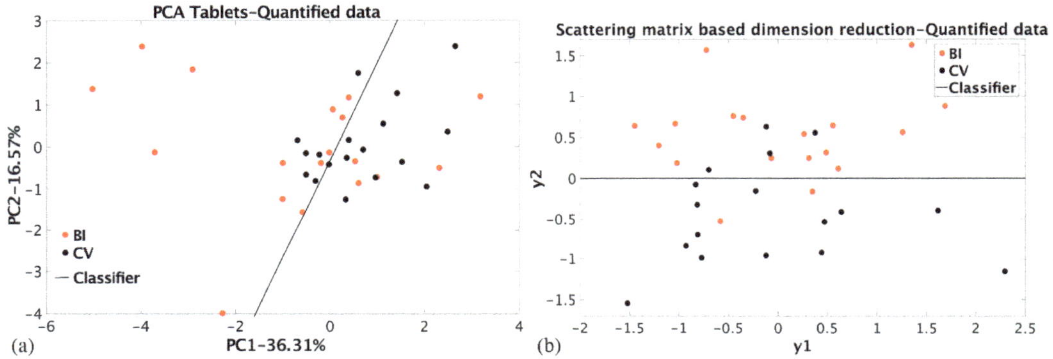

Figure 3. (**a**) PCA dimension reduction of the tablet's quantified elemental composition (QT dataset); (**b**) SMB dimension reduction of the tablet's quantified elemental composition (QT dataset).

Table 2. Dimension reduction and classification parameters.

Material	Dimension Reduction	Dataset	$I_{d/n}$ (%)	μ(1/2)	BI Samples		CV Samples	
					RA *	MC *	RA	MC
Tablets	PCA	Quantified (QT)	52.88 (2.37)	0.31 (0.007)	36	28	33	33
		RBNN feature selection (NT)	43.82 (2.54)	0.08 (0.003)	*	*	*	*
		Full spectra (RT)	14.54 (0.86)	0.33 (0.009)	39	22	36	28
	SMB	Quantified (QT)	100 (0.81)	0.39 (0.022)	44	11	39	22
		RBNN feature selection (NT)	100 (0.72)	0.70 (0.006)	44	11	44	11
Fragments	PCA	RBNN feature selection (NF)	58.24 (2.97)	0.16 (0.008)	42	17	25	50
		Full spectra (RF)	43.12 (1.08)	0.25 (0.007)	39	22	36	28
	SMB	RBNN feature selection (NF)	100 (0.95)	0.35 (0.003)	42	17	39	22

* RA—Recognition ability (%), MC—Misclassification (%).

The results achieved using nine peak maximum values determined for tablets by the RBNN selection procedure (NT dataset, Figure 4) are shown in Table 2. The PCA dimension reduction showed that linear classification is impossible, even though 43.82% of the total variance was preserved. The group coherence is also small, which can be explained by the fact that some grouping information was lost by data selection. The SMB dimension reduction of the same dataset enabled linear classification with the highest value of the Bhattacharya distance (Table 2) and the smallest percentage of the misclassified samples,

11% for both groups. For comparison, the PCA and SMB dimension reduction results for the RBNN-selected data for the ceramic fragments (NF dataset) are shown in Figure 5. For both cases, linear classification is possible. The PCA achieved the highest value of the total variance preserved (58.24%) with a relatively small value of the Bhattacharya distance (0.16).

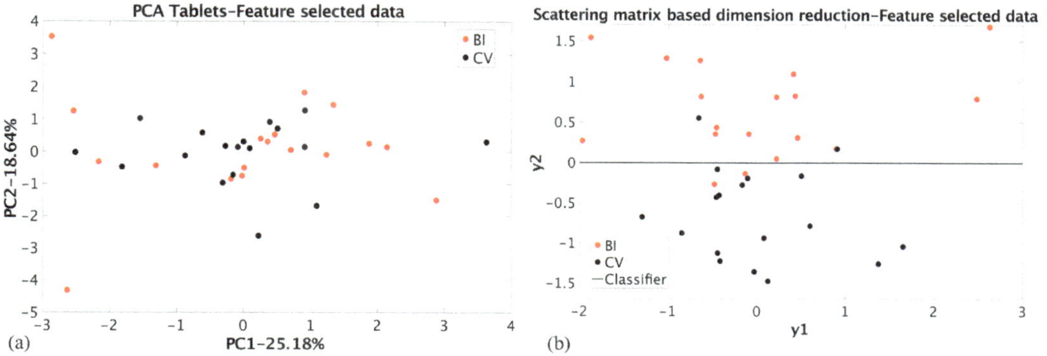

Figure 4. (**a**) PCA dimension reduction of the tablet's RBNN selected elemental composition (NT dataset); (**b**) SMB dimension reduction of the tablet's RBNN selected elemental composition (NT dataset).

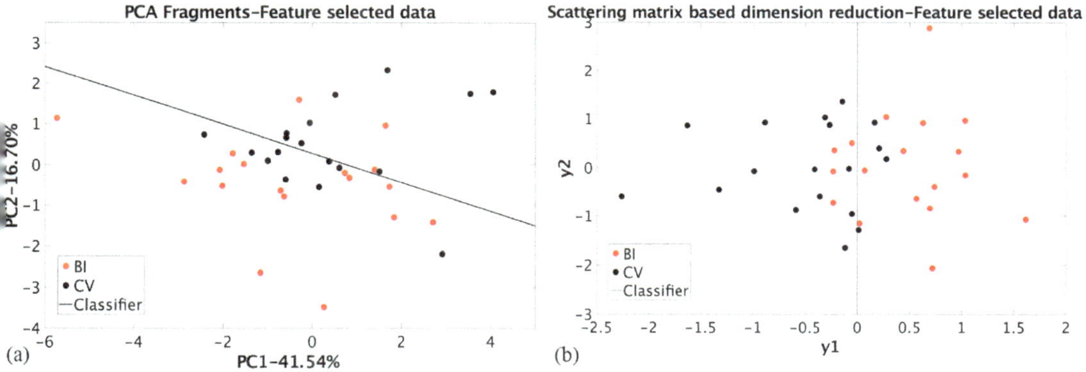

Figure 5. (**a**) PCA dimension reduction of the ceramic fragment's RBNN selected elemental composition (NF dataset); (**b**) SMB dimension reduction of the ceramic fragment's RBNN selected elemental composition (NF dataset).

The designed classifier enabled misclassification of 17% of BI and 50% of CV samples. The SMB dimension reduction achieved a slightly higher misclassification of 17% for BI and 22% for CV samples compared to the results for the tablet's dataset. The comparison of SMB techniques on the two datasets with selected features favors the NT dataset over the QT dataset ($\mu(1/2)$: 0.39 vs. 0.70, RA: 44% for BI and both techniques and 39% vs. 44% for CV, MC: 11% for BI and both techniques and 22% vs. 11% for CV). This is because part of the information about the dataset structure is lost during the quantification process, likely due to feature selection and an ancient porcelain standard. The results of the PCA dimension reduction of the raw EDXRF spectral data for tablets (RT dataset) and ceramic fragments (RF dataset) are shown in Figure 6.

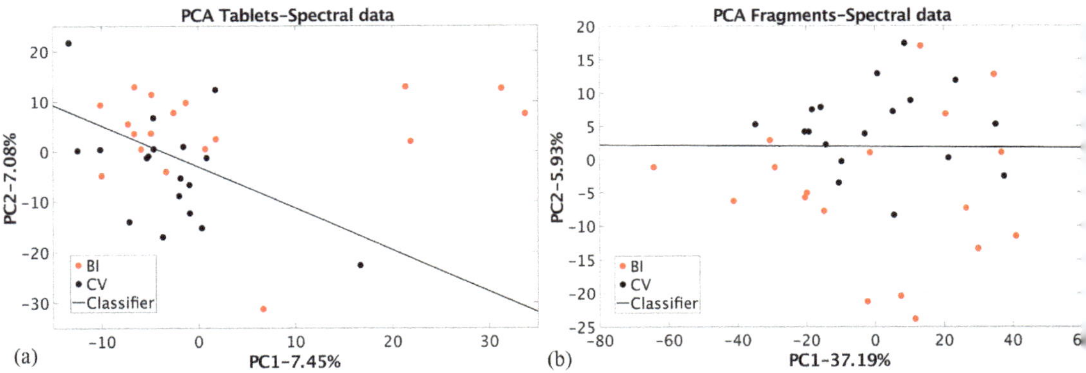

Figure 6. PCA dimension reduction of the raw EDXRF spectral data for (**a**) tablets (RT dataset) and (**b**) ceramic fragments (RF dataset).

The linear classification is possible using raw spectral data even if the $l_{d/n}$ parameter values are small (Table 2). More confident results are achieved for the fragments dataset in the PCA space, where less structural data were lost during the dimension reduction. Group coherence is smaller for the fragments dataset, resulting in a smaller Bhattacharya distance value. The classification parameters have the same values (Table 2). This result is highly significant as it shows that the dimension reduction of the dataset containing the qualitative results of the fragment analysis achieves classification results comparable to those of quantitative results on the homogenized tablets. This implies that a provenance study based on qualitative EDXRF results can be conducted with high reliability.

The small group coherence and the Bhattacharya distance values can be explained by using the same/similar clay sources for ceramic production, resulting in a small variance between data (Figure S1 in Supplementary Materials). The linear classifier was still possible to design, likely due to changes in production technology, such as adding different tempering materials.

The raw EDXRF spectral data obtained for Coţofeni–Kostolac ceramic fragments assemblages were organized in a matrix with a dimension of 13 × 1700. This dataset was subjected to PCA-based dimension reduction with Bubanj Hum I and Cernavoda III fragments. The analysis was performed to compare the results of dimension reduction and classification between the sets of non-similar raw materials (Figures S2 and S3 in Supplementary Materials). The results are shown in Figure 7.

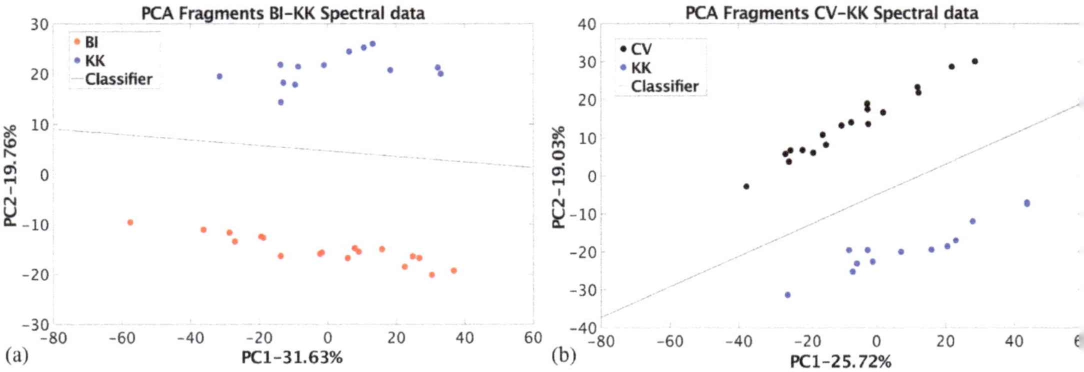

Figure 7. PCA dimension reduction of the raw EDXRF spectral data (**a**) for Bubanj Hum I and Coţofeni–Kostolac dataset; (**b**) for Cernavoda III and Coţofeni–Kostolac dataset.

The percentage of the total variance preserved in the PC1-PC2 space is 51.39% for the BI-KK and 44.75% for the CV-KK datasets. These results are comparable to those presented in Table 2. The Bhattacharya distance values are 20.37 and 27.27, respectively. These values suggest that classification can be performed without any misclassification.

4. Conclusions

The effectiveness of the two most-used dimension reduction techniques, PCA as an unsupervised method and SMB as a supervised method, was evaluated for the archaeological ceramic provenance study. In this evaluation, both quantitative and qualitative EDXRF analytical results were used. The chemical composition of the sampled ceramic materials, homogenized and pressed into tablets, was quantified using CRM and used as referent classification. An RBNN-based procedure was also used to determine the same chemical composition faster and more reliably to test the amount of lost information during feature extraction. Qualitative analysis using EDXRF spectrometry conducted on tablets and ceramic shards aimed to compare their effectiveness in the provenance study among themselves and with quantitative results. The dimension reduction was assessed based on the percentage of preserved total variance and the possibility of linear classification. The Bhattacharya distance was calculated in the reduced space to measure between-class separability as an indicator for effective classification. The recognition ability and percentage of misclassification were calculated for linear classification evaluation.

The preserved total variance ranged from 14.5% to 58%, providing a less representative picture of the initial dataset structure in the reduced space. This result was followed by small group coherence in the reduced spaces and low values of the Bhattacharya distance. Even so, it was possible to design the linear classifier, and the classification results showed a small misclassification. The parameters used to evaluate the success of dimension reduction and classification have similar values for quantitative and qualitative results, which is the most important result ($l_{d/n}$: 52.88% vs. 43.12%, $\mu(1/2)$: 0.31 vs. 0.25, RA: 36% vs. 39% for BI and 33% vs. 36% for CV, MC: 28% vs. 22% for BI and 33% vs. 28% for CV). Although quantification leads to precise chemical composition, it was shown that it might not always be the most informative method for provenance study and comparison with other assemblages, as some important information may be lost during feature selection. The qualitative analysis of the well-prepared shards' surface using EDXRF spectrometry can be utilized in provenance studies, even when the ceramic assemblages were made of similar raw materials. The results indicate that qualitative EDXRF results can be used reliably for provenance studies of archaeological ceramics, similar to the quantitative ones. This method requires easier sample preparation, is much faster, and does not necessitate the use of CRM.

Supplementary Materials: The following supporting information can be downloaded at: https://www.mdpi.com/article/10.3390/ma17153725/s1, Figure S1: PCA dimension reduction of the BI and CV ceramic fragments datasets using raw EDXRF spectral data. The clay material was classified using the linear classifier. Figure S2: PCA dimension reduction of the BI and KK ceramic fragments datasets using raw EDXRF spectral data. The clay material was classified using the linear classifier. Figure S3: PCA dimension reduction of the CV and KK ceramic fragments datasets using raw EDXRF spectral data. The clay material was classified using the linear classifier.

Author Contributions: Conceptualization, M.G.-K.; methodology, M.G.-K. and O.M.; formal analysis, M.G.-K.; resources, O.M. and A.B.; writing—original draft preparation, M.G.-K.; writing—review and editing, V.A., R.J.-H. and A.B.; supervision, R.J.-H. and A.B.; funding acquisition, M.G.-K. and V.A. All authors have read and agreed to the published version of the manuscript.

Funding: This research was funded by the Ministry of Science, Technological Development, and Innovation, Republic of Serbia (Grant No. 451-03-66/2024-03/200017 dated 5 February 2024-Research Program No. 1-Contract No. 110-10/2019-000, T0602403). This paper is the result of the project THE FLOW—Interactions-Transmission-Transformation: Long-distance connections in Copper and

Bronze Age of the Central Balkans, funded by the Science Fund of the Republic of Serbia, Program IDEAS (Grant No. 7750074).

Institutional Review Board Statement: Not applicable.

Informed Consent Statement: Not applicable.

Data Availability Statement: Data are contained within the article and supplementary materials.

Acknowledgments: The authors acknowledge Zeljko Djurovic from the University of Belgrade, School of Electrical Engineering, for valuable comments.

Conflicts of Interest: The authors declare no conflicts of interest.

References

1. Qi, Y.; Qiu, M.-Z.; Jing, H.-Z.; Wang, Z.-Q.; Yu, C.-L.; Zhu, J.-F.; Wang, F.; Wang, T. End-to-end ancient ceramic classifica toolkit based on deep learning: A case study of black glazed wares of Jian kilns (Song Dynasty, Fujian province). *Ceram. Int.* **2** *48*, 34516–34532. [CrossRef]
2. Fanti, L.; Melosu, B.; Cannas, C.; Mameli, V. Pottery vessels and technology of "colouring materials" in the central-west Mediterranean (Sardinia, Italy) during the Middle Neolithic: An interdisciplinary approach combining use-wear and chem physical analysis. *J. Archaeol. Sci. Rep.* **2024**, *53*, 104321. [CrossRef]
3. Vlase, D.; Vlase, G.; Ursut, G.; Sfirloaga, P.; Manea, F.; Budiul, M.; Rotaru, A.; Vlase, T. The in-depth study of Romanian prehist ceramics: Late Neolithic/Eneolithic pottery and clay materials from the Foeni Tell-Orthodox cemetery in Timis county. *Ceram.* **2023**, *49*, 14941–14956. [CrossRef]
4. Merk, V.; Berkh, K.; Rammlmair, D.; Pfeifer, L. Chemical and Mineralogical Analysis of Samples Using Combined LIBS, Rar Spectroscopy and μ-EDXRF. *Minerals* **2023**, *13*, 729. [CrossRef]
5. Orsilli, J.; Galli, A. Angle-Dependent XRF Analyses: Pros and Cons of a Novel Technique in the Field of Cultural Herit *Sustainability* **2024**, *16*, 1460. [CrossRef]
6. Gajić-Kvaščev, M.; Bikić, V.; Wright, V.J.; Radosavljević Evans, I.; Damjanović-Vasilić, L. Archaeometric study of 17th/18th cent painted pottery from the Belgrade Fortress. *J. Cult. Herit.* **2018**, *32*, 9–21. [CrossRef]
7. Emmitt, J.J.; McAlister, A.J.; Phillipps, R.S.; Holdaway, S.J. Sourcing without sources: Measuring ceramic variability with pXR *Archaeol. Sci. Rep.* **2018**, *17*, 422–432. [CrossRef]
8. Ruschioni, G.; Malchiodi, D.; Zanaboni, A.M.; Bonizzoni, L. Supervised learning algorithms as a tool for archaeology: Classi tion of ceramic samples described by chemical element concentrations. *J. Archaeol. Sci. Rep.* **2023**, *49*, 103995. [CrossRef]
9. Pashkova, G.V.; Statkus, M.A.; Mukhamedova, M.M.; Finkelshtein, A.L.; Abdrashitova, I.V.; Belozerova, O.Y.; Chubarov, V Amosova, A.A.; Maltsev, A.S.; Demonterova, E.I.; et al. A Workflow for Uncertainty Assessment in Elemental Analysi Archaeological Ceramics: A Case Study of Neolithic Coarse Pottery from Eastern Siberia. *Heritage* **2023**, *6*, 4434–4450. [Cross
10. Marques, R.; Rodrigues, A.L.; Russo, D.; Gméling, K.; Valera, A.C.; Dias, M.I.; Prudêncio, M.I.; Basílio, A.C.; Fernandes, P.G.; R F. Fingerprinting Ceramics from the Chalcolithic Santa Vitória Enclosure (SW Iberia). *Minerals* **2024**, *14*, 399. [CrossRef]
11. Bulatović, A.; Milanović, D. *Bubanj, The Eneolithic and the Early Bronze Age Tell in Southeastern Serbia*; Mitteilungen der Prä torischen Kommission 90: Wien, Austria, 2020.
12. Bulatović, A.; Vander Linden, M. Absolute Dating of Copper and Early Bronze Age Levels at the Eponymous Archaeolog SITEite Bubanj (Southeastern Serbia). *Radiocarbon* **2017**, *59*, 1047–1065. [CrossRef]
13. Baxter, M.J. A Review of Supervised and Unsupervised Pattern Recognition in Archaeometry. *Archaeometry* **2006**, *48*, 671– [CrossRef]
14. Li, T.; Yao, S.; He, L.; Yu, X.; Shan, S. Compositional study of household ceramic assemblages from a Late Neolithic (5300–450C BP) earthen walled-town in the middle Yangtze River valley of China. *J. Archaeol. Sci. Rep.* **2021**, *39*, 103159. [CrossRef]
15. King, R.H. Provenance of clay material used in the manufacture of archaeological pottery from Cyprus. *Appl. Clay Sci.* **198** 199–213. [CrossRef]
16. Forster, N.; Grave, P.; Vickery, N.; Kealhofer, L. Non-destructive analysis using PXRF: Methodology and application to arch logical ceramics. *X-ray Spectrom.* **2011**, *40*, 389–398. [CrossRef]
17. Shakhashiro, A.; Trinkl, A.; Törvenyi, A.; Zeiller, E.; Benesch, T.; Sansone, U. *Report on the IAEA-CU-2006-06 Proficiency Determination of Major, Minor and Trace Elements in Ancient Chinese Ceramic*; IAEA/AL/168; IAEA: Vienna, Austria, 2006.
18. Chang, Z.; Zhang, Q.; Li, Y.; Xin, X.; Gao, R.; Teng, Y.; Rao, L.; Sun, M. Identification Method for XRF Spectral Analysis Basec an AGA-BP-Attention Neural Network. *Electronics* **2024**, *13*, 507. [CrossRef]
19. Pincé, P.; Braekmans, D.; Lycke, S.; Vandenabeele, P. Ceramic Production in the Kur River Basin (Fars, Iran) During the Middl Late Second Millennium BCE: A Geochemical and Technological Characterization. *Archaeometry* **2019**, *61*, 556–573. [CrossR
20. Barouda, A.; Quinn, P.; Efstratiou, N. Agios Petros and the Neolithic pottery making traditions of the deserted islands, North Sporades, Greece. *Archaeol. Anthrop. Sci.* **2023**, *15*, 16. [CrossRef]

Gajic-Kvascev, M.D.; Maric-Stojanovic, M.D.; Jancic-Heinemann, R.M.; Kvascev, G.S.; Andric, V.D. Non-destructive characterisation and classification of ceramic artefacts using pEDXRF and statistical pattern recognition. *Chem. Cent. J.* **2012**, *6*, 102. [CrossRef] [PubMed]
Andric, V.; Gajic-Kvascev, M.; Korolija Crkvenjakov, D.; Maric-Stojanovic, M.; Gadzuric, S. Evaluation of pattern recognition techniques for the attribution of cultural heritage objects based on the qualitative XRF data. *Microchem. J.* **2021**, *167*, 106267. [CrossRef]
Varmuza, K.; Filzmoser, P. *Introduction to Multivariate Statistical Analysis in Chemometrics*, 1st ed.; CRC Press Taylor & Francis Group: Boca Raton, FL, USA, 2009; p. 63.
Fukunaga, K. *Introduction to Statistical Pattern Recognition*, 2nd ed.; Academic Press Inc.: San Diego, CA, USA, 1990.

claimer/Publisher's Note: The statements, opinions and data contained in all publications are solely those of the individual hor(s) and contributor(s) and not of MDPI and/or the editor(s). MDPI and/or the editor(s) disclaim responsibility for any injury to ple or property resulting from any ideas, methods, instructions or products referred to in the content.

materials

MDPI

Article

Integrating Attenuated Total Reflectance–Fourier Transform Infrared Spectroscopy and Multidetector Computed Tomography for Analysis of Heat-Induced Changes in Bone

Tamara Leskovar [1], Fabio Cavalli [2], Lea Legan [3,4], Dario Innocenti [2], Polonca Ropret [3,5,6] and Matija Črešnar [1,*]

1 Centre for Interdisciplinary Research in Archaeology, Department of Archaeology, Faculty of Arts, University of Ljubljana, Zavetiška 5, 1000 Ljubljana, Slovenia; tamara.leskovar@ff.uni-lj.si
2 Research Unit of Paleoradiology and Allied Sciences, TS—SCIT, Azienda Sanitaria Universitaria Giuliana Isontina, Via della Pietà, 2/1, 34100 Trieste, Italy; fabio.cavalli@asugi.sanita.fvg.it (F.C.); dario.inox@gmail.com (D.I.)
3 Institute for the Protection of Cultural Heritage of Slovenia, Poljanska Cesta 40, 1000 Ljubljana, Slovenia; lea.legan@zvkds.si (L.L.); polona.ropret@zvkds.si (P.R.)
4 National Museum of Slovenia, Prešernova Cesta 20, 1000 Ljubljana, Slovenia
5 Smithsonian Museum Conservation Institute, 4210 Silver Hill Road, Suitland, MD 20746, USA
6 Faculty of Chemistry and Chemical Technology, University of Ljubljana, Večna pot 113, 1000 Ljubljana, Slovenia
* Correspondence: matija.cresnar@ff.uni-lj.si

Abstract: Charred and burned human and animal skeletal remains are frequently found in archaeological records and can also be subjects of forensic investigations. Heat exposure causes specific changes to the physical and chemical characteristics of these remains, offering valuable insights into their taphonomic history. This research combines the commonly used ATR-FTIR (Attenuated Total Reflectance–Fourier Transform Infrared) spectroscopy with the rarely utilised density measurements obtained from Multidetector CT (Computed Tomography) to investigate changes in defleshed bovine cortical bone exposed to different temperatures for varying durations. The inclusion of density measurements is significant because Multidetector CT analysis is non-destructive and can be valuable when remains cannot be removed from their burial context (e.g., urn) or cannot be damaged. The results indicate complex changes in both organic and inorganic components, affecting crystallinity and density. Lower temperatures primarily affect organic matter, while higher temperatures induce significant changes in the mineral lattice and phase transitions. The transformation from β-tricalcium phosphate to α-tricalcium phosphate at high temperatures likely impacts the bone's crystallinity and density. Bone density measured by CT scans provided additional information that complemented the interpretations of FTIR spectroscopy. While CT scans offer important data for planning non-destructive analyses of remains, they present only one layer of information. Therefore, CT scans need to be combined with other techniques to provide comprehensive interpretations of the changes occurring in the bone. Further research is needed on density measurements and other potentially non-destructive analyses to fully unlock the potential of Multidetector CT analyses.

Keywords: skeletal remains; ATR-FTIR spectroscopy; Multidetector CT; FTIR indices; density; cremation

1. Introduction

Be it the result of cooking practices, mortuary and other rituals, destruction of evidence or anything in between, charred and burned human and animal skeletal remains are found in many archaeological and forensic contexts. As exposure to heat leaves a unique pattern in the structural and chemical characteristics of remains, they present a powerful source of understanding regarding the taphonomic history of the remains. In archaeological contexts, they can provide a rich source of information regarding life in the past, while in forensic contexts, they provide means for the reconstruction of events and possible identifications [1–3].

However, bones are a composite material, which makes understanding the effect of heat on their structural and chemical characteristics far from straightforward. Bones can be divided into cancellous and compact types, composed of water (approximately 10%) and closely linked organic (approximately 25%) and inorganic phases (approximately 60–70%). The former is mainly represented by lipids and proteins, 90% of which are type I collagen, formed by amino acid chains in a triple-helical structure [4,5]. Collagen molecules in the bone form elastic fibrils, held together by hydrogen bonds [6]. In the empty spaces between collagen molecules and fibrils are inorganic plate-shaped nanocrystals of non-stoichiometric (highly disordered) apatite with predominantly carbonate (3–8%) but also other (e.g., Na, Mg, Al, Pb, HPO_4) substitutions replacing phosphates and hydroxy groups [5,7–13]. Furthermore, bone mineral is hydrated as it contains water in its structure [14].

With exposure to heat, processes such as combustion and pyrolysis (thermolysis) change the structural and chemical characteristics of bones. To simplify, the changes that occur can be summarised as a loss of water (<250 °C), followed by a decline in organic components (200–600 °C) and changes in the mineral (predominantly > 700 °C). Most free water is lost as the temperature approaches 100 °C and structural water is lost as the temperature approaches 250 °C, with some small, additional changes around 400 °C, when water produced as a byproduct of organic matter combustion evaporates. Lipids completely degrade at temperatures of around 300–340 °C. Between 200 and 350 °C, a loss of collagen is observed. The majority of the organic phase is combusted between 300 and 400 °C in oxidising conditions and 600 and 700 °C in reducing conditions. Collagen is converted into alkylated benzenes and phenols and condensed aromatic compounds, which decline from around 500 °C onwards and almost disappear by 900 °C. At temperatures above 600–700 °C, cyanamide forms due to a reaction between the residual C and N from organic matter and calcium from the inorganic part of the bone [15–18].

Some minor changes in the mineral part of the bone can be observed between 250 and 340 °C, likely correlated with the changes in organic content. Most of the changes in bioapatite occur at temperatures above 600–700 °C, although an increase in crystal size, which is non-linearly correlated with increasing temperature, can start between 300 °C and 500 °C. Due to the increase in crystal size, at temperatures above 500 °C, a slow increase in density occurs, which becomes faster at temperatures above 700 °C. Most alterations in the bone mineral can be seen between 800 °C and 900 °C, when growth of the crystals and ordering of the crystal structure, accompanied by the loss of carbonates and structural water, are most pronounced. Structural carbonate loss starts at around 250–340 °C; however, most of the loss occurs at temperatures above 600 °C. Between 700 °C and 900 °C, in oxidising conditions, an increase in hydroxyl ions can be observed, when carbonate ions reorganise into carbon trioxide to create space for them. Additional carbonate losses occur at temperatures between 700 °C and 1000 °C, with some minor losses reported at temperatures up to 1100 °C [16–18].

Previous studies investigating bone hardness, porosity and density after exposure to high temperatures presented decreasing hardness with exposure to temperatures between

100 °C and 300 °C. The lowest hardness was observed to be between 400 °C and 500 °C, which remained low until exposure to temperatures of 800–900 °C, when it started to rise again [19–22], supposedly due to the increasing density of hydroxyapatite and crystal growth [23–25]. Bone porosity was highest in the samples exposed to temperatures of 600 °C (around 50%), while with increasing temperatures, porosity decreased, with sintering occurring at approximately 1200 °C [26].

Analysing these changes in bones can offer information regarding the reasons and methods for exposing remains to heat. Thus, there are numerous studies focusing on heat-induced changes [27–35]. Although there are various techniques used for this purpose, vibrational spectroscopy, especially Fourier Transform Infrared (FTIR) spectroscopy, is one of the most common choices [28,31,32,34–45]. This is likely because FTIR spectroscopy is a convenient and relatively inexpensive technique for the exploration of molecular structure and material characterisation. Additionally, the Attenuated Total Reflectance (ATR) technique is increasingly being used as it minimises sample preparation, improves signal-to-noise ratios and increases comparability among samples [46,47]. When used on skeletal tissues, FTIR spectroscopy allows researchers to explore the organic (mainly collagen) and inorganic (bioapatite) parts of the tissues. Due to its precision and well-established use in research on skeletal tissues, it also enables detection of small chemical and/or structural variations in the samples [31,33,48,49]. However, the chemical properties of skeletal tissues are heterogeneous, making molecular characterisation using FTIR spectra elusive [31]. To characterise the molecular structure of skeletal samples, calculations of the ratios between peak heights and/or areas are necessary, ensuring that the results present relative and not absolute concentrations of components in the sample [50].

Another less-known technique for studying skeletal tissues is Multidetector CT, which enables the acquisition of three-dimensional (3d) volume images. It is gaining popularity in archaeological research, especially for analysing ancient cinerary, as it can reveal the contents of urns [51,52]. A wide X-ray beam is used in conjunction with multiple rows of detectors to acquire more than one CT slice in a single tube rotation and to simultaneously avoid tube overloading. In general, a Multidetector CT slice consists of a matrix of the smallest elements called "voxels". The coefficient of absorption at each voxel is converted into a CT value, reported in Hounsfield units (HUs), which is defined by Equation (1) as follows:

$$\text{CT value} = \text{k} \times \frac{\mu_t - \mu_w}{\mu_w} \tag{1}$$

where μ_t is the coefficient of absorption at the scanning volume, μ_w is the coefficient of absorption with respect to water and k is a proportionality constant, which is fixed at a value of 1000.

Hounsfield unit (HU) values are defined as a measurement of the linear attenuation coefficient of a tissue, based on a defined scale of 0 for the radiodensity of distilled water at standard pressure and temperature and −1000 for air at standard pressure and temperature. In CT scans, the reconstruction software calculates HU values for each volume unit (voxel) of the scan [53].

The density of cortical bone varies depending on whether it is measured in living bone or in bone remains, including historical/archaeological ones. However, ranges vary between 1150 and 1550 HU. In skeletal remains, there is some loss of the mineral component of the bone due to taphonomic processes. Cremated bones instead seem to have a higher density, generally over 2000 HU and up to more than 2500 HU, probably due to changes in the composition of the mineral component and alterations in the morphology and size of crystallites of bone hydroxyapatite [54]. Preliminary experiments seem to express a certain correlation between the density of the cortex of the cremated bone and the cremation temperature [55].

Aims of the Research

The vast amount of research using FTIR spectroscopy to address changes in bones exposed to heat has allowed researchers to establish FTIR indices with relatively well-known meanings (see Table 1). However, there are still some gaps in our understanding of FTIR results and heat-induced changes, one being the effect of time of exposure. Although it was previously observed that time of exposure does play a role [40,56], it is not commonly and systematically included in research designs. Thus, the first aim of this study was to address the influence of time of exposure on changes in bones observed with FTIR spectroscopy.

Contrarily to FTIR spectroscopy, bone density calculated from CT scans is not commonly used to understand heat-induced changes in bones. As it could provide a completely non-destructive technique for assessing information about bones, this study expands the commonly applied FTIR indices with CT density. First, a set of baseline changes was established using calculated indices from the FTIR spectra. In the next step, density calculations were performed and compared against the FTIR indices. The second aim of this study was to establish a correlation between FTIR indices and density to better understand what information density can tell us about the bone. Since Multidetector CT is a non-destructive method, which can be performed without disturbing the burial context, the aim was to determine if it could be used as a standalone technique for studying and interpreting taphonomic changes in remains or if other methods need to be incorporated.

Research was conducted on a single bone with known taphonomic history to eliminate the influence of individual variability and the unknown complexities of bone diagenesis. This approach was chosen to isolate the effects of time and temperature of exposure, as FTIR spectra and density are known to be significantly influenced by extrinsic factors (e.g., post-depositional environment) and intrinsic factors (e.g., age, sex, skeletal element) [30,33,57–59]. By focusing on a single specimen, the study effectively controlled for these variables, ensuring that the observed changes were directly attributable to the experimental conditions. This method provides a clearer understanding of how time and temperature impact bone properties, free from the confounding effects of natural variability.

Table 1. Included FTIR indices.

Index	~Peaks	Characterisation	References
IRSF	(560 + 600)/585	crystallinity, indicating crystal size and order in the matrix	[60,61]
H1640	1640 height	relative amount of Amide I	[62,63]
H1415	1415 height	relative amount of type B carbonates + organic matter	[28,64,65]
OH	630 height	relative amount of hydroxyl groups	[32]
CN	2015 height	relative amount of cyanamide	[32,66]
H960	960 * height	relative amount of well-crystallised apatite	[67]
H872	872 * height	relative amount of B-type carbonates	[68]
H878	878 * height	relative amount of A-type carbonates	[68]

* Peaks measured in second-derivative spectra.

2. Materials and Methods

2.1. Samples

The mid part of the femur diaphysis of a single adult bovine was manually defleshed and cut into rectangular samples of 4 × 4.5 cm. One sample was used as a control while sixteen others were exposed to low, mid and high temperatures in a vented muffle furnace. Each sample was placed into the furnace individually once the set temperature was reached.

Samples were exposed to temperatures of 300 °C, 600 °C, 900 °C and 1200 °C for 30, 60, 90 and 120 min each. After exposure, samples were naturally cooled to room temperature and stored inside perforated polyethylene bags.

2.2. ATR-FTIR Spectroscopy

Each sample was analysed using a Perkin Elmer Spectrum 100 FTIR spectrophotometer in Attenuated Total Reflectance (ATR) mode, equipped with a thallium–bromoiodide (KRS-5) crystal. To ensure full physical contact between the samples and the ATR setup, firm and consistent pressure was applied during measurements to optimise the interaction between the sample surface and the ATR crystal. The samples were measured in their original form, with full pieces of bone used; they were not crushed or altered. ATR spectra were collected in the range between 4000 cm^{-1} and 380 cm^{-1}. For each sample, measurements were taken at five distinct spots, with 32 scans collected at a spectral resolution of 4 cm^{-1} per spot. All spectra underwent ATR correction using Spectrum software (Perkin Elmer, Norwalk, CT, USA). Additionally, the spectra were baseline-corrected, normalised (Min-Max normalisation) to the highest peak at ~1010 cm^{-1} and averaged using the OPUS 7.0 software package (Bruker, Bremen, Germany).

As most of the peaks used to understand the changes in the bone due to heat exposure were found to be between 550 cm^{-1} and 2020 cm^{-1}, scanned spectra were reduced to wavelengths between 480 cm^{-1} and 2100 cm^{-1}. Due to overlapping, three peaks were measured in second-derivative spectra using the Savitzky–Golay filter with 7 smoothing points (Supplementary Materials Figures S1 and S2).

FTIR indices were measured/calculated (Supplementary Materials, Figures S1 and S2) to help better understand the molecular structure of the bones. Indices were chosen based on the published research studying bones exposed to high temperatures (Table 1). Although there are more indices mentioned in various other studies, only the ones where peaks were clearly measurable for all the samples were included. Due to the normalisation of the spectra to the PO_4 peak at ~1010 cm^{-1}, some of the peaks, namely OH and CN, were slightly adjusted and a peak height of ~1010 cm^{-1} was used for the ratio calculation instead of ~605 cm^{-1}.

2.3. CT Scanning

After cremation, the samples were scanned with a 16-slice Multidetector CT scanner (Toshiba Aquilion 16, 129 KVp, 200 mA, isotropic 0.5 mm voxel). The HU value of each voxel of the sample was registered and classified by frequency. In order to exclude the contribution of the partial-volume effect due to the micro-cracking of the bone structure caused by the heat, the histogram was deconvoluted to obtain the mean HU value of the higher component of the curve (CT).

2.4. Statistical Analyses

Statistical analyses were performed using IBM SPSS Statistics. First, normality of the data was checked using the Shapiro–Wilk test. Based on the results, parametric tests for normally distributed and non-parametric tests for not-normally distributed data were conducted. Spearman correlation was used to observe correlations between different FTIR indices and the CT density index. A confidence interval of 95% was used for means or medians to explore the degree of change in the bones due to exposure to heat and to evaluate their significance.

3. Results

All the obtained FTIR indices and densities are presented in the Supplementary Materials Table S1 and averaged measured/calculated FTIR and CT density indices are presented in Table 2.

Table 2. Measured/calculated FTIR indices and average CT density (CT).

	Temperature (°C)				
Index	0	300	600	900	1200
H1640	**1**	0.16 ± 0.096	0.035 ± 0.014	0.019 ± 0.004	0.012 ± 0.005
CN	**0.136**	0.034 ± 0.044	0.043 ± 0.028	0.163 ± 0.118	0.061 ± 0.078
H1415	**0.70**	0.19 ± 0.07	0.098 ± 0.021	0.028 ± 0.007	0.009 ± 0.003
H872	**0.0052**	0.0029 ± 0.0003	0.0027 ± 0.0005	0.0008 ± 0.0003	0.00006 ± 0.00001
H878	**0.0002**	0.0025 ± 0.0007	0.0023 ± 0.0005	0.0020 ± 0.0009	0.0010 ± 0.0015
SF	**3.17**	3.55 ± 0.17	4.29 ± 0.36	5.40 ± 0.35	4.81 ± 0.45
H960	**0.0003**	0.0016 ± 0.0001	0.0033 ± 0.0012	0.0122 ± 0.0032	0.0108 ± 0.0020
OH	**0.022**	0.078 ± 0.027	0.115 ± 0.022	0.183 ± 0.086	0.248 ± 0.080
CT	2139.8	1934.3 ± 195.1	1711.5 ± 89.1	2656.6 ± 294.8	3344.9 ± 216.5

3.1. Changes with Time and Temperature

The obtained FTIR results (Figure 1) showed that with exposure to temperatures of 300 °C, significant changes had already occurred in the bone. Significant decreases in H1640, H1415 and H872, as well as a significant increase in H878, had already occurred after 30 min of exposure. An increase in OH also occurred but was only clear after 60 min of exposure at 300 °C and there were no further significant changes until prolonged (90 min) exposure at 900 °C. IRSF slightly increased at 300 °C, continuing to gradually increase with prolonged exposure, as did stoichiometric phosphate (H960), but the latter presented no differences with prolonged exposure.

At 600 °C, H1640 continued to slowly decrease and was lost after 90–120 min of exposure. Similarly, H1415 continued to decrease; however, the changes were small and barely significant in the first 30 min of exposure and again after 90 min of exposure. H872 and H878 presented non-significant but observable decreases after at least 90 min of exposure, while at the same time, OH presented a similar pattern of increase. IRSF continued to increase slowly and gradually, as did H960; however, the latter presented a more pronounced increase upon 90 min and 120 min of exposure.

At 900 °C, H1640, H1415 and H872 were almost completely lost. In some cases, some remnants might still be detectable in the first 30 min. On the other hand, H878 and OH both initially decreased and then significantly increased after 90 min and 120 min of exposure. CN significantly increased after 30 min and 60 min of exposure and significantly decreased after 90 min and 120 min of exposure. The increase in IRSF was still present but stopped after 90 min and 120 min of exposure, while H960 showed an extreme increase after 60 min of exposure with no further significant change.

At 1200 °C, H1640, H1415, H872 and H878 has been largely removed. In some cases, some leftovers of H878 might still be visible after 30 min of exposure, and the same goes for CN, which seemed to persist for a bit longer, with peaks sometimes still detected after 60 min and 90 min of exposure. IRSF and H960 both experienced a decrease at 1200 °C exposure. IRSF decreased non-significantly, without any obvious differences caused by different time of exposure. H960 decreased significantly with some noticeable variation due to time of exposure, presenting significant differences between 30 min and 120 min of exposure, as it started to increase again with prolonged time of exposure. The OH peak seemed to resemble the peaks at 900 °C, although an increase was seen with exposure at 1200 °C for 120 min.

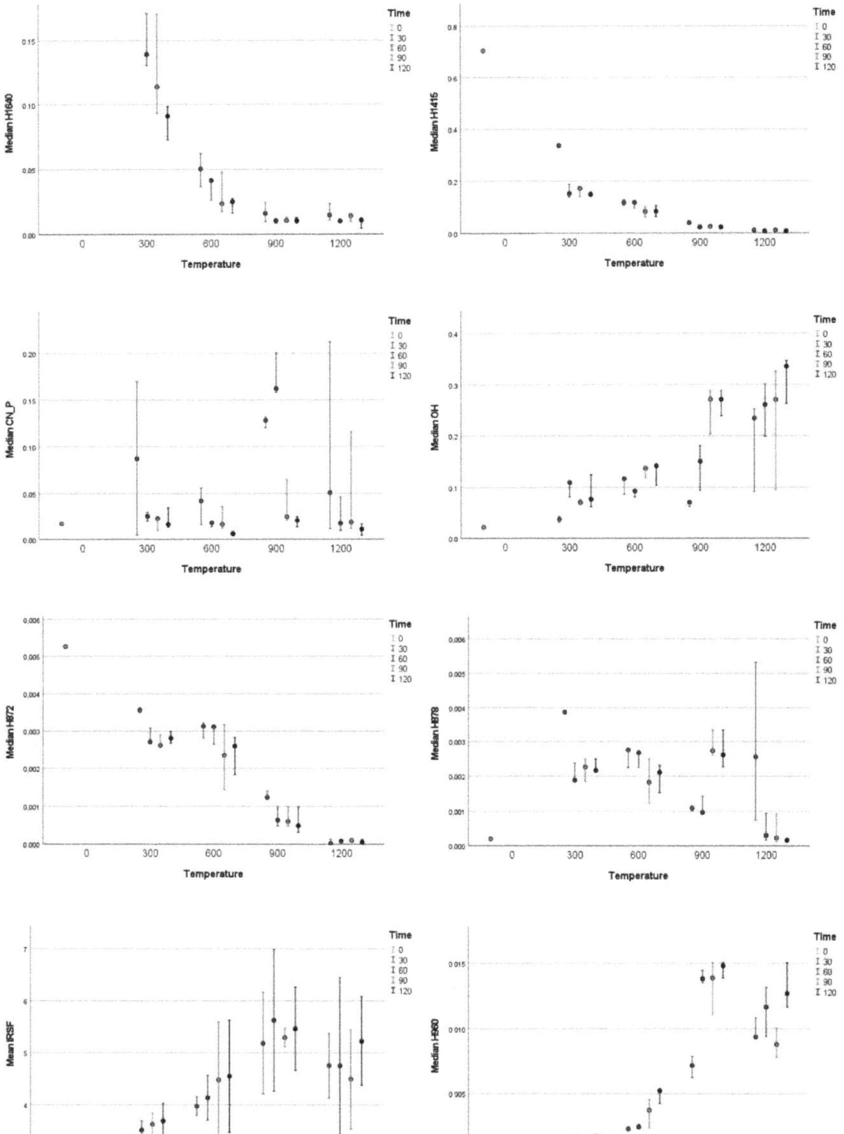

Figure 1. Confidence intervals of FTIR indices for different temperatures and durations of exposure.

CT density only significantly changed after 120 min of exposure at 300 °C, when it decreased. The next significant change only occurred after 90–120 min of exposure at 600 °C; however, a more substantial increase was observed after exposure at 900 °C, which continued until 30 min of exposure at 1200 °C. With 60–120 min of exposure at 1200 °C, CT density decreased again (Figure 2).

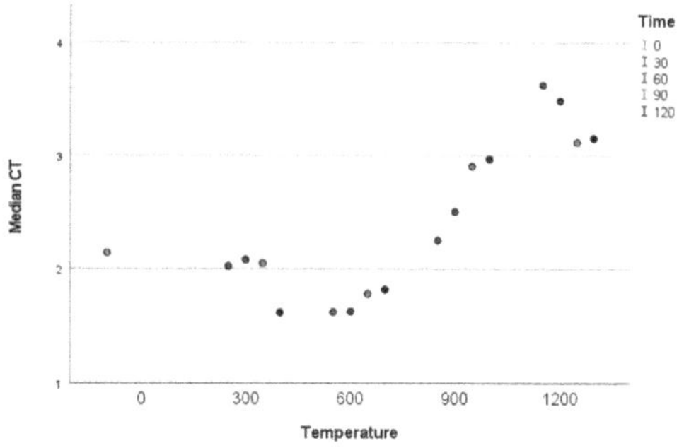

Figure 2. CT density for different temperatures and durations of exposure.

3.2. Correlations

Considering the observed changes in CT density and FTIR spectroscopy measurements, the correlation analysis was adjusted as follows:

(1) First decrease in density (300 °C/120 min–600 °C/60 min): strong negative correlation with H1640 (0.84) and H1415 (−0.74); strong positive correlation with IRSF (0.84) and H960 (0.79).
(2) First small increase in density (600 °C/90–120 min): strong negative correlation with CN (−0.88); moderate positive correlation with H960 (0.68).
(3) First high increase in density (900 °C/30–60 min): strong negative correlation with H872 and H1415 (−0.88); strong positive correlation with H960, CN and OH (0.88).
(4) Second high increase in density (900 °C/90 min–1200 °C/30 min): strong negative correlation with H872 and H1415 (−0.80); moderate negative correlation with H960 and IRSF (−0.64).
(5) Final decrease in density (1200 °C/60–120 min): strong negative correlation with H1415 (−0.79); moderate negative correlation with H872 (−0.53); moderate positive correlation with H960 (0.53).

4. Discussion

4.1. FTIR Spectroscopy

Changes at 300 °C indicated loss of organic matter (H1640, partially H1415) following a short period (30 min) of exposure. Changes in the mineral lattice also took place in these early stages, as losses of B-type carbonates (H872) were accompanied by increases in A-type carbonates (H878), crystallinity (IRSF) and stoichiometric phosphates (H960). With prolonged exposure, further losses of collagen occurred (H1640), hydroxyl ions (OH) were incorporated into the lattice and crystallinity further increased (IRSF), while both types of carbonates (H1415, H872 and H878) and stoichiometric phosphates (H960) were not affected by the additional time of exposure. Significant changes in the organic part are to be expected [16–18], while the observed changes in the mineral part might predominantly be a consequence of changes in the organic part, as some of the spectral locations are shared and calculated ratios only present relative concentrations of specific molecules in the bone.

The presence of the Amide I peak (H1640) at 600 °C and its small, gradual decrease were visible due to the conversion of collagen into aromatic compounds [17]. Decreases in all type of carbonates (H1415, H872 and H878), especially with prolonged (90–120 min) exposure, indicated the non-selective loss of carbonates, which predominantly occurred only after prolonged exposure. On the other hand, a gradual increase in crystallinity (IRSF) was accompanied by an increase in stoichiometric phosphates (H960) and the incorporation of hydroxyl ions (OH) into the mineral lattice. These changes were consistent with reported reorganisation of the mineral lattice due to carbonate losses and crystal growth [16,17,31,69].

Although some significant changes in the mineral already occurred with prolonged exposure at 600 °C, they became more pronounced at higher temperatures. Furthermore, with prolonged exposure at 900 °C, a clear distinction between continuingly decreasing B-type carbonates (H872) and increasing A-type carbonates (H878) occurred, which was accompanied by an increase in hydroxyl ions (OH) and a decrease in cyanamides (CN). The latter presented a tangible increase during short periods of exposure (30 min and 60 min) and decreased significantly with prolonged (90 min and 120 min) exposure. The predominant loss of B-type carbonates observed agrees with the findings of previous studies [31,32,48,68], although one should take into account the very low relative amounts of A-type carbonates in the bone [70,71]. When acknowledging hydroxyl groups, cyanamides and A-type carbonates together, an interesting interplay can be seen. Short-term exposure resulted in an increase in cyanamides, accompanied by a decrease in A-type carbonates and hydroxyl groups, while with prolonged exposure, the opposite occurred: there was a decrease in cyanamides, accompanied by an increase in A-type carbonates and hydroxyl groups. When considering only samples exposed at 900 °C (hydroxyl peaks are known to increase only after exposure at 700 °C) [26,31], there was a strong negative correlation between A-type carbonates and cyanamides ($r = -0.74$), a moderate negative one between hydroxyl groups and cyanamides ($r = -0.64$) and a strong positive one between A-type carbonates and hydroxyl groups ($r = 0.76$). Thus, at 900 °C, cyanamides increased while A-type carbonates and OH decreased. Previous observations indicate that during heating, A-type carbonates are lost and replaced by hydroxyl groups [32]. When also considering cyanamides, Snoeck et al. [40] proposed that, when present, cyanamides substitute for hydroxyl groups and replace A-type carbonates. However, based on our results, at 900 °C, A-type carbonates and hydroxyl groups seemed to act together, with cyanamides on the other side. This only changed with complete loss of both A-type carbonates and cyanamides following prolonged exposure at 1200 °C, when only hydroxyl groups were left and were further incorporated into the lattice. The presence of cyanamides mostly correlated to the combustion of organics, namely to the heating of ammonia [72], in the case of bone possibly deriving from organic matter (collagen) [40]. Thus, it is surprising that they were still detectable at 1200 °C. As some leftovers of aromatic compounds were also present, they might be "feeding" the peak of cyanamides. A strong positive correlation was also seen between CN and H1640 (0.83) following short-term exposure at 1200 °C, almost completely disappearing (0.26) with prolonged time of exposure. With complete loss of cyanamides and A-type carbonates, only the hydroxyl group remained as substitutes in the lattice and were thus able to further increase. At this point, IRSF did not seem to change significantly, even decreasing slightly. This decrease had been previously noted in [31,32], but the reasons behind it are unclear, generally being attributed to the rearrangement of the apatite structure. Legan et al. [36] noticed additional bands characteristic of α-tricalcium phosphate (TCP) after exposure at 1200 °C, which agrees with a thermal transformation of β-TCP to α-TCP, occurring at approximately 1150 °C [73]. Since β-TCP has a rhombohedral, more stable structure with uniformly distributed Ca vacancies when compared to monoclinic α-TCP [74,75], transformation to α-TCP might be associated with the decreasing crystallinity.

4.2. CT-Obtained Density

The presented results showed an initial significant decrease in density with prolonged exposure at 300 °C, followed by a gradual increase until the period of prolonged exposure at 1200 °C. A decrease in density following exposure at 600 °C, followed by an increase upon exposure at 900 °C and 1200 °C, was also observed by Figueiredo et al. [26]. It was explained by changes in bone porosity caused by a loss of collagen at lower temperatures and complete removal of carbonate from the mineral lattice at higher temperatures. Based on the FTIR indices and correlations, our results partially agree with these interpretations. It seems like density only starts to change with prolonged exposure to low temperatures (300 °C) or after a short period of exposure to medium temperatures (600 °C). Correlations indicate that changes in the density occur due to losses of organic matter with the initial changes in crystallinity. However, according to FTIR indices, the latter are significantly less substantial, leading to lower density. This agrees with previous research showing that most of the water and organic matter is lost at temperatures between 300 °C and 800 °C when bone losses equate to approximately 30–45% of its weight [21,69,76–78]. Since the presence of organic matter protects inorganic components of the bone [69,78] and bone hydroxyapatite remains unchanged up to high temperatures [21,79], the results are understandable. Significant increases in density only occurred with exposure to high temperatures (900 °C), a phenomenon also observed by other researchers [26]. However, the reasons for the increase were more complex than just the loss of carbonates from the matrix. B-type carbonates were removed, but they were replaced, first by cyanamides and hydroxyl groups, and with prolonged exposure, following the removal of cyanamides, by A-type carbonates and hydroxyl groups. Although organic matter and B-type carbonates were removed, other molecules were being incorporated into the mineral matrix. Furthermore, at these temperatures, increases in stoichiometric apatite and crystallinity were observed, altogether likely causing density to increase. With longer exposure to very high temperatures (1200 °C), cyanamides and A-type carbonates were also removed and only hydroxyl groups persisted as substitutions. Although the mineral matrix was becoming "cleaner", density and crystallinity both decreased. The latter is possibly due to the rearrangement of the apatite structure [31,32], while the decrease in density is unusual, if we assume crystal growth occurred and a closely interlocked structure was produced [69]. It might be due to the observed transition of β-TCP to α-TCP, as the former has higher density than α-TCP (3.07 g/cm^3 vs. 2.86 g/cm^3) [80–82].

5. Conclusions

This study aimed to improve our knowledge of heat-induced changes in bones observable with FTIR spectroscopy, also acknowledging the time of exposure and exposure to very high temperatures (1200 °C). Furthermore, it correlates relatively well-established FTIR indices with the simultaneous changes in bone density measured with CT.

The exposure of bone to varying temperatures results in complex changes in both organic and inorganic components, influencing crystallinity and density. Lower temperatures primarily affected organic matter, while higher temperatures induced significant mineral lattice changes and phase transitions. The transformation from β-TCP to α-TCP at high temperatures likely impacts the crystallinity and density of the bone. The changes at different temperatures can be summarised as follows:

Low-Temperature Changes (300 °C):

- Initial exposure leads to a loss of organic matter and changes in the mineral lattice.
- Losses of B-type carbonates coincided with increases in A-type carbonates, crystallinity and stoichiometric phosphates.
- Prolonged exposure resulted in further collagen loss, the incorporation of hydroxyl ions into the lattice and increased crystallinity.

Medium-Temperature Changes (600 °C):

- Collagen converted into aromatic compounds.

- Crystallinity and stoichiometric phosphates increased, accompanied by hydroxyl ion incorporation.
- There was a non-selective loss of all carbonate types with prolonged exposure.
- There was an initial significant density decrease due to the loss of organic matter.

High-Temperature Changes (900 °C):

- There was interplay between hydroxyl groups, cyanamides and A-type carbonates; short-term exposure resulted in an increase in cyanamides, accompanied by a decrease in A-type carbonates and hydroxyl groups, while with prolonged exposure, the opposite occurred.
- There was a significant density increase due to interplay between hydroxyl groups, cyanamides and A-type carbonates, and an increase in stoichiometric apatite and crystallinity.

Very-High-Temperature Changes (1200 °C):

- Complete loss of cyanamides and A-type carbonates.
- Hydroxyl groups continued to increase.
- Density and crystallinity decreased with prolonged exposure, possibly attributed to the transition from β-TCP to α-TCP.

The observed changes in the FTIR indices agreed with previous studies, while also offering a possible explanation—the transition from β-TCP to α-TCP—for the decreasing crystallinity previously observed with exposure to very high temperatures. Additionally, the study revealed significant differences between the short- and long-term exposure of the bones to different temperatures, indicating that time, although often neglected, is also a crucial factor in heat-induced changes. Because samples originated from a singular bone with known taphonomic history, the observed changes can directly be correlated to the exposure conditions. As such, the results provided a solid baseline for understanding the components of temperature and time, and the correlations between changes observed in FTIR indices and CT density. This is particularly important when working with valuable heritage materials. Firstly, FTIR analyses were performed on a piece of bone and not on powder, limiting the destructiveness of the method. Furthermore, Multidetector CT is a non-destructive technique. It can be, for instance, carried out without disturbing the burial context (e.g., urn) and/or damaging the remains (e.g., human bones). When results of Multidetector CT were analysed alongside FTIR indices, which offer analysis of several layers of data (changes in different molecules attributed to the organic and mineral phases of the bone), CT density only provided one value. Although there was a pattern of initial decrease, followed by a significant, gradual increase and another decrease in CT density, it would be hard to classify most of the samples in terms of the explicit (time and) temperature to which they were exposed. CT density could be a useful parameter to study bones exposed to heat but, for now, not as a standalone technique. Future research should include other parameters, also obtainable in a non-destructive way (e.g., porosity, microstructure), to interpret the taphonomic changes in bones exposed to heat.

The results offer meaningful insights and thus a sort of baseline for investigating the chemical and structural changes occurring in bones exposed to different temperatures and for different times. This could be useful when trying to decipher mortuary and other rituals, and also in forensic cases of combustion of the body. It was relatively easy to distinguish between samples exposed to different temperatures and, to an extent, times, acknowledging all the FTIR indices and density measurements using non-linear dimensionality reduction (Supplementary Materials Figure S4). However, it is well known that various intrinsic and extrinsic factors will significantly affect the FTIR and density results. For the results to be transferable to archaeological and forensic contexts, the effect of the biological profile, the skeletal element and post-depositional bone diagenesis first need to be evaluated.

Supplementary Materials: The following supporting information can be downloaded at https://www.mdpi.com/article/10.3390/ma18040742/s1, Figure S1: Process of spectra manipulation; Figure S2: Peak measurement; Figure S3: Non-linear dimensionality reduction using Manifold Learning, a technique which finds a non-linear manifold within the higher-dimensional space and outputs new coordinates which correspond to a two-dimensional space (Orange Data Mining). Data is visualized with Scatter Plot; Table S1: Obtained indices.

Author Contributions: T.L. prepared the manuscript and performed ATR-FTIR spectroscopy data analyses. L.L. and P.R. carried out the ATR-FTIR spectroscopy part of the analyses and reviewed the manuscript. F.C. and D.I. prepared all the samples. F.C. performed CT data analyses. M.Č. organised all the work steps, was the main communication link and reviewed the article. All authors have read and agreed to the published version of the manuscript.

Funding: The study was conducted within the framework of E-RIHS PP (H2020-INFRADEV-2016-2017, Grant Agreement No. 739503) and Centre for Interdisciplinary Research in Archaeology. It was supported by the Slovenian Research Agency (ARIS) through research core funding P6-0247 (Archaeology), P6-0282 ("Objects and prestige: taste, status, and power (Researches of the material culture in Slovenia)"), P1-0447 ("N-DAD—Non-destructive analysis and diagnostics") and the research project "Next generation analytical tools for heritage science" (J7-50226).

Institutional Review Board Statement: Not applicable.

Informed Consent Statement: Not applicable.

Data Availability Statement: The datasets generated and/or analysed during the current study are available from the corresponding author on reasonable request.

Acknowledgments: We would like to thank Omega D.O.O. (Slovenia) for collaboration and the loan of the PerkinElmer Universal ATR accessory.

Conflicts of Interest: The authors declare that they have no known competing financial interests or personal relationships that could have appeared to influence the work reported in this paper.

ferences

Fairgrieve, S. Burned Remains in Forensic Contexts. In *Encyclopedia of Global Archaeology*; Smith, C., Ed.; Springer: New York, NY, USA, 2014; pp. 1072–1077, ISBN 978-1-4419-0465-2.
Ubelaker, D.H. The Forensic Evaluation of Burned Skeletal Remains: A Synthesis. *Forensic Sci. Int.* **2009**, *183*, 1–5. [CrossRef] [PubMed]
Symes, S.A.; Rainwater, C.W.; Chapman, E.N.; Gipson, D.R.; Piper, A.L. Patterned Thermal Destruction of Human Remains in a Forensic Setting. In *The Analysis of Burned Human Remains*; Schmidt, C.W., Symes, S.A., Eds.; Elsevier: London, UK, 2008; pp. 15–54.
White, T.D.; Folkens, P.A. *The Human Bone Manual*; Elsevier: Oxford, UK, 2005; ISBN 0120884674/9780120884674.
Boskey, A.L. Mineralization of Bones and Teeth. *Elements* **2007**, *3*, 385–391. [CrossRef]
Adamiano, A.; Fabbri, D.; Falini, G.; Giovanna Belcastro, M. A Complementary Approach Using Analytical Pyrolysis to Evaluate Collagen Degradation and Mineral Fossilisation in Archaeological Bones: The Case Study of Vicenne-Campochiaro Necropolis (Italy). *J. Anal. Appl. Pyrolysis* **2013**, *100*, 173–180. [CrossRef]
Termine, J.D.; Posner, A.S. Amorphous/Crystalline Interrelationships in Bone Mineral. *Calcif. Tissue Res.* **1967**, *1*, 8–23. [CrossRef]
Posner, A.S. Crystal Chemistry of Bone Mineral. *Physiol. Rev.* **1969**, *49*, 760–792. [CrossRef]
Gokhale, J.A.; Boskey, A.L.; Robey, P.G. Chapter 4—The Biochemistry of Bone. In *Osteoporosis*, 2nd ed.; Marcus, R., Feldman, D., Kelsey, J., Eds.; Academic Press: San Diego, CA, USA, 2001; pp. 107–188, ISBN 978-0-12-470862-4.
Wopenka, B.; Pasteris, J.D. A Mineralogical Perspective on the Apatite in Bone. *Mater. Sci. Eng. C* **2005**, *25*, 131–143. [CrossRef]
Lebon, M.; Reiche, I.; Fröhlich, F.; Bahain, J.J.; Falguères, C. Characterization of Archaeological Burnt Bones: Contribution of a New Analytical Protocol Based on Derivative FTIR Spectroscopy and Curve Fitting of the $\nu_1 \nu_3$ PO_4 Domain. *Anal. Bioanal. Chem.* **2008**, *392*, 1479–1488. [CrossRef]

12. Rey, C.; Combes, C.; Drouet, C.; Glimcher, M.J. Bone Mineral: Update on Chemical Composition and Structure. *Osteoporos.* **2009**, *20*, 1013–1021. [CrossRef]
13. Skinner, H.C.W. *Mineralogy of Bones BT—Essentials of Medical Geology: Revised Edition*; Selinus, O., Ed.; Springer: Dordre The Netherlands, 2013; pp. 665–687, ISBN 978-94-007-4375-5.
14. Pasteris, J.D.; Yoder, C.H.; Wopenka, B. Minerals in the Human Body: Molecular Water in Nominally Unhydrated Carbona Hydroxylapatite: The Key to a Better Understanding of Bone Mineral. *Am. Mineral.* **2014**, *99*, 16–27. [CrossRef]
15. Dowker, S.E.P.; Elliott, J.C. Infrared Absorption Bands from NCO^- and NCN^{2-} in Heated Carbonate-Containing Apat Prepared in the Presence of $NH_4{}^+$ Ions. *Calcif. Tissue Int.* **1979**, *29*, 177–178. [CrossRef]
16. Van Hoesel, A.; Reidsma, F.H.; van Os, B.J.H.; Megens, L.; Braadbaart, F. Combusted Bone: Physical and Chemical Change Bone during Laboratory Simulated Heating under Oxidising Conditions and Their Relevance for the Study of Ancient Fire Us *Archaeol. Sci. Rep.* **2019**, *28*, 102033. [CrossRef]
17. Reidsma, F.H.; van Hoesel, A.; van Os, B.J.H.; Megens, L.; Braadbaart, F. Charred Bone: Physical and Chemical Changes du Laboratory Simulated Heating under Reducing Conditions and Its Relevance for the Study of Fire Use in Archaeology. *J. Arch. Sci. Rep.* **2016**, *10*, 282–292. [CrossRef]
18. Shehata, T.P.; Krap, T. An Overview of the Heat-Induced Changes of the Chemical Composition of Bone from Fresh to Calci *Int. J. Legal Med.* **2024**, *138*, 1039–1053. [CrossRef] [PubMed]
19. Amprino, R. Investigations on Some Physical Properties of Bone Tissue. *Acta Anat.* **1958**, *34*, 161–186. [CrossRef]
20. Holck, P. *Cremated Bones: A Medical-Anthropological Study of an Archaeological Material on Cremation Burials*; Oslo University: C Norway, 1987.
21. Kalsbeek, N.; Richter, J. Preservation of Burned Bones: An Investigation of the Effects of Temperature and PH on Hardness. S *Conserv.* **2006**, *51*, 123–138. [CrossRef]
22. Imaizumi, K.; Taniguchi, K.; Ogawa, Y. DNA Survival and Physical and Histological Properties of Heat-Induced Alteration Burnt Bones. *Int. J. Legal Med.* **2014**, *128*, 439–446. [CrossRef]
23. Hoepfner, T.P.; Case, E.D. The Influence of the Microstructure on the Hardness of Sintered Hydroxyapatite. *Ceram. Int.* **2003** 699–706. [CrossRef]
24. Ruys, A.J.; Wei, M.; Sorrell, C.C.; Dickson, M.R.; Brandwood, A.; Milthorpe, B.K. Sintering Effects on the Strength of Hydrox atite. *Biomaterials* **1995**, *16*, 409–415. [CrossRef]
25. Muralithran, G.; Ramesh, S. The Effects of Sintering Temperature on the Properties of Hydroxyapatite. *Ceram. Int.* **2000** 221–230. [CrossRef]
26. Figueiredo, M.; Fernando, A.; Martins, G.; Freitas, J.; Judas, F.; Figueiredo, H. Effect of the Calcination Temperature on Composition and Microstructure of Hydroxyapatite Derived from Human and Animal Bone. *Ceram. Int.* **2010**, *36*, 2383–2 [CrossRef]
27. Piga, G.; Malgosa, A.; Thompson, T.J.U.; Enzo, S. A New Calibration of the XRD Technique for the Study of Archaeolog Burned Human Remains. *J. Archaeol. Sci.* **2008**, *35*, 2171–2178. [CrossRef]
28. Thompson, T.J.U.; Gauthier, M.; Islam, M. The Application of a New Method of Fourier Transform Infrared Spectroscopy to Analysis of Burned Bone. *J. Archaeol. Sci.* **2009**, *36*, 910–914. [CrossRef]
29. Gonçalves, D.; Thompson, T.J.U.; Cunha, E. Implications of Heat-Induced Changes in Bone on the Interpretation of Fune Behaviour and Practice. *J. Archaeol. Sci.* **2011**, *38*, 1308–1313. [CrossRef]
30. Figueiredo, M.M.; Gamelas, J.A.F.; Martins, A.G. Characterization of Bone and Bone-Based Graft Materials Using FTIR S troscopy. In *Infrared Spectroscopy—Life and Biomedical Sciences*; Theophile, T., Ed.; InTech: Rijeka, Croatia, 2012; pp. 315–338.
31. Thompson, T.J.U.; Islam, M.; Bonniere, M. A New Statistical Approach for Determining the Crystallinity of Heat-Altered B Mineral from FTIR Spectra. *J. Archaeol. Sci.* **2013**, *40*, 416–422. [CrossRef]
32. Snoeck, C.; Lee-Thorp, J.A.; Schulting, R.J. From Bone to Ash: Compositional and Structural Changes in Burned Modern Archaeological Bone. *Palaeogeogr. Palaeoclimatol. Palaeoecol.* **2014**, *416*, 55–68. [CrossRef]
33. Snoeck, C.; Schulting, R.J.; Lee-Thorp, J.A.; Lebon, M.; Zazzo, A. Impact of Heating Conditions on the Carbon and Oxy Isotope Composition of Calcined Bone. *J. Archaeol. Sci.* **2016**, *65*, 32–43. [CrossRef]
34. Gonçalves, D.; Vassalo, A.R.; Mamede, A.P.; Makhoul, C.; Piga, G.; Cunha, E.; Marques, M.P.M.; Batista de Carvalho, L. Crystal Clear: Vibrational Spectroscopy Reveals Intrabone, Intraskeleton, and Interskeleton Variation in Human Bones. *A Phys. Anthropol.* **2018**, *166*, 296–312. [CrossRef]
35. Lemmers, S.A.M.; Gonçalves, D.; Cunha, E.; Vassalo, A.R.; Appleby, J. Burned Fleshed or Dry? The Potential of Bioerosio Determine the Pre-Burning Condition of Human Remains. *J. Archaeol. Method Theory* **2020**, *27*, 972–991. [CrossRef]
36. Legan, L.; Leskovar, T.; Črešnar, M.; Cavalli, F.; Innocenti, D.; Ropret, P. Non-Invasive Reflection FTIR Characterizatio Archaeological Burnt Bones: Reference Database and Case Studies. *J. Cult. Herit.* **2019**, *41*, 13–26. [CrossRef]
37. Mamede, A.P.; Vassalo, A.R.; Piga, G.; Cunha, E.; Parker, S.F.; Marques, M.P.M.; Batista De Carvalho, L.A.E.; Gonçalves Potential of Bioapatite Hydroxyls for Research on Archeological Burned Bone. *Anal. Chem.* **2018**, *90*, 11556–11563. [CrossRe

Mamede, A.P.; Gonçalves, D.; Marques, M.P.M.; Batista de Carvalho, L.A.E. Burned Bones Tell Their Own Stories: A Review of Methodological Approaches to Assess Heat-Induced Diagenesis. *Appl. Spectrosc. Rev.* **2018**, *53*, 603–635. [CrossRef]
Marques, M.P.M.; Mamede, A.P.; Vassalo, A.R.; Makhoul, C.; Cunha, E.; Gonçalves, D.; Parker, S.F.; Batista de Carvalho, L.A.E. Heat-Induced Bone Diagenesis Probed by Vibrational Spectroscopy. *Sci. Rep.* **2018**, *8*, 15935. [CrossRef] [PubMed]
Squires, K.E.; Thompson, T.J.U.; Islam, M.; Chamberlain, A. The Application of Histomorphometry and Fourier Transform Infrared Spectroscopy to the Analysis of Early Anglo-Saxon Burned Bone. *J. Archaeol. Sci.* **2011**, *38*, 2399–2409. [CrossRef]
Thompson, T.J.U.; Islam, M.; Piduru, K.; Marcel, A. An Investigation into the Internal and External Variables Acting on Crystallinity Index Using Fourier Transform Infrared Spectroscopy on Unaltered and Burned Bone. *Palaeogeogr. Palaeoclimatol. Palaeoecol.* **2011**, *299*, 168–174. [CrossRef]
Ellingham, S.T.D.; Thompson, T.J.U.; Islam, M. The Effect of Soft Tissue on Temperature Estimation from Burnt Bone Using Fourier Transform Infrared Spectroscopy. *J. Forensic. Sci.* **2016**, *61*, 153–159. [CrossRef]
Salesse, K.; Stamataki, E.; Kontopoulos, I.; Verly, G.; Annaert, R.; Boudin, M.; Capuzzo, G.; Claeys, P.; Dalle, S.; Hlad, M.; et al. These Boots Are Made for Burnin': Inferring the Position of the Corpse and the Presence of Leather Footwears during Cremation through Isotope ($\delta^{13}C$, $\delta^{18}O$) and Infrared (FTIR) Analyses of Experimentally Burnt Skeletal Remains. *PLoS ONE* **2021**, *16*, e0257199. [CrossRef]
Stamataki, E.; Kontopoulos, I.; Salesse, K.; McMillan, R.; Veselka, B.; Sabaux, C.; Annaert, R.; Boudin, M.; Capuzzo, G.; Claeys, P.; et al. Is It Hot Enough? A Multi-Proxy Approach Shows Variations in Cremation Conditions during the Metal Ages in Belgium. *J. Archaeol. Sci.* **2021**, *136*, 105509. [CrossRef]
Rosa, J.; Vassalo, A.R.; Amarante, A.; Batista de Carvalho, L.A.E.; Marques, M.P.M.; Ferreira, M.T.; Gonçalves, D. Burned and Buried: A Vibrational Spectroscopy Analysis of Burial-Related Diagenetic Changes of Heat-Altered Human Bones. *Am. J. Biol. Anthropol.* **2023**, *180*, 534–547. [CrossRef]
Beasley, M.M.; Bartelink, E.J.; Taylor, L.; Miller, R.M. Comparison of Transmission FTIR, ATR, and DRIFT Spectra: Implications for Assessment of Bone Bioapatite Diagenesis. *J. Archaeol. Sci.* **2014**, *46*, 16–22. [CrossRef]
De Carvalho Almança Lopes, C.; Limirio, P.H.J.O.; Novais, V.R.; Dechichi, P. Fourier Transform Infrared Spectroscopy (FTIR) Application Chemical Characterization of Enamel, Dentin and Bone. *Appl. Spectrosc. Rev.* **2018**, *53*, 747–769. [CrossRef]
Lebon, M.; Reiche, I.; Bahain, J.J.; Chadefaux, C.; Moigne, A.M.; Fröhlich, F.; Sémah, F.; Schwarcz, H.P.; Falguères, C. New Parameters for the Characterization of Diagenetic Alterations and Heat-Induced Changes of Fossil Bone Mineral Using Fourier Transform Infrared Spectrometry. *J. Archaeol. Sci.* **2010**, *37*, 2265–2276. [CrossRef]
Querido, W.; Ailavajhala, R.; Padalkar, M.; Pleshko, N. Validated Approaches for Quantification of Bone Mineral Crystallinity Using Transmission Fourier Transform Infrared (FT-IR), Attenuated Total Reflection (ATR) FT-IR, and Raman Spectroscopy. *Appl. Spectrosc.* **2018**, *72*, 1581–1593. [CrossRef] [PubMed]
Ou-Yang, H.; Paschalis, E.P.; Mayo, W.E.; Boskey, A.L.; Mendelsohn, R. Infrared Microscopic Imaging of Bone: Spatial Distribution of CO_3^{2-}. *J. Bone Miner. Res.* **2001**, *16*, 893–900. [CrossRef] [PubMed]
Cavalli, F.; Innocenti, D.; Cresnar, M.; Vinazza, M. Multidetector Computed Tomography and Micro-Excavation of Prehistoric Urn from Novine/Hoarachkogel (Slovenia/Austria). *Schild Steier* **2015**, *6*, 238–244.
Waltenberger, L.; Bosch, M.D.; Fritzl, M.; Gahleitner, A.; Kurzmann, C.; Piniel, M.; Salisbury, R.B.; Strnad, L.; Skerjanz, H.; Verdianu, D.; et al. More than Urns: A Multi-Method Pipeline for Analyzing Cremation Burials. *PLoS ONE* **2023**, *18*, e0289140. [CrossRef]
Schreiber, J.J.; Anderson, P.A.; Hsu, W.K. Use of Computed Tomography for Assessing Bone Mineral Density. *Neurosurg. Focus* **2014**, *37*, E4. [CrossRef]
Piga, G.; Baró, M.D.; Escobal, I.G.; Gonçalves, D.; Makhoul, C.; Amarante, A.; Malgosa, A.; Enzo, S.; Garroni, S. A Structural Approach in the Study of Bones: Fossil and Burnt Bones at Nanosize Scale. *Appl. Phys. A Mater. Sci. Process.* **2016**, *122*, 1031. [CrossRef]
Walker, P.L.; Miller, K.W.P.; Richman, R. Time, Temperature, and Oxygen Availability: An Experimental Study of the Effects of Environmental Conditions on the Color and Organic Content of Cremated Bone. In *The Analysis of Burned Human Remains*; Schmidt, W.C., Symes, S., Eds.; Elsevier: London, UK, 2008; pp. 129–135.
Shin, J.Y.; Hedges, R.E.M. Diagenesis in Bone and Enamel Apatite Carbonate; the Potential of Density Separation to Assess the Original Composition. *J. Archaeol. Sci.* **2012**, *39*, 1123–1130. [CrossRef]
Scaggion, C.; Dal Sasso, G.; Nodari, L.; Pagani, L.; Carrara, N.; Zotti, A.; Banzato, T.; Usai, D.; Pasqualetto, L.; Gadioli, G.; et al. An FTIR-Based Model for the Diagenetic Alteration of Archaeological Bones. *J. Archaeol. Sci.* **2024**, *161*, 105900. [CrossRef]
Tătar, A.; Ponta, O.; Kelemen, B.; Tatar, A.S.; Oana, P.; Kelemen, B.S.; Tătar, A.; Ponta, O.; Kelemen, B.S. Bone Diagenesis and FTIR Indices: A Correlation. *Stud. Univ. Babes Bolyai Biol.* **2014**, *59*, 101–113.
Nielsen-Marsh, C.M.; Hedges, R.E.M. Patterns of Diagenesis in Bone I: The Effects of Site Environments. *J. Archaeol. Sci.* **2000**, *27*, 1139–1150. [CrossRef]

60. Surovell, T.A.; Stiner, M.C. Standardizing Infra-Red Measures of Bone Mineral Crystallinity: An Experimental Approac Archaeol. Sci. **2001**, *28*, 633–642. [CrossRef]
61. Weiner, S.; Bar-Yosef, O. States of Preservation of Bones from Prehistoric Sites in the Near East: A Survey. *J. Archaeol. Sci.* **1990** 187–196. [CrossRef]
62. Lebon, M.; Reiche, I.; Gallet, X.; Bellot-Gurlet, L.; Zazzo, A. Rapid Quantification of Bone Collagen Content by ATR-F Spectroscopy. *Radiocarbon* **2016**, *58*, 131–145. [CrossRef]
63. Trueman, C.N.G.; Behrensmeyer, A.K.; Tuross, N.; Weiner, S. Mineralogical and Compositional Changes in Bones Exposed on Surfaces in Amboseli National Park, Kenya: Diagenetic Mechanisms and the Role of Sediment Pore Fluids. *J. Archaeol. Sci.* **2** *31*, 721–739. [CrossRef]
64. Olsen, J.; Heinemeier, J.; Bennike, P.; Krause, C.; Margrethe Hornstrup, K.; Thrane, H. Characterisation and Blind Testin Radiocarbon Dating of Cremated Bone. *J. Archaeol. Sci.* **2008**, *35*, 791–800. [CrossRef]
65. Wright, L.E.; Schwarcz, H.P. Infrared and Isotopic Evidence for Diagenesis of Bone Apatite at Dos Pilas, Guatemala: Palaeodie Implications. *J. Archaeol. Sci.* **1996**, *23*, 933–944. [CrossRef]
66. Zazzo, A.; Lebon, M.; Chiotti, L.; Comby, C.; Delqué-Količ, E.; Nespoulet, R.; Reiche, I. Can We Use Calcined Bones for Dating the Paleolithic? *Radiocarbon* **2013**, *55*, 1409–1421. [CrossRef]
67. Lebon, M.; Zazzo, A.; Reiche, I. Screening in Situ Bone and Teeth Preservation by ATR-FTIR Mapping. *Palaeogeogr. Palaeoclim Palaeoecol.* **2014**, *416*, 110–119. [CrossRef]
68. Madupalli, H.; Pavan, B.; Tecklenburg, M.M.J. Carbonate Substitution in the Mineral Component of Bone: Discriminating Structural Changes, Simultaneously Imposed by Carbonate in A and B Sites of Apatite. *J. Solid State Chem.* **2017**, *255*, 27 [CrossRef]
69. Etok, S.E.; Valsami-Jones, E.; Wess, T.J.; Hiller, J.C.; Maxwell, C.A.; Rogers, K.D.; Manning, D.A.C.; White, M.L.; Lopez-Cape Collins, M.J.; et al. Structural and Chemical Changes of Thermally Treated Bone Apatite. *J. Mater. Sci.* **2007**, *42*, 9807–9 [CrossRef]
70. Rey, C.; Collins, B.; Goehl, T.; Dickson, I.R.; Glimcher, M.J. The Carbonate Environment in Bone Mineral: A Resolution-Enhan Fourier Transform Infrared Spectroscopy Study. *Calcif. Tissue Int.* **1989**, *45*, 157–164. [CrossRef] [PubMed]
71. LeGeros, R.Z.; Trautz, O.R.; Klein, E.; LeGeros, J.P. Two Types of Carbonate Substitution in the Apatite Structure. *Experientia* **1** *25*, 5–7. [CrossRef] [PubMed]
72. Habelitz, S.; Pascual, L.; Durán, A. Transformation of Tricalcium Phosphate into Apatite by Ammonia Treatment. *J. Mater.* **2001**, *36*, 4131–4135. [CrossRef]
73. Destainville, A.; Champion, E.; Bernache-Assollant, D.; Laborde, E. Synthesis, Characterization and Thermal Behavior of Apa Tricalcium Phosphate. *Mater. Chem. Phys.* **2003**, *80*, 269–277. [CrossRef]
74. Yin, X.; Stott, M.J.; Rubio, A. α- and β-Tricalcium Phosphate: A Density Functional Study. *Phys. Rev. B* **2003**, *68*, 205205. [Cross
75. Mathew, M.; Schroeder, L.W.; Dickens, B.; Brown, W.E. The Crystal Structure of α-$Ca_3(PO_4)_2$. *Acta Crystallogr. Sect. B* **1977** 1325–1333. [CrossRef]
76. Trueman, C.N.; Privat, K.; Field, J. Why Do Crystallinity Values Fail to Predict the Extent of Diagenetic Alteration of Bone Min *Palaeogeogr. Palaeoclimatol. Palaeoecol.* **2008**, *266*, 160–167. [CrossRef]
77. Person, A.; Bocherens, H.; Mariotti, A.; Renard, M. Biogenic Phosphates as Palaeoenvironmental IndicatorsDiagenetic Evolu and Experimental Heating of Bone Phosphate. *Palaeogeogr. Palaeoclimatol. Palaeoecol.* **1996**, *126*, 135–149. [CrossRef]
78. Thompson, T.J.U. Recent Advances in the Study of Burned Bone and Their Implications for Forensic Anthropology. *Forensic. Int.* **2004**, *146*, S203–S205. [CrossRef]
79. Shipman, P.; Foster, G.; Schoeninger, M. Burnt Bones and Teeth: An Experimental Study of Color, Morphology, Crystal Struc and Shrinkage. *J. Archaeol. Sci.* **1984**, *11*, 307–325. [CrossRef]
80. Miranda, P.; Pajares, A.; Saiz, E.; Tomsia, A.P.; Guiberteau, F. Mechanical Properties of Calcium Phosphate Scaffolds Fabricate Robocasting. *J. Biomed. Mater. Res. A* **2008**, *85A*, 218–227. [CrossRef]
81. Ryu, H.-S.; Youn, H.-J.; Sun Hong, K.; Chang, B.-S.; Lee, C.-K.; Chung, S.-S. An Improvement in Sintering Property of β-Tricalc Phosphate by Addition of Calcium Pyrophosphate. *Biomaterials* **2002**, *23*, 909–914. [CrossRef] [PubMed]
82. Lin, K.; Chang, J.; Lu, J.; Wu, W.; Zeng, Y. Properties of β-$Ca_3(PO_4)_2$ Bioceramics Prepared Using Nano-Size Powders. *Ceram.* **2007**, *33*, 979–985. [CrossRef]

Disclaimer/Publisher's Note: The statements, opinions and data contained in all publications are solely those of the indivic author(s) and contributor(s) and not of MDPI and/or the editor(s). MDPI and/or the editor(s) disclaim responsibility for any injur people or property resulting from any ideas, methods, instructions or products referred to in the content.

materials

MDPI

:icle

nalysis of Late Antique and Medieval Glass from Koper :apodistria, SI): Insights into Glass Consumption and roduction at the Turn of the First Millennium CE

;a Šmit [1,2,*] and Tina Milavec [3]

[1] Faculty of Mathematics and Physics, University of Ljubljana, Jadranska 19, SI-1000 Ljubljana, Slovenia
[2] Jožef Stefan Institute, Jamova 39, SI-1000 Ljubljana, Slovenia
[3] Department of Archaeology, Faculty of Arts, University of Ljubljana, Aškerčeva 2, SI-1000 Ljubljana, Slovenia; tina.milavec@ff.uni-lj.si
* Correspondence: ziga.smit@fmf.uni-lj.si

Abstract: A series (n = 22) of glasses from the site Kapucinski vrt (garden of the Capuchin monastery, 5th–17th c. CE) in Koper (Capodistria), a port town in the northern Adriatic, was measured using a combined PIXE and PIGE method. Koper has been continuously populated since the late Roman period, with a rich medieval history, thus offering an opportunity to study Early Medieval glass. Stemmed goblet fragments, in the original publication dated between the 6th–9th centuries CE, and several other vessel types (beakers and flasks or bottles and lamps) were selected for analysis. The measurements were expected to show the trends in glass production and consumption from Late Antiquity until the Middle Ages, notably the transition between natron to plant ash glass and the supply of fresh glass. Among the set of 22 glass vessel fragments, both natron and plant ash glass were identified. For finer classification, we relied on a newly developed method of Euclidean distances with respect to major concentrations. Natron glass of the types Foy 2.1 (9 examples), Magby (2 examples), and Levantine I (Apollonia; 2 examples) was found. Two glasses remain undetermined but testify to an Egyptian origin. Most natron glasses show signs of recycling. Among the three unrecycled glasses (about 20% of the whole set), there are two examples of Levantine glass and a Magby glass lamp; this may indicate a modest supply of fresh glass during the period. Plant ash glass may be attributed to the Early or High Middle Ages, exploiting the purified alkalis of the Levantine coasts (known as *alume catino* in later Venetian glassmaking), and the admixture of impurities in the siliceous sands suggests the circulation and consumption of glass that was produced and traded in the eastern Mediterranean since the 10th century CE.

Keywords: glass analysis; natron glass; plant-ash glass; northen Adriatic; PIXE; PIGE

1. Introduction

In glass studies, the 8th–10th centuries CE represent the transition period between the use of natron and plant ash glass compositions. In a broader historical context, this reflects the availability of material supplies and the intensity of trade routes; especially important is the ratio between the old, recycled material and freshly supplied raw glass. In Egypt and Mesopotamia, the production of plant ash glass never completely disappeared. In the Roman world outside this region, it is documented already in the 1st c. CE, though limited to intensively colored blue or green glass [1,2]. A few examples of plant ash glass during Late Antiquity are mentioned in the eastern part of the Roman Empire, such as Crete [3] or Moesia [4,5]. In Italy, the first examples of plant ash or mixed natron-plant ash glass are dated to the 8th c. CE, in Lombardy and around Venice [6–8]. An earlier occurrence, such as in the 7th c. CE from Comacchio [9], is opposed in [10]: high MgO values were hypothetically explained as contamination in the crucible. Most of the secondary glass production still used recycled natron glass, at least until the 12th–13th c. CE [7,9,11].

At the same time as the compositions, vessel forms were also changing, especially the stemmed goblet. There is a gap in their development between the ubiquitous Isings 111 form of the 6th–8th c. CE and the appearance of tall-stemmed variants in the 13th–14th c. CE. An opportunity to study the mechanisms of survival of the stemmed goblet during these dark ages is given in the examples from central Italy and the Caput Adriae region. Here, a small group of rare and unique goblet types was documented, usually linked to high-status sites where the demand for drinking vessels and consumer power continued after the 8th c. CE [12,13].

To deepen our understanding of the trends in glass production from Late Antiquity to the Middle Ages, we selected 22 samples of vessels from the excavations of the garden of the Capuchin monastery in the port town of Koper/Capodistria in the Slovenian part of Istria. Situated on the Adriatic coast, the archaeology of Koper, previously an island, displays Byzantine and Carolingian/Ottonian influence and, later, a Venetian influence. The garden of the Capuchin monastery revealed a complex stratigraphy of stone buildings from the 5th to the 17th c. CE, when the monastery was built. Until now, only the first two phases, the Late Antique and the Early Medieval phases, have been published, dated from the 5th–10th c. CE (Figure 1). Among the small finds, a 10th c. CE Byzantine belt buckle was discovered in the same building as a coin of Charlemagne, as well as a fitting of a Carolingian spur set [14]. Two illegible dirhems have also been discovered, dated to the beginning of the 9th c. CE, as well as a coin of Constantine VII and Zoe from the beginning of the 10th c. CE [15]. Historical sources report a bishop in Koper (Caprae) in 599 CE. In 908 CE, the Italian king Berengar I promised protection to Adlegida, the abbess of a female cenobium in Koper, named civitas Justinopolitana [16]. A detailed study of the social and political situation of the time revealed that the abbess might have originated from the highest Italian noble families. Caught in the conflicts between the Istrian margrave and the Venetians in the 10th c. CE, the town signed multiple agreements with Venice. Between the 12th and the 13th c. CE, the city prospered as an independent commune, and in 1279, Koper eventually came under Venetian dominance. Historical analysis of the recently available archival sources for 13th–14th c. CE shows the town was comparable to the most important Late Medieval cities on the Adriatic coast, Zadar and Dubrovnik, and that it represented one of Venice's most important supplying areas. The beginnings of the town's elevated status can be traced back to the 10th c. CE [17,18].

Excavations in the garden of the Capuchin monastery (45°32′53″, 13°44′03″) in the 1980s uncovered settlement remains dated from the 5th c. CE onwards [14]. They are represented by several houses built in local stone bound with clay and additional buildings built using the post-hole technique (Figure 1). Stone-paved hearths, water channels, and graves of small children were also discovered among the walls. During the excavations, several Isings 111 goblets, lamps, and window glass were found, but also some exceptional Early Medieval goblet types (Cunja types 2 and 4; Figure 4: 1, 3 in ref. [19]). As the site archive is currently under re-evaluation, only a limited amount of glass was offered for analysis by the Regional Museum in Koper. Mainly, goblet feet and stems and parts thereof were chosen for analysis because, among the heavily fragmented material, they could most reliably be assigned to the vessel forms. One sample (22) can be identified as Cunja 2 type. The vessels were given an approximate age estimation (Late Antique: 6th–7th c. CE; Early Medieval: 8th–11th c. CE) according to their stratigraphic position, the phasing of the site published by Cunja [14], and excavation documentation, kept in the Regional Museum in Koper. At the time, the ceramic and especially glass typo-chronologies of the period between the 10th and 13th c. CE were sketchy, and representative objects were not recognized [20]; thus, the end of the Early Medieval phase in the 10th c. CE may have been set too early. A more confident dating of the ceramics from the 13th c. CE onwards allowed the authors to define the Late Medieval phase [21]. The site stratigraphy remains to be published in detail.

Figure 1. Drawing of the excavation area (after [14]).

Goblet feet with a small diameter (ca. 3 cm) belong to the Late Antique phase, and the large ones (ca. 4 cm in diameter) to the Early Medieval. This increase in foot size over time is visible in the published goblet types from Koper [14,19] and elsewhere [22,23]. Among the glass materials not published in 1996, a part of a long and thin aqua-colored stem with a disc was discovered and selected for analysis (without context, Figure 2). Comparable tall-stemmed goblets are relatively rare, and in Italy, they are usually dated from 9th–11th c. CE [24–26]. Apart from goblets, we also sampled a lamp handle, concave beaker bases, and bottles. We also sampled five pushed-in bases and a bottle or flask rim with a bulge found during the excavation of one of the Early Medieval houses. Still, at the time of sampling, it was estimated to be post-Early Medieval on account of their typological similarity with European Medieval goblet and bottle forms (Figure 2, Table 1).

Table 1. Description of the samples with approximate typological (C2—type Cunja 2) and available stratigraphic dating. Small letters distinguish samples with the same temporary inventory number.

No.	ID	Description	Color	Archaeological Dating
1	626	vessel base	aqua	Middle Ages?
2	429 a	bottle neck	aqua	Middle Ages?
3	429 b	vessel base	brown	indeterminable
4	429 c	vessel base	aqua	Middle Ages?
5	429 d	vessel base	aqua	Middle Ages?
6	624	goblet foot	aqua (patina)	Early Middle Ages?
7	646	goblet foot	greenish (patina)	Early Middle Ages?
8	633	goblet foot	aqua	Late Antiquity
9	695	vessel fragment	aqua	Early Middle Ages?
10	601	goblet foot	aqua	Early Middle Ages?
11	447	beaker base	indeterm. (patina)	Antiquity/Late Antiquity
12	619	lamp/balsamarium	aqua	LA/EMA
13	151	goblet foot	aqua	Late Antiquity
14	594	goblet stem	aqua	EMA/MA
15	715	goblet foot	aqua	Early Middle Ages?
16	417 a	rim of a small bottle	indeterm. (patina)	Middle Ages?
17	417 b	vessel fragment	aqua	Early Middle Ages?
18	875	goblet foot	aqua	Late Antiquity?
19	647	goblet foot	greenish (patina)	Early Middle Ages?
20	186	lamp handle	aqua (patina)	Late Antiquity?
21	170 a	goblet rim	aqua	Early Middle Ages
22	170 b	goblet stem and foot	greenish (patina)	Early Middle Ages (C2)

Our main research question was whether the putative Early and High Medieval goblet types and other vessels were made using natron or plant ash and how our results compare to the other analyses of Early Medieval glass in northern Italy conducted so far.

Figure 2. Photos of the analyzed glass. The numbers are the same as in Tables 1 and 2.

Table 2. Oxide concentrations in the analyzed glasses. The first 11 columns report the oxide concentrations in mass %, and the last 8 columns report the oxide concentrations in μg/g. Single zeros denote values below the detection limit.

	Na_2O	MgO	Al_2O_3	SiO_2	SO_3	Cl	K_2O	CaO	TiO_2	MnO	Fe_2O_3	CuO	ZnO	Br	Rb_2O	SrO	ZrO_2	BaO	PbO
1	10.7	3.75	0.93	66.2	0.43	0.77	2.44	13.6	0.106	0.27	0.65	57	127	33	14	841	241	0	95
2	12.1	3.03	0.79	67.3	0.43	0.84	2.34	12.2	0.087	0.18	0.63	29	83	61	19	866	271	0	12
3	9.65	2.44	3.21	67.3	0.33	0.83	2.27	11.8	0.070	1.47	0.47	16	34	55	27	937	32	446	0
4	11.7	3.61	0.91	66.1	0.48	0.86	2.33	13.2	0.119	0.08	0.52	39	157	59	12	837	299	0	58
5	11.2	2.99	0.90	67.3	0.63	0.92	1.79	13.4	0.098	0.03	0.64	19	348	49	12	791	209	0	15
6	17.3	1.62	2.37	66.3	0.81	1.00	0.83	6.94	0.183	0.98	1.02	1300	74	7	8	645	117	0	2230
7	16.7	1.78	2.51	66.8	0.63	0.79	0.79	6.63	0.203	1.13	1.17	1950	96	10	9	590	125	0	4830
8	14.8	0.84	3.08	67.6	0.37	0.91	1.52	10.1	0.108	0.02	0.60	13	11	5	11	609	53	0	30
9	18.5	0.98	2.56	65.0	0.71	0.95	0.78	7.32	0.218	1.22	1.19	1290	62	13	5	697	136	0	2470
10	14.0	1.57	2.32	68.4	0.45	0.96	0.90	9.85	0.267	0.20	0.95	430	35	7	5	319	198	0	172
11	11.7	1.09	7.21	68.5	0.26	0.24	0.95	7.24	0.215	0.67	1.42	1050	69	10	7	588	145	0	2770
12	18.6	0.85	2.26	66.4	0.58	1.09	0.87	7.61	0.141	0.69	0.81	355	36	7	11	550	72	0	438
13	17.2	0.96	2.97	69.1	0.56	0.69	0.55	7.40	0.083	0.02	0.39	4	6	3	8	567	58	0	8
14	17.5	0.86	2.43	67.4	0.55	1.03	0.80	7.49	0.132	0.70	0.82	649	55	7	9	624	79	0	703
15	17.6	1.27	2.48	66.2	0.74	1.09	0.71	7.21	0.190	0.99	0.94	1070	54	10	9	580	102	0	2620
16	11.9	3.27	0.78	66.2	0.61	0.99	1.82	13.3	0.125	0.18	0.61	14	594	53	10	930	326	0	63
17	17.6	0.85	2.48	66.2	0.88	0.87	0.84	7.10	0.153	0.82	0.95	3030	177	15	0	649	39	0	6000
18	15.2	1.21	2.32	67.7	0.52	0.97	0.75	9.23	0.259	0.53	1.05	399	78	4	11	399	187	0	1150
19	16.7	1.21	2.16	65.8	1.14	1.07	0.79	7.58	0.149	0.90	1.05	3190	102	11	19	656	49	0	7160
20	18.3	2.10	1.65	64.8	0.64	0.70	1.69	8.85	0.171	0.13	0.83	12	17	4	9	845	138	0	19
21	16.6	1.52	2.41	66.7	0.66	0.92	1.11	8.04	0.173	0.55	0.91	895	96	4	7	522	123	0	1980
22	17.5	1.39	2.51	68.0	0.72	0.98	0.66	5.88	0.157	0.80	0.76	576	36	9	2	566	94	0	2110

2. The Analytical Method

2.1. Experimental

The glasses were analyzed using a proton-induced X-ray (PIXE) and gamma ray (PIGE) method, applying the in-air beam of the Tandetron accelerator at the Jožef Stefan Institute in Ljubljana (HVE, Amersfoort, The Netherlands). The cleanest part of the samples (where oxide layers peeled off upon washing with alcohol) was used for the measurements. The nominal energy of the beam was 3 MeV, but after passing a 200 nm exit window of Si_3N_4 and a 7 mm air gap, the impact energy at the target was about 2.94 MeV. The beam intensity was a few nA, and the proton current was measured by an RBS signal from a gold foil on a chopper, periodically intersecting the beam in a vacuum. The measurement of the proton number was checked according to the intensity of the argon signal induced in the fixed air gap between the exit window and the target. As the fluctuations of the signal ratio were below 3%, the chopper measurements were recognized as reliable. Typical measurements were about 30 min per sample. The induced X-rays were detected by a Si(Li) detector (PGT) of 160 eV resolution at 5.89 keV, positioned about 6 cm from the target. The exact distance was determined by measuring a series of elemental and simple chemical compound targets. The detector was further equipped with a pinhole filter made of 0.05 mm thick aluminum foil with a relative opening of 9%; the pinhole transmission function was carefully measured and modeled. The combination of the air gap as an X-ray absorber and the pinhole filter allowed for the detection of X-rays from silicon until antimony in a single spectrum (including, however, the L-lines of heavier elements). The lighter elements, Na, Mg, and Al, were then detected according to their gamma rays, induced by inelastic scattering of protons on the nuclei. The gamma lines used in the analysis were 440 keV for Na, 585 keV for Mg, and 844 and 1014 keV for Al; they were detected by an intrinsic germanium detector (Ortec) of 40% relative efficiency positioned about 10 cm from the target. The concentrations were determined according to the method of fundamental physical parameters for X-rays and according to the surface approximation based on the NIST 620 glass standard for gamma rays, considering the effects of proton stopping and photon absorption simultaneously for both sets of data. The sum of elemental concentrations in oxide form was set to unity, yet for control purposes, it was also compared to the calculated virtual concentration of argon induced in the air gap between the exit window and the target. Departure from the argon nominal value signaled sample mis-orientation or its roughness, which was then considered as a correction in the calculation. The detection limit for Na was about 50 μg/g, for Mg about 0.2%, and for Al about 0.1%. Here, the most critical was the measurement of Mg, on account of low counting statistics of its 585 keV line; obtaining a satisfactory result, thus, regulated the measuring time. The detection limits for X-ray-based elements were about 10 μg/g for mid-Z elements until $Z = 30$ but worsened to about 50 μg/g around $Z = 50$ on account of a smaller ionization cross-section. The accuracy of the method, measured according to the NIST 620 and 621 glass standards, was about 5% for major elements but worsened to 10–15% for minor and trace elements.

2.2. Determination of Glass Types According to the Euclidean Distance

For the designation of glass types, we designed a numerical method that calculates the Euclidean distance of an unknown glass sample n from the multi-dimensional ellipsoid of a specific glass type. The significance of the methods is intended to go beyond the current case; we expected to develop it into a more general tool for the designation of glass types. In the present stage, the Euclidean distances replaced the method of score numbers, tentatively introduced in [27]; this approach often produced undeterminable results, as several glass samples could achieve the same number of scores. A stricter criterion is the Euclidean distance, defined as

$$d^2 = \sum_{i=1}^{N} \frac{1}{N} \left(\frac{x_i - m_i}{t\sigma_i} \right)^2 \tag{1}$$

where x_i represents the oxide concentrations in the unknown glass, and m_i and σ_i are the mean concentrations and their standard deviations in the specific glass group or series. For a 95% agreement with the specific group, we take $t = 2$. Identification with a specific type is successful if $d < 1$. The ellipsoid has nine dimensions ($N = 9$), considering the major and minor glass composition with the oxides of Na, Mg, Al, Si, K, Ca, Ti, Mn, and Fe, which are reported as most analytical results obtained by different methods. The elements with concentrations around 0.05% and lower, such as Sr and Zr, or trace elements like Li and B, are presently not considered in the calculation but are rather studied in graphs or considered as complementary criteria. It is also important that the number of variables N in (1) is not too large, as a disagreement for one selected concentration can be screened by good agreement of the remaining $N - 1$.

The advantage and convenience of the present method is avoiding large sets of individual experimental data that are indispensable in other analytical methods, such as PCA. Instead, the user relies on the elemental averages m_i and standard deviations σ_i of the recognized glass groups that are published by several authors. However, the uncertainty of the method lies in the distribution of individual elements that may deviate from Gaussian values and in incomplete databases.

In the following calculation, we tried to collect databases of different authors that distinguish in the number of collected data and the precision of calculation; especially, the standard deviations σ_i need to be calculated with the same precision as the averages m_i since they define the length scale. Priority was, however, given to the authors who calculated their means and standard deviations from the most complete datasets.

3. Results

3.1. Elemental Concentrations and Broad Distribution into Groups

The list of samples (Figure 2) with their description is given in Table 1, together with the available stratigraphic/typological dating. Sample selection relied heavily on availability. We are aware of the limitations, but nevertheless, we trust the results are interesting enough to publish them at this stage.

The glasses were first distributed according to the type of flux, which can be resolved from the MgO-K_2O diagram (Figure 3); natron glass typically has an MgO value below 1.6%, with plant ash above 2.2% [28]. K_2O values are below 1.5% in natron glass and above 2.2% in plant ash glass [28]. In this and the following graphs, we use different symbols for different glass vessel forms and different colors for the historical periods, as suggested by available typological designations and stratigraphic data. For the vessel shapes, we used diamonds for lamps, circles for bottles, squares for bases, left triangles for goblets, and right triangles for goblets of type Cunja 2. The color scales for the time periods are blue for Late Antiquity, green for the Early Middle Ages, yellow for the 9th–11th centuries, red for the Middle Ages, and white for indeterminable.

Two groups are evident from Figure 3: the natron glass type, which involves both Late Antique and Early Medieval glass forms, as well as the sample dated to the 9th–11th centuries (no. 14), and glass made from the ash of halophytes, which involves presumably post-Early Medieval glass and one chronologically undeterminable example. Only two glasses largely stand apart from the two groups. One is Late Antique goblet no. 8, which shows a somewhat higher concentration of potassium (1.5% K_2O); however, this value is not exceptionally high and might have been caused by pollution during remelting, where additional potassium might have resulted from contamination from the crucible wall and/or from fuel fume [29,30]. The second sample is Late Antique lamp no. 20, which is found in the intermediate region between the natron and plant-ash glass. This indicates that it was produced either from mixed alkalis or from a mixture of natron and plant ash glass. Its position is also close to the so-called Byzantine Magby glass, as specified in [31].

For a broad distribution into groups, we studied the glasses using principal component analysis, considering 11 elemental oxides; to overcome the influence of very large and very small concentrations, we used the logarithmic transform $x' = \ln(1 + x)$ [32]. Figure 4, again, shows that the presumably earlier glasses (up to the 11th century) form a rather compact group with slightly different samples (nos. 8, 10, and 20). There are two rather different samples: undeterminable no. 3 and beaker no. 11; it is located far in the direction of the aluminum eigenvector, which is due to its high Al_2O_3 content. On the other hand, the plant ash glass of the presumably post-Early Medieval group forms an independent compact group.

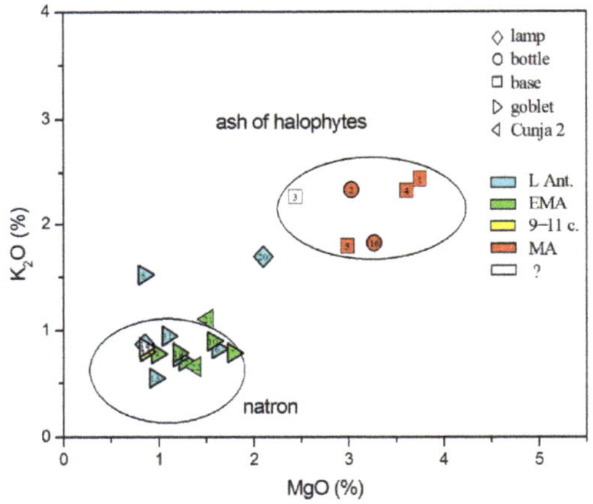

Figure 3. Distribution of measured glasses according to MgO and K_2O oxides reveals the source of alkalis: natron or halophytic plant ash.

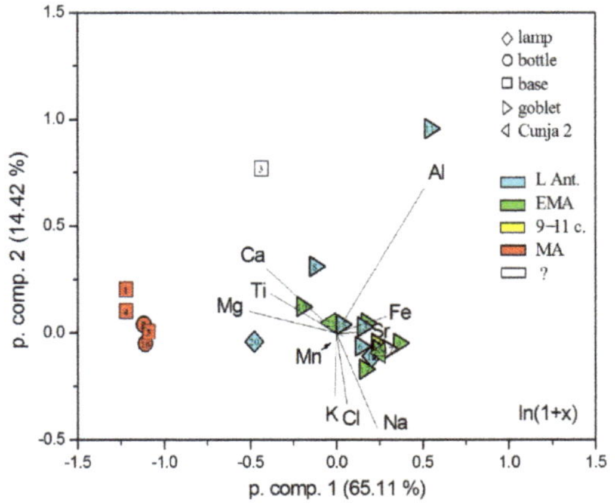

Figure 4. Distribution of the analyzed glasses according to principal component analysis (PCA). The concentrations of 10 elemental oxides plus Cl were logarithmically transformed. The eigenvector of SO_3 is too small to be shown.

3.2. Natron-Type Glass

The distribution of natron-type glass into groups or series has a complex history; in this work, we will operate with the following terms:

Roman Mn and *Roman Sb* [33] will designate pre-4th c. AD Roman glass discolored with Mn and Sb, respectively. These two glasses approximately agree with *Foy Série 3.1 non-tardif* and *Foy Groupe 4* glasses [34] or *RBGY2* and *RBGY1* [35]. The first glass type is of Levantine origin, and the second is of Egyptian origin, respectively. We will not specifically consider naturally colored blue-green-yellow glass [36]. Glass with a higher content of impurities encountered after the 4th c. AD was identified as *Foy Groupe 1* or as *HIMT* (high iron, manganese, and titanium) by Freestone [37], though it was experimentally detected earlier [38]. Several subspecies were identified by several authors, though only *HIMT1* or *strong HIMT* is now recognized as true HIMT. According to its iron content, it is divided into *HIMTa* and *HIMTb* [39]. The other derivatives of HIMT glasses are then rather related to Late Antique glasses of the Foy scheme. Glass *Foy Série 3.2* also involves *HIMT2*, and *Foy Série 2.1* includes *weak HIMT*, *HLIMT* (high lime), and *Ca-rich HIMT* [40]. All HIMT glasses are now considered of Egyptian origin; their Levantine counterparts were designated as *Levantine I* by Freestone and include 4th c. glasses from Jalame and 6th c. glasses from Apollonia [41]. Of the glasses that appear after the 6th century, we considered *Egypt I* (7th to 8th century) and *Egypt II* (8th to 10th century), as well as *Levantine II* (or Bet Elie'zer, 6th to 8th century) [42]. *Egypt II* was split by Schibille into *Egypt 2 (<815 CE)* and *Egypt 2 (>815 CE) [43]*. We further added High Al glasses produced from the evaporitic source of alkalis in Asia Minor [44] and a mixed-alkali Magby glass [45].

The data of the mean elemental concentrations and their standard deviations (m_i and σ_i) are given in Table 3. *For Roman Sb*, they were taken from [46] (Table 1: 269–680 data from refs. [47–53]. For Roman Mn, we used a compilation [46] (Table 4: 138–239 data from refs. [48,49,52–57]. Two subgroups of *Roman Mn* glass are taken from [40]: *Roman Mn—Britain* (7 glasses from [50]) and *Roman Mn—Italy* (12 glasses from [53]). The data for *HIMTa* (14 glasses) and *HIMTb* (5 glasses) are from [58]. The compilation from [46] is used for *Foy Série 3.2* (Table 1: 65–99 data from [34,59–62], *Foy Série 2.1* (Table 1: 157–180 data from [31,34,58], and *Jalame* (*Levantine I*) (Table 4: 50 data from [63]). Balvanović [64] distinguished two subgroups: *Jalame Mn* (14 glasses from [65]) and *Jalame no Mn* (28 glasses from [65]). Schibille [46] further provides data for *Apollonia* (*Levantine I*) (Table 4: 30 data from [66,67]) and *Bet Eli'ezer* (Table 4, 27–79 data from [41,66,68,69]. Phelps [69] summarizes data for *Egypt I* (24 glasses from [70]) and *Egypt II* (17 glasses from [70]). Data for both Egypt 2 groups are taken from [43] (12 and 24 glasses, respectively, data from [34,41,71], and for Magby, they were taken from [46] (Table 1, 55–65 data from [31,72,73].

The relation between *Egypt II* and *Egypt 2* glasses was inspected from the perspective of Euclidean distances. There is a close relation between *Egypt II* and *Egypt 2 (<815 CE)* (d = 0.383), whereas *Egypt 2 (>815 CE)* differs from *Egypt II* (d = 4.581) and is closer to *Magby* glass (d = 1.336).

As natron is chemically a relatively pure agent, the distinction between different glass groups is based on the impurities of the siliceous sand, including aluminum, titanium, iron, and zirconium. The main distinction is between the Egyptian sands, rich in heavy elements brought by the Nile, and the Levantine sands, rich in feldspars, composed of lighter elements such as aluminum. In 2005, Freestone proposed distinguishing glass types according to the Al_2O_3-CaO diagrams [42]. In Figure 5, we can distinguish post-Early Medieval plant ash glasses as a separate group, and among the Early Medieval glasses, there are two in the Levantine I area (nos. 8 and 13), one or possibly two are *HIMT* (nos. 22 and 7), and two (nos. 10 and 18 and possibly no. 20 also) are in the *Egypt II* region. All other glasses form a compact group between these groups. As such characterization is now regarded as insufficient, Freestone later [33] presented another diagram, which is based on the Al_2O_3/SiO_2 vs. TiO_2/Al_2O_3 diagram initially proposed by Schibille [57]. In Figure 6, we display our data against the shaded areas of glass types, which Freestone plotted as individual points; additionally, we added the data points of Magby glass [31,45,67]. We can, again, see that the post-Early Medieval plant ash glasses form an individual group, whereas most of the glasses up to the 11th c. form a compact group of *Foy Série 2.1*. There are three glasses in the region of Levantine glass (nos. 3, 8, and 13). Two glasses are in the boundary region between *Egypt II* and *HIMT* glass (nos. 10 and 18), and one (no. 20) seems to be at the other edge of *Egypt II*. There is also an outsider at the high Al side (no. 11).

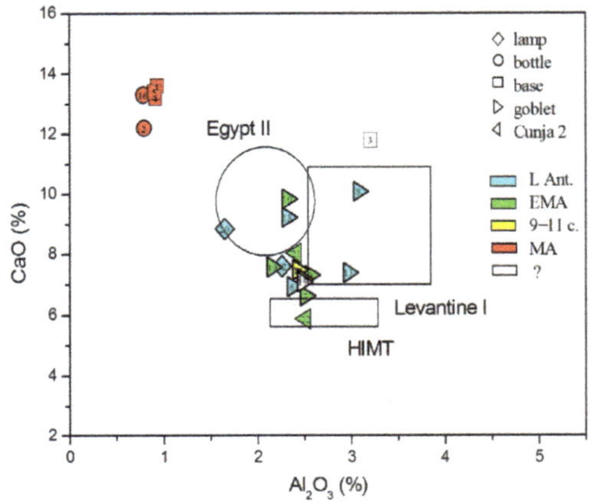

Figure 5. Al_2O_3 vs. CaO concentrations approximately distinguish between Levantine and Egyptian sands.

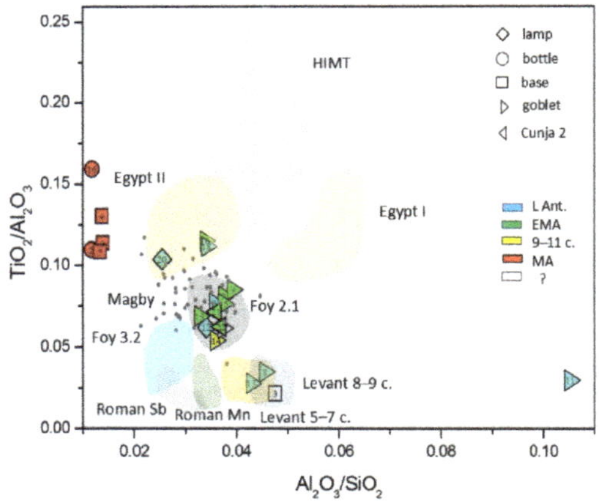

Figure 6. Distribution of measured glasses according to titanium and aluminum oxides. Regions of individual glass types according to the data collected in [33]. Magby data (see references in the text) are added using points.

The conclusions from Figures 3 and 6 are summarized in the second column of Table 4. As a more thorough test, we performed the calculations of Euclidean distances according to Equation (1). The glass type or group with the smallest Euclidean distance is given in the third column of Table 4, and the numerical values of the distances from the group centroids are listed in the fifth column. The respective dating according to the glass type is given in the sixth column of Table 4. The table also marks glasses with apparent signs of recycling.

Table 3. Glass types of natron glass applied in the numerical classification of the measured glasses according to Equation (1), with their reference mean concentrations and standard deviations. Data were taken from the compilations [46] (1, 4, 7, 8, 11–13, 16, 17, 19), [5] (2, 3), [67] (9, 10), [58] (5, 6), [69] (14, 15), and [44] (18; averaging 11 individual data).

		Na_2O	MgO	Al_2O_3	SiO_2	K_2O	CaO	TiO_2	MnO	Fe_2O_3
1	Roman Sb	18.7 ± 1.3	0.41 ± 0.11	1.91 ± 0.21	71.4 ± 1.8	0.45 ± 0.09	5.53 ± 0.84	0.06 ± 0.02	0.01 ± 0.01	0.36 ± 0.1
2	Roman Mn (Britain)	18.31 ± 2.09	0.67 ± 0.14	2.32 ± 0.17	69.62 ± 2.62	0.74 ± 0.14	6.66 ± 1.06	0.10 ± 0.03	0.99 ± 0.12	0.59 ± 0.17
3	Roman Mn (Italy)	15.18 ± 0.84	0.57 ± 0.10	2.59 ± 0.13	70.29 ± 1.08	0.51 ± 0.07	7.83 ± 0.3	0.07 ± 0.01	1.39 ± 0.21	0.20 ± 0.16
4	Roman Mn	16.1 ± 1.3	0.54 ± 0.10	2.62 ± 0.24	69.6 ± 2.3	0.65 ± 0.23	7.92 ± 0.76	0.07 ± 0.02	0.74 ± 0.56	0.4 ± 0.15
5	HIMTa	18.33 ± 1.21	1.05 ± 0.18	2.99 ± 0.33	65.43 ± 1.44	0.47 ± 0.14	6.30 ± 1.02	0.43 ± 0.15	1.92 ± 0.57	1.79 ± 0.38
6	HIMTb	18.25 ± 0.11	1.17 ± 0.12	3.31 ± 0.25	63.8 1± 0.55	0.40 ± 0.03	5.70 ± 0.24	0.54 ± 0.07	1.69 ± 0.16	3.81 ± 0.22
7	Foy Série 3.2	19.0 ± 1.1	0.64 ± 0.21	1.94 ± 0.19	68.1 ± 1.7	0.47 ± 0.16	6.61 ± 0.86	0.10 ± 0.03	0.83 ± 0.27	0.68 ± 0.16
8	Foy Série 2.1	17.7 ± 1.3	1.12 ± 0.25	2.53± 0.23	65.7 ± 1.7	0.75 ± 0.19	8.12 ± 0.92	0.15 ± 0.02	1.41 ± 0.44	1.16 ± 0.5
9	Jalame Mn	15.89 ± 0.85	0.59 ± 0.10	2.69 ± 0.15	68.4 ±1.36	0.80 ± 0.08	8.77 ± 0.46	0.08 ± 0.02	1.93 ± 1.11	0.47 ± 0.08
10	Jalame no Mn	15.74 ± 0.81	0.60 ± 0.15	2.70 ± 0.13	70.55 ± 1.18	0.76 ± 0.12	8.77 ± 0.71	0.08 ± 0.02	0.11 ± 0.09	0.38 ± 0.06
11	Jalame	15.7 ± 0.9	0.59 ± 0.12	2.73 ± 0.17	69.9 ± 1.6	0.78 ± 0.13	8.74 ± 0.67	0.09 ± 0.02	0.65 ± 0.94	0.44 ± 0.19
12	Apollonia (Lev. I)	14.2 ± 1.1	0.68 ± 0.28	3.25 ± 0.18	71.2 ± 1.4	0.62 ± 0.19	8.43 ± 0.79	0.09 ± 0.02	0.02 ± 0.005	0.50 ± 0.11
13	Bet Eli'ezer (Lev. II)	12.3 ± 1.2	0.59 ± 0.12	3.38 ± 0.3	74.4 ± 1.5	0.48 ± 0.08	7.35 ± 0.7	0.11 ± 0.03	0.02 ± 0.004	0.69 ± 0.24
14	Egypt I	18.25 ± 1.38	0.93 ± 0.14	4.05 ± 0.29	70.05 ± 1.21	0.40 ± 0.11	3.03 ± 0.23	0.50 ± 0.12	0.051 ± 0.007	1.74 ± 0.28
15	Egypt II	17.26 ± 1.96	0.58 ± 0.13	2.19 ± 0.35	67.85 ± 1.90	0.32 ± 0.24	9.34 ± 1.27	0.27 ± 0.06	0.03 ± 0.015	0.98 ± 0.23
16	Egypt 2 (<815)	16.5 ± 1.0	0.47 ± 0.09	2.00 ± 0.31	69.7 ± 1.9	0.33 ± 0.09	8.51 ± 1.32	0.20 ± 0.03	0.045 ± 0.083	0.84 ± 0.31
17	Egypt 2 (>815)	13.4 ± 0.6	0.70 ± 0.15	2.52 ± 0.20	70.1 ± 1.4	0.51 ± 0.25	9.57 ± 0.54	0.27 ± 0.03	0.44 ± 0.47	1.18 ± 0.32
18	High Al	16.34 ± 1.74	1.14 ± 0.22	6.08 ± 2.30	62.38 ± 3.67	1.57 ± 0.37	8.38 ± 2.31	0.29 ± 0.25	1.22 ± 0.69	1.02 ± 0.52
19	Magby	16.3 ± 1.3	1.87 ± 0.25	2.03 ± 0.29	65.1 ± 1.7	1.54 ± 0.28	9.09 ± 0.78	0.17 ± 0.03	1.25 ± 0.92	1.27 ± 0.41

Table 4. Characterization of the measured samples according to the Al_2O_3/SiO_2 vs. TiO_2/Al_2O_3 diagram (Figure 6) and according to the calculation regarding Equation (1). In columns 3 and 5, the nearest type and the respective Euclidean distance to the group centroids are given. Foy 2.1 is a shorthand notation for *Foy Série* 2.1. The determined glass type implies approximate dating.

No.	Type (Figure 6)	Type (Calculated)	Recycling Markers	d (Equation (1))	Compositional a Da Dating
1	plant ash	Tyre		0.803	10th–11th c.
2	plant ash	Raqqa 1		0.638	10th–11th c.
3	plant ash	Tyre/Nishapur		1.227/1.317	10th–11th c.
4	plant ash	Tyre		0.785	10th–11th c.
5	plant ash	Raqqa 1		0.912	10th–11th c.
6	Foy 2.1	Foy 2.1	x	0.534	6th–7th c.
7	Foy 2.1	Foy 2.1	x	0.709	6th–7th c.
8	Levantine I	Apollonia (Lev. I)		1.009	6th c.
9	Foy 2.1	Foy 2.1	x	0.610	6th–7th c.
10	Egypt (?)	Egypt (?)/Maby	x	0.888	?
11	?	Indet./High Al		0.636	?
12	Foy 2.1	Foy 2.1	x	0.449	6th–7th c.
13	Levantine I	Apollonia (Lev. I)		0.668	6th c.
14	Foy 2.1	Foy 2.1	x	0.432	6th–7th c.
15	Foy 2.1	Foy 2.1	x	0.429	6th–7th c.
16	plant ash	Tyre		0.990	10th–11th c.
17	Foy 2.1	Foy 2.1	x	0.364	6th–7th c.
18	Egypt (?)	Magby	x	0.852	?
19	Foy 2.1	Foy 2.1	x	0.376	6th–7th c.
20	Magby	Magby		0.471	late 6th–7th c.
21	Foy 2.1	Magby	x	0.531	late 6th–7th c.
22	Foy 2.1	Foy 2.1	x	0.574	6th–7th c.

The results of the calculation of Euclidean distances (Equation (1)) largely agree with Figure 6. Among the natron-type glass, nine glasses were identified as *Foy Série 2.1* and two as *Magby*. The two glasses (nos. 8 and 13) that appear among the Levantine glasses in Figures 4 and 5 are also Levantine, according to the calculation: both are closest to the glass from Apollonia.

The characterization of two glasses (nos. 10 and 18) that, in Figure 6, lie in the region intersecting the areas of *HIMT*, *Egypt II*, and *Magby* glasses is problematic. *Magby* glasses are characterized by the mean values of A_2O_3/SiO_2 = 0.0314 ± 0.0052 and TiO_2/Al_2O_3 = 0.0832 ± 0.0150 (calculated from 53 data points in [31,45,67], while the individual points spread between the upper region of Foy 2.1 and lower region of *Egypt II* (Figure 6). For no. 10, we calculated the following distances with respect to *HIMTa* (d = 1.355), *Egypt 2 (d = 1.062)*, and *Magby* (d = 0.880). For no. 18, we obtain *HIMTa* (d = 1.017), *Egypt 2* (d = 0.852), and *Magby* (d = 0.893). For a distinction between the three types, we further inspect SrO and ZrO_2 concentrations, which, in both glasses, amount to 300–400 μg/g and about 200 μg/g, respectively. The mean values for *HIMTa* are 519 μg/g and 276 μg/g (from the data of [58], for *Egypt 2 (>815 CE)*, they are 221 μg/g and 244 μg/g, respectively, and for *Magby*, 890 μg/g and 118 μg/g, respectively [46]. Nos. 10 and 18 lie somewhere in between these values and, therefore, cannot be assigned to any definite type. The common property of the three glass types considered is their Egyptian origin. Therefore, we will use the notation *Egypt (?)* for nos. 10 and 18.

3.3. Plant Ash Glass

According to the MgO-K_2O diagram in Figure 3, all post-Early Medieval glasses (nos. 1, 2, 4, 5, and 16) appear to be made of alkalis obtained from the ash of halophytic plants, and among them, the undeterminable base no. 3. Glass no. 20, with its mixed alkali composition, has been determined as Magby glass and is studied among the natron glasses.

Plant ash glass was also subject to the calculation of Euclidean distances (Table 4). For the database, we used the data compilations and measurements from [74] (Tables 11.4 and 11.6), [43] (Table 1), [46] (Table 3), [75] (Supplementary Tables S4 and S5), and [76] (Table 1). Here, the most consistent results were obtained from the data compiled by Phelps, as the plant ash glass nos. 1–5 and 16 were characterized as Tyre (10th–11th c.) or Raqqa (8th–11th c.)—see Table 4; experimental data were taken from [41,77] (8 glasses) and [78] (90 glasses). The distances calculated according to his own data for Ramla (*P1*, *P3*, and *P4*) were greater than unity. According to [43], the classification was Levantine plant ash (data from [78]; 40 glasses)—except for no. 16, which resulted, here, as Mesopotamian due to a slightly smaller distance (d = 1.158) in comparison with d = 1.166 for the Levantine plant ash. The distances for Egyptian plant ash glasses *E1–E4* [46] were greater than unity. According to the data collected in [75], the plant ash samples also appeared to be made in Tyre (13 glasses 8th–12th c.) according to the data in [77]; however, this classification was a consequence of a very large data spread of values [77], as a large σ make the distances smaller. If we consider as a potential source all distances to be smaller than unity, possible proveniences also include Ctesifon (9 glasses from [78]), Raqqa 1 (91 glasses from [78,79]), Bayreuth (7 glasses from [78]), Raqqa 4 (74 glasses from [78,79]), and Siraf Main A 9th–12th c. (15 glasses from [80]—for no. 2 only). According to the data compiled and calculated in [76], glasses 1–3 were determined as Raqqa 1 (database of 103 glasses from [79]). For glasses 4, 5, and 16, Raqqa 1 remained the second closest, though smaller distances were obtained for Khirbet al Minya (no. 4, d = 0.651; database of 6 glasses from [78]) and Sagalassos (no. 5, d = 0.612; no. 16, d = 0.611; database of 11 glasses from [81]).

Though these locations are quite diverse, most of them are on the Mediterranean coast or its close background, with three exceptions: Sagalassos in Asia Minor (glass could have traveled there by trade), Siraf in Iran (encountered as a modest possibility for glass no. 2), and Ctesifon near Baghdad (attribution to this site may be due to the large standard deviation from the reference concentrations).

4. Discussion

4.1. Natron Glass

Natron-type glass of Late Antiquity was produced in two regions sufficiently close to exploit the dry deposits of Egyptian lakes, Egypt itself, and the Levantine coast. The two regions producing primary raw glass differ according to the impurities in the siliceous sand. Figure 7, showing Al_2O_3 vs. Fe_2O_3, clearly distinguishes between Levantine and Egyptian sands: Levantine (nos. 8 and 13) is characterized by higher aluminum values and smaller iron content. The situation is similar in the TiO_2 vs. ZrO_2 plot (Figure 8). Glasses of Levantine sands (nos. 8 and 13, as well as the plant ash glass no. 3) show both low titanium and zirconium values. Higher values of both elements, showing a linear correlation, are perceived in Egyptian sands, with the highest values in the undetermined Egyptian glasses of nos. 10 and 18. Glass 20, which is made of mixed natron and plant ash alkalis, and glass 21 are then among the Egyptian glass, in accordance with their characterization as Magby glass.

Strontium can be used to distinguish mineral sources of calcium from their source in mollusks or plant ash [42]. Figure 9 (showing SrO vs. CaO) reveals that SrO concentrations are typically larger than 300 μg/g, which excludes a mineral source of calcium. The lowest SrO values are observed in the two Egyptian glasses of nos. 10 and 18. Though these values are closest to Sr concentrations in HIMT glass, such classification can be excluded on account of major composition. Juan de Ares [72] noted the structured distribution of *HIMTa* and *HIMTb* glass in the eastern and western Mediterranean, with the absence of temporally later (beginning of the 5th c. CE) *HIMTb* in certain regions, including the northeastern Mediterranean. This may indicate a limited supply of HIMT glasses since the beginning of the 5th c. CE, caused by specific political or economic events. In the region of Koper, this overlaps with changed supply routes, as reflected in amphorae imports. Until mid-5th c. CE, North Africa predominates as the export region, while in the 6th c. CE, most ceramic imports arrive from the Eastern Mediterranean. This change follows a decline in Tunisian pottery workshops, most probably linked to wine and oil production and circulation dynamics [82].

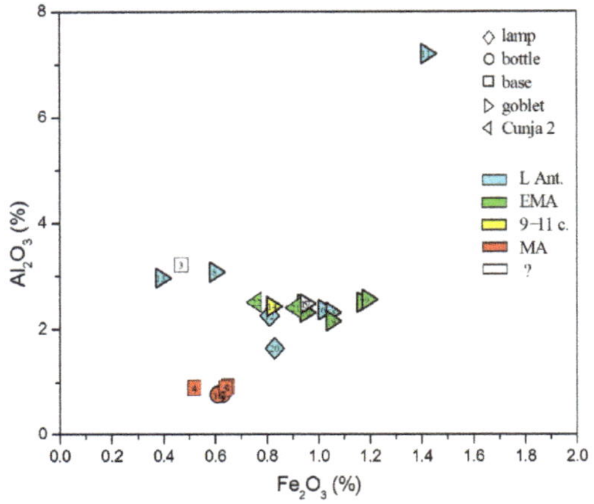

Figure 7. Distribution of glasses according to iron and aluminum oxides.

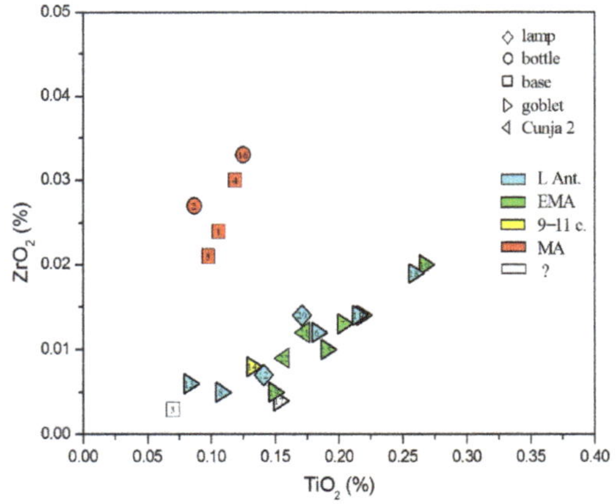

Figure 8. Distribution of glasses according to titanium and zirconium oxides.

Manganese can enter glass either as an impurity or as a decolorizer added intentionally, for example, in the form of pyrolusite (MnO_2). In Figure 10 (showing MnO vs. SrO), it is evident that two Levantine glasses (8 and 13) were made of glass that was not discolored with MnO, while glass no. 3 exhibits the highest MnO level of 1.47%. Of the two Egyptian (?) glasses, one (no. 18) exhibits 0.53% MnO, while for no. 10, its content is only 0.2%; for this reason, no. 10 departs from *HIMTa* more than no. 18.

It is further important to consider the percentage of recycled/non-recycled glass. There are several criteria for the distinction of recycled glass: the content of antimony below the level that ensures discoloration [52] and the admixture of heavy elements that enter the glass batch through the colored glass. In our glasses, we did not detect antimony and tin (the detection limit for both elements was about 50 μg/g), which means that the recycling process did not involve a significant amount of glass discolored with antimony and glass opacified with tin. Therefore, we could only rely on the admixture of heavy elements Cu, Zn, and Pb, the values of which in the recycled glass are typically above 100 μg/g, though Zn values may be slightly below this value even in the recycled glass [9]. According to these criteria, all natron glasses in our set are recycled, except for both Levantine glasses (8 and 13). Another example includes lamp 20 (Magby), produced using mixed alkalis; the siliceous component points to this being non-recycled; this finding makes the possibility of mixing natron and plant ash glass less probable.

Compared to vessel typology, both non-recycled Levantine glasses are goblet feet with a diameter of ca. 3 cm, found in the Late Antique (6th–7th c. CE) phase layers at the site. The typology, stratigraphy, and composition of these two samples fit very well. One Magby glass is a lamp handle (no. 20); here, again, the stratigraphy, typology, and glass composition fit. The second Magby (no. 21) is the rim of a further undetermined vessel, quite possibly a goblet or a lamp. It was found together with sample no. 22, a Cunja 2 goblet of recycled Foy 2.1 composition, on a stone floor of the Early Medieval phase. Recycled Foy 2.1 glasses are represented mainly by goblet feet with a diameter of ca. 4 cm or by parts of goblets (nos. 9 and 17) and one lamp or balsamarium. They belong to the Early Medieval phase of the site. Among them are the Cunja 2 type goblet (no. 22) and the thin-stemmed goblet stem (no. 14). As to where they were produced, it is, of course, not possible to give a definitive answer. Still, it seems worth stressing that Cunja type 2 goblets are very similar to a goblet from the workshop at Torcello, now dated to the 9th c. CE or even slightly later [22]; Figure 46. They are both composed of two parts, with a hollow stem and a narrow knob at the top of the stem. The thin-stemmed goblet (no. 14) confirms the reuse of old glass until the 10th or 11th c. CE in the workshop where it was made. To our knowledge, these types of goblets are not found in the same layers as the sturdier Cunja 2 and 4 types or similar vessels. They are also more widely found in the Italian Peninsula, and their development continues into the Middle Ages. It may well be that they were produced in different workshops and, more importantly, in different social and political contexts of the 10th c. CE and later. The second half of the 10th c. marks the beginning of the consolidation of the Ottonian Empire and the rise of the power of Venice. More political stability allowed the now firmly established elites to also pursue their demand for luxury tableware. It also allowed the workshops to meet this demand with a slightly more constant supply than in the two centuries before, yet still relying on the circulation of recycled material. A high percentage of recycled earlier glass was also detected in medieval sites from Italy, such as in Nogara [83], Vetricella [84], or the medieval castle of San Giuliano [85], while fresh Levantine glass still reached Islamic Sicily [86].

Glass no. 11 is a fragment of a concave beaker base, which was found on a layer of fired clay and ash, probably a hearth of the Early Medieval phase. It contains a low level of MgO, which points to a mineral source of alkalis, but it contains a high amount of aluminum. The reason for the high aluminum content is not clear; if we exclude surface pollution, high-Al glasses may be associated with the production in Asia Minor exploiting evaporitic mineral sources of soda [44]. Their characteristic also involves boron and lithium contents, the presence of which we could detect as gamma lines at 429 keV and 478 keV, respectively [87]; however, the two lines in this object could not be discerned from the background in our measured spectra. Using another set of data ([76] and unpublished results), the detection limits for Li and B were estimated to be 10 μg/g and 300 μg/g, respectively. The range of both elements in 11 glasses from Asia Minor is between 16 and 438 μg/g for Li and between 657 and 1810 μg/g, so both elements could have been observed. On the other hand, agreement with the major composition of high-Al glasses is quite good (d = 0.636). Another possible origin is glass produced in central Asia and used for beads [88], which is characterized by a high K_2O concentration and low CaO and SrO concentrations. No. 11 does not match these properties, so the question of its origin remains open.

4.2. Plant Ash Glass

Plant ash glasses are produced from cleaner silica sources, as the vegetal ash contains oxides of both alkali and earth-alkaline elements, as well as several impurities. Investigations of the silica matrix usually determine the cleanliness of the silica source [89] or connections with the geological background according to neodymium isotopes [90]. Significant differences are then sought according to the plant ash component.

The Euclidean distances show little differences between the plant ash glasses of our samples and do not allow distinction within quite a broad region involving present-day Lebanon and Syria, with a small probability including Mesopotamia and Asia Minor. As glass was a trading material within the Islamic-Byzantine world, we explore the relation between our glasses and the glass cullet from the shipwreck of Serçe Limani, sunk around 1025 CE. The Euclidean distances according to the mean values of Serçe Limani glass ([76] using 99 data from [63]) calculated for plant ash glasses nos. 1–5 and 16 vary between 0.637 and 0.837. These values show a high similarity between our samples and the traded glass samples.

As Venice was also a renowned trading city in the Mediterranean world, closely interacting with Constantinople and North Africa [91], we compared our glasses (nos.. 1–5, 16) with the Venetian soda glass: greenish-brown and uncolored, from the 11th–14th centuries, and later 15th and 16th c. *Venetian commune* and *vitrum blanchum* glass according to the data compiled by Verità ([92], Tables 6.2.3 and 6.2.4). The smallest distances in the range 0.500–0.658 were obtained for the uncolored glass of the 11th–14th centuries (except no. 3, which was closest to the green-blue glass of the same period; d = 0.345). The distances for the *commune* glass were in the range of 0.655–0.834, and the distances for *vitrum blanchum* were in the range of 0.738–0.988. For glass no. 3, both distances were greater than unity (1.047 and 1.528, respectively). Summarily, the mean distances were 0.563 for the glass of the 11th–14th centuries, 0.797 for the later *vitrum commune*, and 1.017 for *vitrum blanchum*, respectively. These values argue strongly towards earlier dating of our plant ash samples and point to the type of glass that was produced in the Levantine area and the matter of extensive trade since the 10th c. CE.

Next, we inspected the properties of the alkali component. The quality of the plant ash can be monitored from the diagram that shows the relative fraction of sodium and potassium oxides in the total sum of alkaline and earth–alkaline oxides (Figure 11). All samples are sorted within an area that, in our previous works, embraced certain fractions of glass from Ljubljana and Celje in Slovenia, Antwerp in Belgium, and Lezha in Albania [93,94]. Within this group, there are also glasses with original Venetian provenance, and according to the conclusion of de Raedt [95], they are made of the finest plant ash, named *allume catino* in the 15th and 16th c. CE, in Venetian glass making, which was harvested along the Levantine coasts. All our post-Early Medieval glasses were also made of alkalis matching *allume catino*; we do not encounter any glass made of lower quality alkalis harvested elsewhere. Measurements of strontium isotopes suggest two production areas for harvesting halophytic plants in the Levant: coastal areas and the interior around the Euphrates River [96]. In Figure 11, we also plotted data for Banias (a representative of the coastal region; data from [41]), Raqqa (a region around Euphrates; data from [79]), and Samarra (a region around Tigris, towards the Zagros mountains; data from [78]). The plot shows differences between the coastal alkalis and those harvested inland (the Raqqa and

Banias datasets contain some natron-type glass as well). Our data agrees better with the centroid of the Banias values, confirming that alkalis from the coastal region were more frequently used in the maritime trade.

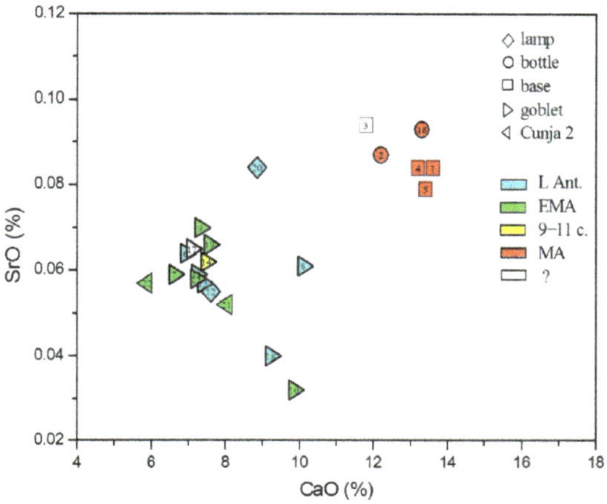

Figure 9. Strontium oxide content with respect to calcium oxide.

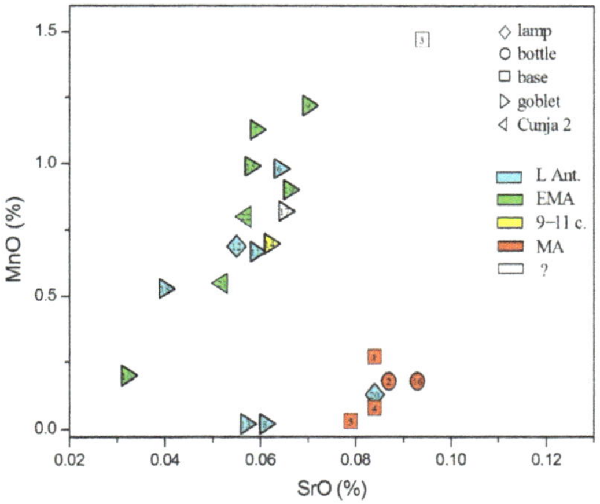

Figure 10. Glasses according to manganese and strontium oxides.

Considering the origin of siliceous sand material, we, again, rely on the admixtures of light and heavy elements, as presented in Figures 5–10. In all these figures, all post-Early Medieval glasses appear as a compact group, suggesting their common origin. In Figure 5 (CaO vs. Al_2O_3), the post-Early Medieval glasses exhibit the lowest Al_2O_3 (below 1%) and highest CaO values (above 12%). High CaO implies high SrO values, between 700 and 930 μg/g (Figure 9). Sample no. 3 departs the group due to its high Al_2O_3 concentration, which, in Figure 6, puts it among Levantine glass. The local origin of plant ash glass is also evident in the MgO/CaO vs. Al_2O_3 diagram, according to [74]. Figure 12 shows that all post-Early Medieval glasses are made of Eastern Mediterranean ingredients, while the indeterminable no. 3 is rather *Mesopotamian Type I*. The Euclidean distance for no. 3 is inconclusive and is larger than unity, both for Levantine and Mesopotamian glass (d = 1.317 for Nishapur). It also bears a resemblance to the Egyptian *E4* glass (d = 1.297), though details about this type of glass, dated to 1035–1149 CE, are not yet clear [46]; in Figure 12, *E4* would be characterized as *Mesopotamian Type I*.

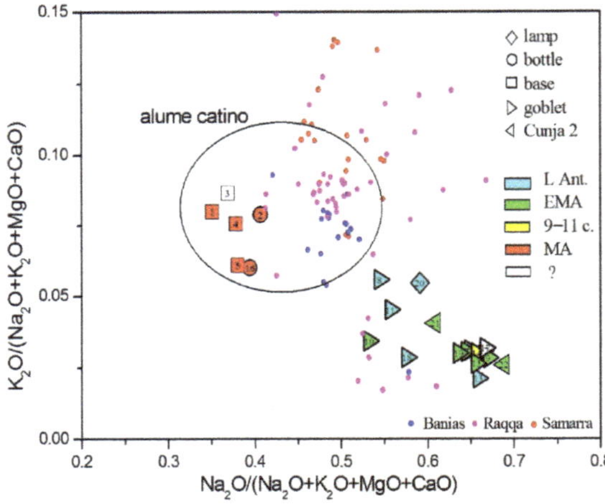

Figure 11. Relative fraction of sodium and potassium oxides reveals the source of plant ash alkalis. Smaller symbols show published data for Banias [41], Raqqa [79], and Samarra [78].

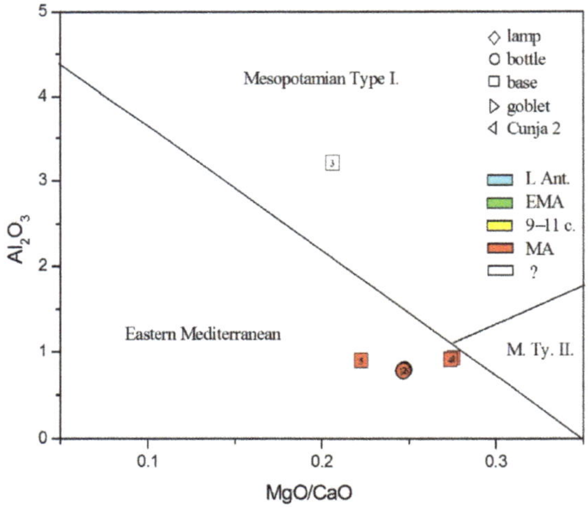

Figure 12. Distinction between Mediterranean and Mesopotamian glass according to Al, Mg, and Ca oxides from [74].

All post-Early Medieval samples have an MgO/CaO ratio between 0.2 and 0.3. This ratio was also studied in [96]: MgO/CaO values around and below 0.3 refer to glasses from Banias, Ramla, Beirut, and Damascus, i.e., to the glasses from the Levantine coast. This confirms the hypothesis that the exported glass mainly came from the coastal area [96].

We also consider the possibility that the post-Early Medieval plant ash glasses (nos.. 1, 2, 4, 5, and 16) were imported to Koper from Venice, which mastered the Mediterranean glass market in later centuries. For this, we inspected the minor and trace elements. The content of TiO_2 varies between 0.087% and 0.125%, and ZrO_2 between 210 and 330 μg/g. Both elements are then present at concentrations that highly exceed the values for original Venetian production. De Raedt [95] set the zirconium limit at 30 μg/g for a distinction between the imported Venetian glass in Antwerp and domestic production. Similarly, original Venetian glass, according to [97], should not exceed the concentrations of 2% Al_2O_3, 0.07% TiO_2, and 40 μg/g ZrO_2. Since our titanium values exceed these limits moderately, while the zirconium values considerably, our glasses were not produced from glass using siliceous sands of Venetian glassworks. This would imply they were imported to Koper from a site different from Venice. However, as Venice also imported large amounts of raw glass or its raw materials from the Levant and elsewhere [92,98], it is also possible that the objects were manufactured in the Venetian or other workshops from the imported raw glass or cullet.

High Zr values reopen the question of imports from Egypt, as its prehistoric glass contains significantly high values of zirconium. However, high Zr values are also found in glass from Islamic Ramla, where raw glass chunks were imported from Mesopotamia and Iran during the Abbasid period [74]. This complies with our previous statement that the plant ash from Koper relates to the locations of present-day Palestine, Lebanon, and Syria, as well as possibly Mesopotamia and Asia Minor, in contrast to Egypt.

The presence of six plant ash glass samples in Koper also fits well into the picture sketched by Italian colleagues, especially for Lombardy and the area around Venice [7,8,92]. It is not possible to determine with any certainty whether the glasses are the result of trade with finished products (e.g., the Cape Stoba shipwreck [99]), raw materials, or cullet (e.g., the Serçe Limani shipwreck [100]). However, in line with the rising mercantile role of Venice between major forces such as the Ottonian Empire, the Byzantines, and the Arabs [101], the major pull of the North Adriatic for long-distance traders is hardly surprising. Perhaps Istrian ports and sites in the hinterland benefited from this success indirectly, trading with Venice or other large centers. However, it is also possible that some of the maritime trade supplied the eastern Adriatic coast directly. Moreover, the more widely known Cape Stoba case and the shipwrecks near Savudrija, Umag, and Poreč carrying Byzantine amphorae indicate that some of the merchants sailed along the Istrian coast as well [102]. The dating of the plant ash glass compositions from Koper is not precise but revolves around the 10th or 11th c. In a wide sense, then, the plant ash glasses arrived in the same time window as the much more strongly represented reused old natron glass. The typology of plant ash glasses is more difficult. The pushed-in bases (nos. 1 and 3–5) most probably represent beakers. They are incompletely preserved, so the base diameters cannot be determined with certainty. The bases could also belong to small flasks or bottles, like the rim of sample no. 16. In this case, if they arrived as complete vessels, they may have represented containers, perhaps for precious liquids or perfumes. The shape of the partially preserved neck with a bulge (no. 2) is analogous to the bottles in Syro-Palestine since the Umayyad period and became particularly popular in the 10th and 11th centuries [103]. In Europe, bottles with bulges on the neck appear from the 13th c. onwards [26,104]. The pushed-in bases (in contrast to the merely concave earlier form) appear in Europe around the 12th–13th c. [26], but they are already present in Umayyad contexts in Jerusalem [105] and continued into the following centuries. Based on typology alone, then, an occasional import of vessels, perhaps even as containers, seems more likely than a very early production of these vessel forms.

5. Conclusions

Glass from Koper exhibits two major groups. One is natron-type glass according to the Roman tradition, though it has properties specific to Late Antiquity and Early Middle Ages. The most numerous is type *Foy Série 2.1* glass (nine samples: eight goblet feet and one lamp or balsamarium). This type of glass includes a Cunja 2 type and a 9th–11th c. CE thin-stemmed goblet. Two Late Antique goblet feet of small dimensions are made of Levantine sands, attributed to the 6th c. CE Apollonia. Two glasses, a rim, and a lamp handle are of the late 6th–7th c. CE-type Magby, and the composition of the two goblets cannot be determined but points to Egyptian origin. Such a composition shows the predominance of Egyptian glass, which has also recently been confirmed for Italy [106] and the Balkans [64].

It is interesting to note an absence of unaltered HIMT glass, which is normally more frequent in western Europe [107], and a small percentage of the high-quality *Levantine* glass, which is unusual in the Adriatic cities. Glass *Foy Série 2.1* was imported to the Balkans mostly along the major rivers Danube and Sava and from the Aegean ports [40,64]. This is certainly not an optimal way for Koper, and we may rather imagine maritime trade through the Adriatic, as in the case of Croatian islands [108]. This trade might have also continued inland for a certain distance until Korinjski hrib, for instance, which also contains a considerable fraction of *Foy Série 2.1* glass [109].

The second group of six samples is composed of halophytic plant ash of the type *alume catino*, harvested in the Levant. The siliceous component shows rather uniform properties, yet its titanium and zirconium concentrations exclude its Venetian origin. The present calculation also excludes imports from Egypt but points towards Lebanon and Syria, with a small probability also to Iraq and Iran. The glass might have been imported from there as vessels or containers; the other possibility is that the glass vessels were made in (Venetian?) workshops from imported raw glass or cullet. In any case, they seem to predate the ubiquitous glass of the Renaissance period and give an important insight into the very rare presence of Islamic glass on the Adriatic coast.

The role of Egyptian glassworks in the period of Islamic glassmaking is unclear. A shipwreck on the Israel coast loaded with glass cullet testifies to contact with Egypt [110]. According to [46], the Egyptian glassworks were overloaded by the production of architectural glass for the monumental mosques since the end of the 7th c. CE.

Three glasses are out of this scheme. A beaker base was made of plant ash alkalis. Still, its aluminum content suggests Mesopotamian origin, yet its specifics include high manganese content (no. 3). Though its precise attribution is not so clear, it matches very well with traded glass, such as that found in the shipwreck of Serçe Limani. One lamp handle was made of mixed alkalis (composed of natron and plant ash), while its siliceous component is likely of Egyptian origin (no. 20), in accordance with the Magby glass. The third sample, a beaker base, was made of natron glass, but it shows a high aluminum content, pointing towards some other, not yet determinable provenience (no. 11).

Our analysis confirmed the reuse of old natron glass for Early Medieval vessels and for the 9th–11th c. CE thin-stemmed goblet, as has previously been observed in northern, central, and southern Italy [7,9,11,12,23,111]. It is characteristic that all natron glass of Egyptian provenience is recycled. A small presence of non-recycled glass (three samples or roughly 20%) is of Levantine or (in one case) Mesopotamian origin and suggests a modest supply of fresh glass from this region during Late Antiquity.

The goblet feet of larger dimensions were made of the following glass types: Foy 2.1 (6), Levantine I (Apollonia; 2), Magby (1), and an undeterminable glass of Egyptian origin (2). All these types represent natron glass, which, in the literature, is dated from the end of the 5th c. CE (*Foy 2.1*) to the 7th century (*Foy 2.1* and Magby) until the 9th century (*Egypt 2 > 815 CE*). In Koper, this type of glass circulated for two centuries longer, as the last object made of (recycled) natron glass can be dated to the 10th–11th century CE. The glass market re-intensifies after the 10th c. CE, with the influx of plant ash glass from the Syrian and neighboring glassworks.

From a methodology point of view, the Euclidean distance method proved effective and discriminative enough for natron-type glass. For plant ash glass, it seems less selective, probably on account of the larger dispersion of the plant ash elemental concentrations, which also partly overshadow the elements of the siliceous component. A solution may be an improved database with recalculated standard deviations based on critically evaluated experimental data.

Author Contributions: Ž.Š. executed the measurements, conceived mathematical models, performed calculations, and created the graphs. T.M. initiated the study, described the archaeological part of the paper, and selected the samples. Both authors discussed and interpreted the results and contributed to the final conclusions. All authors have read and agreed to the published version of the manuscript.

Funding: This research received no external funding.

Institutional Review Board Statement: Not applicable.

Informed Consent Statement: Not applicable.

Data Availability Statement: The original contributions presented in this study are included in the article. Further inquiries can be directed to the corresponding author.

Acknowledgments: We are grateful to Maša Saccara and Špela Prunk from Pokrajinski muzej Koper for the permission to analyze the material of the Kapucinski vrt excavations. The source of funding that has supported the work of one of us (T.M.) is Slovenian Research and Innovation Agency grant number P6-0247 and the MATRES project, approved as part of the first (preparatory) phase for the preparation of large interdisciplinary projects of the University of Ljubljana, funded by the Slovenian Research Agency development pillar (ARIS, RSF-A). Measurements were enabled through the Slovenian node of the European Research Infrastructure for Heritage Science (E-RIHS.si; Infrastructure Programme I0-E012 founded by the Slovenian Research and Innovation Agency).

Conflicts of Interest: The authors declare no conflicts of interest.

Abbreviations

The following abbreviations are used in this manuscript:

PIXE	Particle- induced X-ray emission
PIGE	Proton-induced gamma-ray emission
PCA	Principle Component Analysis

ferences

Nenna, M.-D.; Gratuze, B. Étude diachronique des compositions de verre employés dans les vases mosaïqués antiques: Résultats préliminaire. In *Annales du 17e de l'Association Internationale pour l'Histoire du Verre*; Janssens, K., Degryse, P., Cosyns, P., Caen, J., Van't dack, L., Eds.; AIHV: Antwerp, Belgium, 2006; pp. 199–205.
Jackson, C.M.; Cottam, S. 'A green thought in a green shade'; Compositional and typological observations concerning the production of emerald green glass vessels in the 1st century A.D. *J. Archaeol. Sci.* **2015**, *61*, 139–148. [CrossRef]
Oikonomou, A.; Rehren, T.; Fiolotaki, A. An Early Byzantine glass workshop at Argyroupolis, Crete: Insights into complex glass supply networks. *J. Archaeol. Sci. Rep.* **2021**, *35*, 102766. [CrossRef]
Drauschke, J.; Greiff, S. Early Byzantine glass from Caričin grad/Iustiniana prima (Serbia): First results concerning the composition of raw glass chunks. In *Glass Along the Silk Road from 200 BC to AD 100*; Zorn, B., Hilgner, A., Eds.; RGZM Mainz: Mainz, Germany, 2010; pp. 53–67.
Balvanović, R.; Šmit, Ž.; Marić-Stojanović, M.; Spasić-Đurić, D.; Špehar, P.; Milović, O. Late Roman glass from Viminacium and Egeta (Serbia): Glass making patterns on Iron Gates Danubian Limes. *Archaeol. Anthropol. Sci.* **2022**, *14*, 79. [CrossRef]
Silvestri, A.; Pescarin Volpato, M.; Marcante, A. A review of medieval glass composition from northern and central Italy: A statistical approach. In *Le Verre du VIIIe du XVIe Siècle en Europe Occidentale. Actes du 8e Colloque International de l'AFAV, Collection les Cahiers de la MSHE Ledoux. Série Dynamiques Territoriales, Besançon, France, 5–7 December 2016*; Pactat, I., Munier, C., Eds.; Presses Universitaires de Franche-Comté: Besançon, France, 2020; pp. 47–79.
Neri, E. Produzione e circolazione del vetro nell'alto medioevo: Une entrée en matière. In *Il Vetro in Transizione (IV–XII Secolo) Produzione e Commercio in Italia Meridionale e nell'Adriatico*; Coscarella, A., Noyé, G., Neri, E., Eds.; Firenze University Press: Florence, Italy, 2021; pp. 19–31.
Uboldi, M.; Verità, M. Composizione chimica e processi produttivi del vetro tra tarda antichità e medioevo in Lombardia. In *Il Vetro in Transizione (IV–XII Secolo) Produzione e Commercio in Italia Meridionale e nell'Adriatico*; Coscarella, A., Noyé, G., Neri, E., Eds.; Firenze University Press: Florence, Italy, 2021; pp. 235–244.
Bertini, C.; Henderson, J.; Chenery, S. Seventh to eleventh century CE glass from Northern Italy: Between continuity and innovation. *Archaeol. Anthropol. Sci.* **2020**, *12*, 120. [CrossRef]

10. Gliozzo, E.; Braschi, E.; Ferri, M. New data and insights on the secondary glass workshop in Comacchio (Italy): MgO conte steatite crucibles and alternatives to recycling. *Archaeol. Anthropol. Sci.* **2024**, *16*, 114. [CrossRef]
11. Boschetti, C.; Kindberg Jacobsen, J.; Parisi Presicce, C.; Raja, R.; Schibille, N.; Vitti, M. Glass in Rome during the transition f late antiquity to the early Middle Ages: Materials from the Forum of Caesar. *Herit. Sci.* **2022**, *10*, 95. [CrossRef]
12. Sannazaro, M.; Guglielmetti, A.; Uboldi, M. Manufatti del quotidiano: Pietra ollare, ceramiche e vetri tra VIII e XIII secolo *1287 e Dintorni. Ricerche su Castelseprio a 730 Anni Dalla Distruzione*; Sannazaro, M., Lusuardi Siena, S., Giostra, C., Eds.; S Società Archeologica: Quingentole, Italy, 2017; pp. 129–159.
13. Uboldi, M. San Tomé di Carvico: I vetri. In *San Tomé di Carvico. Archeologia di Una Chiesa Altomedievale*; Brogiolo, G.P., Ed.; Com di Carvico: Carvico, Italy, 2016; pp. 197–204.
14. Cunja, R. Poznoantični in srednjeveški Koper. Arheološka izkopavanja na bivšem Kapucinskem vrtu v letih 1986-1987 v drobnih najdb od 5. do 9. stoletja stoletja/Capodistria tardoromana e altomedievale. Lo scavo archeologico nell'ex orto Cappuccini negli anni 1986–1987 alla luce dei reperti dal V al IX secolo d. C. In *Založba Annales*; ZRS Koper: Koper, Slovenia, 1
15. Kos, P. *Die Fundmünzen der Römischen Zeit in Slowenien*; Mann: Berlin, Germany, 1988.
16. Štih, P. Il diploma del re Berengario I del 908 e il monastero femminile di Capodistria. *Atti Del Cent. Di Ric. Storiche Rovigno* **2** *40*, 67–98.
17. Mlacović, D. Koper v poznem srednjem veku: Opažanja o mestu in njegovih portah po pregledu knjig koprskih vicedomin konca 14. stoletja (Late Medieval Koper: Observations about the Town and Its Portae from a Survey of the Books of the Kc Vicedomini from the End of the 14th Century). *Acta Histriae* **2022**, *30*, 819–854.
18. Mlacović, D. Kartuzija Bistra in Koper v 14. stoletju (The Bistra Carthusian Monastery and Koper in the 14th Century). Z *Časopis* **2023**, *77*, 98–346.
19. Milavec, T. The elusive early medieval glass. Remarks on vessels from the Nin—Ždrijac cemetery, Croatia. *Pril. Instituta Ar Zagrebu* **2018**, *35*, 239–250. [CrossRef]
20. Cunja, R. Archeologia urbana in Slovenia. Alcuni risultati e considerazioni degli scavi a Capodistria. *Archeol. Mediev.* **1** *25*, 199–212.
21. Cunja, R. *Koper med Rimom in Benetkami. Izkopavanje na Vrtu Kapucinskega Samostana: [Razstava] = Capodistria Tra Roma e Venezia. Scavi Nel Convento dei Cappuccini: [Mostra]*; Medobčinski zavod za varstvo naravne in kulturne dediščine = Istituto intercomu per la tutela dei beni naturali e culturali: Piran, Slovenia, 1989.
22. Gelichi, S.; Negrelli, C.; Ferri, M.; Cadamuro, S.; Cianciosi, A.; Grandi, E. Importare, produrre e consumare nella lagun Venezia dal IV al XII secolo Anfore, vetri e ceramiche. In *Adriatico Altomedievale (VI–XI Secolo) Scambi, Porti, Produzioni*; Gelich Negrelli, C., Eds.; Studi e Ricerche 4: Venezia, Italy, 2017; pp. 23–114.
23. Ferri, M. *Il Vetro Nell'alto Adriatico fra V e XV Secolo*; All'insegna del Giglio: Sesto Fiorentino, Italy, 2022.
24. Arena, M.S.; Delogu, P.; Paroli, L.; Ricci, M.; Saguì, L.; Vendittelli, R. (Eds.) *Roma Dall'antichità al Medioevo, Archeologia e St* Museo Nazionale Romano Crypta Balbi: Milano, Italy, 2004.
25. Del Vecchio, F. I vetri di IX-XII secolo dalla domus porticata del foro di Nerva. In *Il Vetro Nell'alto Medioevo. Atti VII Giorna Studio A.I.H.V.*; Ferrari, D., Ed.; Comitato Nazionale Italiano: Imola, Italy, 2005; pp. 45–48.
26. Uboldi, M.; Lerma, S.; Marcante, A.; Medici, T.; Mendera, M. Le verre au Moyen Âge en Italie (VIIIe-XVIe siècle): État connaissances et mise à jour. In *Le Verre du VIIIe au XVIe Siècle en Europe Occidentale*; Pactat, I., Munier, C., Eds.; Pre universitaires de Franche-Comté: Besançon, France, 2020; pp. 31–47. [CrossRef]
27. Šmit, Ž. Glass analysis in relation to historical questions. In *Bridging Science and Heritage in the Balkans: Studies in Archaeom Cultural Heritage Restoration and Conservation*; Palincaş, N., Ponta, C.C., Eds.; Archaeopress: Oxford, UK, 2019; pp. 103–109.
28. Lilyquist, C.; Brill, R.H. *Studies in Early Egyptian Glass*; The Metropolitan Museum of Art: New York, NY, USA, 1993; p. 56.
29. Paynter, S.; Jackson, C. Re-used Roman rubbish: A thousand years of recycling glass. *Eur. J. Postclassical Archaeol.* **2016**, *6*, 31
30. Al-Bashaireh, K.; Al-Mustafa, S.; Freestone, I.C.; Al Housan, A.Q. Composition of Byzantine glasses from Umm el Jimal, north Jordan: Insights into glass origins and recycling. *J. Cult. Herit.* **2016**, *21*, 809–818. [CrossRef]
31. Schibille, N.; Meek, A.; Tobias, B.; Entwistle, C.; Avisseau-Broustet, M.; Da Mota, H.; Gratuze, B. Comprehensive chem characterization of Byzantine glass weights. *PLoS ONE* **2016**, *11*, e0168289. [CrossRef]
32. Duewer, D.L.; Kowalski, B.R. Source identification of oil spills by pattern recognition analysis of natural elemental composi *Anal. Chem.* **1975**, *47*, 1573–1583. [CrossRef]
33. Freestone, I.C.; Degryse, P.; Lankton, J.; Gratuze, B.; Schneider, J. HIMT, glass composition and commodity branding in primary glass industry. In *Things That Travelled. Mediterranean Glass in the First Millennium CE*; Rosenow, D., Phelps, M., Meek Freestone, I., Eds.; UCL Press: London, UK, 2018; pp. 159–190.
34. Foy, D.; Picon, M.; Vichy, M.; Thirion-Merle, V. Caractérisation des verres de la fin de l'Antiquité en Méditerranée occiden L'émergence de nouveaux courants commerciaux. In *Échange et Commerce du Verre Dans le Monde Antique. Actes du Coll de l'Association Française pour l'Archéologie du Verre, Aix-en-Provence et Marseille*; Foy, D., Nenna, M.D., Eds.; Éditions Mon Mergoil: Montagnac, France, 2003; pp. 41–85.

Gliozzo, E.; Santagostino, A.; D'Acapito, F. Waste glass, vessels and window-panes from Thamusida (Morocco): Grouping natron-based blue-green and colorless Roman glasses. *Archaeometry* **2013**, *55*, 609–639. [CrossRef]
Gliozzo, E. The composition of colourless glass: A review. *Archaeol. Anthropol. Sci.* **2017**, *9*, 455–483. [CrossRef]
Freestone, I.C. Appendix: Chemical analysis of 'raw' glass fragments. In *Excavations of Carthage Vol. II 1. The Circular Harbour North Side. The Site and Finds Other than Pottery. British Academy Monographs in Archaeology No. 4*; Hurst, H.R., Ed.; Oxford University Press: Oxford, UK, 1994; p. 290.
Mirti, P.; Casoli, A.; Appolonia, L. Scientific analysis of Roman glass from Augusta Praetoria. *Archaeometry* **1993**, *35*, 225–240. [CrossRef]
Ceglia, A.; Cosyns, P.; Nys, K.; Terryn, H.; Thienpont, H.; Meulenbroeck, W. Late antique glass distribution and consumption in Cyprus: A chemical study. *J. Archaeol. Sci.* **2015**, *61*, 213–222. [CrossRef]
Balvanović, R.; Šmit, Ž. Emerging Glass Industry Patterns in Late Antiquity Balkans and Beyond: New Analytical Findings on Foy 3.2 and Foy 2.1 Glass Types. *Materials* **2022**, *15*, 1086. [CrossRef]
Freestone, I.C.; Gorin-Rosen, Y.; Hughes, M.J. Primary glass from Israel and the production of glass in Late Antiquity and the Early Islamic period. In *La Route du Verre*; Nenna, M.D., Ed.; Maison de l'Orient et de la Méditerranée Jean Pouilloux: Lyon, France, 2000; pp. 64–83.
Freestone, I.C. The provenance of ancient glass through compositional analysis. *Mater. Res. Soc. Proceeding* **2005**, *852*, OO8.1.1-13. [CrossRef]
Schibille, N.; Gratuze, B.; Ollivier, E.; Blondeau, E. Chronology of early Islamic glass compositions from Egypt. *J. Archaeol. Sci.* **2019**, *104*, 10–18. [CrossRef]
Schibille, N. Late Byzantine mineral soda high alumina glasses from Asia Minor: A new primary glass production group. *PLoS ONE* **2011**, *6*, e1897. [CrossRef] [PubMed]
De Juan Ares, J.; Vigil-Escalera Guirado, A.; Cáceres Gutiérez, Y.; Schibille, N. Change in the supply of eastern Mediterranean glasses to Visigothic Spain. *J. Archaeol. Sci.* **2019**, *107*, 23–31. [CrossRef]
Schibille, N. *Islamic Glass in the Making—Chronological and Geographic Dimensions*; Leuven University Press: Leuven, Belgium, 2022.
Baxter, M.J.; Cool, H.E.M.; Jackson, C.M. Further studies in the compositional variability of colourless Romano-British vessel glass. *Archaeometry* **2005**, *47*, 47–68. [CrossRef]
Degryse, P. *Glass Making in the Greco-Roman World: Results of the ARCHGLASS Project*; Leuven University Press: Leuven, Belgium, 2014.
Gratuze, B. Contribution à l'étude des verres décolorés à l'antimoine produits le 1er s. et la fin du IIIe s. de notre ère: Nouvelles données analytiques. In *Verres Incolores de l'Antiquité Romaine en Gaule at Aux Marges de la Gaule. Vol. 2: Typologie et Analyses*; Foy, D., Labaune-Jean, F., Leblond, C., Pruvot, C.M., Marty, M.-T., Massart, C., Munier, C., Robin, L., Roussel-Ode, J., Adrian, Y.-M., et al., Eds.; Archaeopress Roman Archaeology 42: Oxford, UK, 2018; pp. 682–714.
Jackson, C.M. Making colourless glass in the Roman period. *Archaeometry* **2005**, *47*, 763–780. [CrossRef]
Paynter, S. Analyses of colourless Roman glass from Binchester, County Durham. *J. Archaeol. Sci.* **2006**, *33*, 1037–1057. [CrossRef]
Silvestri, A. The coloured glass of Iulia Felix. *J. Archaeol. Sci.* **2008**, *35*, 1489–1501. [CrossRef]
Silvestri, A.; Molin, G.; Salviulo, G. The colourless glass of Iulia Felix. *J. Archeol. Sci.* **2008**, *35*, 331–341. [CrossRef]
Silvestri, A.; Molin, G.; Salviulo, G. Roman and medieval glass from the Italian area: Bulk characterization and relationships with production technologies. *Archaeometry* **2005**, *47*, 797–816. [CrossRef]
Jackson, C.M.; Paynter, S. A great big melting pot: Exploring patterns of glass supply, consumption and recycling in Roman Coppergate, York. *Archaeometry* **2016**, *58*, 68–95. [CrossRef]
Schibille, N.; Degryse, P.; O'Hea, M.; Izmer, A.; Vanhaecke, F.; McKenzie, J. Late Roman glass from the 'Great Temple' at Petra and Khirbet et-Tannur, Jordan—Technology and Provenance. *Archaeometry* **2012**, *54*, 997–1022. [CrossRef]
Schibille, N.; Sterrett-Krause, A.; Freestone, I.C. Glass groups, glass supply and recycling in late Roman Carthage. *Archaeol. Anthropol. Sci.* **2017**, *9*, 1223–1241. [CrossRef]
Ceglia, A.; Cosyns, P.; Schibille, N.; Meulebroeck, W. Unravelling provenance and recycling of late antique glass from Cyprus with trace elements. *Archaeol. Antropol. Sci.* **2019**, *11*, 279–291. [CrossRef]
Balvanović, R.; Stojanović Marić, M.; Šmit, Ž. Exploring the unknown Balkans: Early Byzantine glass from Jelica Mt. in Serbia and its contemporary neighbours. *J. Radioanal. Nucl. Chem.* **2018**, *317*, 1175–1189. [CrossRef]
Cholakova, A.; Rehren, T.h. A late antique manganese decolourised glass composition: Interpreting patterns and mechanisms of distribution. In *Things That Travelled: Mediterranean Glass in the First Millenium CE*; Rosenow, D., Phelps, M., Meeks, A., Freestone, I., Eds.; UCL Press: London, UK, 2018; pp. 46–71.
Gallo, F.; Marcante, A.; Silvestri, A.; Molin, G. The glass of the "Casa delle Bestie Ferite": A first systematic archaeometric study on late Roman vessels from Aquileia. *J. Archaeol. Sci.* **2014**, *41*, 7–20. [CrossRef]

62. Maltoni, S.; Chinni, T.; Vandini, M.; Cirelli, E.; Silvestri, A.; Molin, G. Archaeological and archaeometric study of the glass fi from the ancient harbour of Classe (Ravenna-Italy): New evidence. *Herit. Sci.* **2015**, *3*, 13. [CrossRef]
63. Brill, R.H.; Rising, B.A. *Chemical Analyses of Early Glasses, Volume 2*; The Corning Museum of Glass: Corning, NY, USA, 1 pp. 178–187.
64. Balvanović, R.; Šmit, Ž.; Marić Stojanović, M.; Špehar, P.; Milović, O. Sixth-century Byzantine glass form Limes Fortifications Serbian Danube. *Archaeol. Anthrop. Sci.* **2023**, *15*, 166. [CrossRef]
65. Brill, R.H. Scientific investigations of Jalame glass and related finds. In *Excavations in Jalame: Site of a Glass Factory in Late Ro Palestine: Excavations Conducted by a Joint Expedition of the University of Missouri and the Corning Museum of Glass*; Weinberg, C Ed.; University of Missouri: Columbia, MO, USA, 1988; pp. 257–294.
66. Brems, D.; Freestone, I.C.; Gorin-Rosen, Y.; Scott, R.; Devulder, V.; Vanhaecke, F.; Degryse, P. Characterisation of Byzantine early Islamic primary tank furnace glass. *J. Archaeol. Sci. Rep.* **2018**, *20*, 722–735. [CrossRef]
67. Freestone, I.C.; Hughes, M.J.; Stapleton, C.P. The composition and production of Anglo-Saxon glass. In *Catalogue of Anglo-Sa Glass in the British Museum*; Evison, V.I., Ed.; British Museum: London, UK, 2008; pp. 29–46.
68. Freestone, I.C. The recycling and reuse of Roman glass: Analytical approaches. *J. Glass Stud.* **2015**, *57*, 29–40.
69. Phelps, M.; Freestone, I.C.; Gorin-Rosen, Y.; Gratuze, B. Natron glass production and supply in the late antique and early medi Near East: The effect of the Byzantine-Islamic transition. *J. Archaeol. Sci.* **2016**, *75*, 57–71. [CrossRef]
70. Gratuze, B.; Barrandon, J.N. Islamic glass weights and stamps: Analysis using nuclear techniques. *Archaeometry* **1990**, *32*, 155– [CrossRef]
71. Kato, N.; Nakai, I.; Shindo, Y. Change in chemical composition of early Islamic glass excavated in Raya, Sinai Peninsula, Eg On-site analysis using a portable X-ray fluorescence spectrometer. *J. Archaeol. Sci.* **2009**, *36*, 1698–1707. [CrossRef]
72. De Juan Ares, J.; Schibille, N.; Molina Vidal, J.; Sánchez de Prado, M.D. The supply of glass at Portus Ilicitanus (Alicante, Sp A meta-analysis of HIMT glasses. *Archaeometry* **2019**, *61*, 647–662. [CrossRef] [PubMed]
73. Freestone, I.C.; Jackson-Tal, R.E.; Tal, O. Raw glass and the production of glass vessels at late Byzantine Apollonia-Arsuf, Is *J. Glass Stud.* **2008**, *50*, 67–80.
74. Phelps, M. Glass supply and trade in early Islamic Ramla: An investigation of the plant ash glass. In *Things That Trave Mediterranean Glass in the First Millenium CE*; Rosenow, D., Phelps, M., Meeks, A., Freestone, I., Eds.; UCL Press: London, 2018; pp. 236–282.
75. Gliozzo, E.; Braschi, E.; Langone, A.; Ignelzi, A.; Favia, P.; Giuliani, R. New geochemical and Sr-Nd isotopic data on medi plant ash-based glass: The glass collection from San Lorenzo in Carmignano (12th–14th centuries AD, Italy). *Microchem. J.* **2** *168*, 106371. [CrossRef]
76. Balvanović, R.; Šmit, Ž.; Marić Stojanović, M.; Spasić-Đurić, D.; Branković, T. Colored glass bracelets from Middle Byzan (11th–12th century CE) Morava and Braničevo (Serbia). *J. Arch. Sci. Rep.* **2025**, *61*, 104950.
77. Freestone, I.C. Composition and affinities of glass from the furnaces on the island site, Tyre. *J. Glass Stud.* **2002**, *44*, 67–76.
78. Henderson, J.; Chenery, S.; Faber, E.; Kröger, J. The use of electron probe microanalysis and laser ablation-inductively cou plasma-mass spectrometry for the investigation of 8th–14th century plant ash glasses from the Middle East. *Microchem. J.* **2** *128*, 134–152. [CrossRef]
79. Henderson, J.; McLoughlin, S.D.; McPhal, D.S. Radical changes in Islamic glass technology: Evidence for conservatism experimentation with new glass recipes from early and middle Islamic Raqqa, Syria. *Archaeometry* **2004**, *46*, 439–468. [Cross
80. Swan, C.M.; Rehren, T.; Lankton, J.; Gratuze, B.; Brill, R.H. Compositional observations for Islamic Glass from Sirāf, Iran, in Corning Museum of Glass collection. *J. Archaeol. Sci. Rep.* **2017**, *16*, 102–116. [CrossRef]
81. Lauwers, V.; Degryse, P.; Waelkens, M. Middle Byzantine (10–13th century A.D.) glass bracelets at Sagalassos (SW Turk In *Glass in Byzantium—Production, Usage, Analyses. RGZM Tagungen, Band 8*; Drauschke, J., Keller, D., Eds.; Verlag des Römis Germanischen Zentralmuseum: Mainz, Germany, 2010; pp. 145–152.
82. Modrijan, Z.; Milavec, T. *Late Antique Fortified Settlement Tonovcov Grad near Kobarid*; Opera Instituti Archaeologici Slover Ljubljana, Slovenia, 2011; Volume 24, pp. 121–219.
83. Silvestri, A.; Marcante, A. The glass of Nogara (Verona): A "window" on production technology of mid-Medieval time Northern Italy. *J. Archaeol. Sci.* **2011**, *38*, 2509–2522. [CrossRef]
84. Gratuze, B.; Castelli, L.M.; Bianchi, G. The glass finds from the Vetricella site (9th–12th c.) Introduction. *Mélanges L'école Fran Rome* **2023**, *135*, 349–359. [CrossRef]
85. Zori, C.; Fulton, J.; Tropper, P.; Zori, D. Glass from the 11th–13th century medieval castle of San Giuliano (Lazio Province, Cer Italy). *J. Archaeol. Sci. Rep.* **2023**, *47*, 103731. [CrossRef]
86. Colangeli, F.; Schibille, N. Glass from Islamic Sicily: Typology and composition from an urban and a rural site. *Mélanges L'é Française Rome* **2023**, *135*, 321–331. [CrossRef]
87. Neri, E.; Biron, I.; Verità, M. New insights into Byzantine glass technology from loose mosaic tesserae from Hierapolis (Turk PIXE/PIGE and EPMA analyses. *Archaeol. Anthropol. Sci.* **2018**, *10*, 1751–1768. [CrossRef]

Siu, I.; Henderson, J.; Dashu, Q.; Ding, Y.; Ciu, J. A study of 11th–15th century AD glass beads from Mambrui, Kenya: An archaeological and chemical approach. *J. Archaeol. Sci. Rep.* **2021**, *36*, 102750.

Cagno, S.; Mendera, M.; Jeffries, T.; Janssens, K. Raw materials for medieval to post-medieval Tuscan glassmaking: New insight from LA-ICP-MS analyses. *J. Archaeol. Sci.* **2010**, *37*, 3030–3036. [CrossRef]

Henderson, J.; Ma, H.; Evans, J. Glass production of the Silk Road? Provenance and trade of Islamic glasses using isotopic and chemical analyses in a geological context. *J. Archaeol. Sci.* **2020**, *119*, 105164. [CrossRef]

Lane, F.C. *Venice, A Maritime Republic*; John Hopkins University Press: Baltimore, MD, USA, 1973.

Verità, M. Venetian Soda Glass. In *Modern Methods for Analysing Archaeological and Historical Glass*; Janssens, K., Ed.; John Wiley & Sons: Chichester, UK, 2013; pp. 515–536.

Šmit, Ž.; Janssens, K.; Schalm, O.; Kos, M. Spread of façon-de-Venise glassmaking through central and western Europe. *Nucl. Instrum. Methods Phys. Res. Sect. B* **2004**, *213*, 717–722. [CrossRef]

Šmit, Ž.; Stamati, F.; Civici, N.; Vevecka-Priftaj, A.; Kos, M.; Jezeršek, D. Analysis of Venetian-type glass fragments from the ancient city of Lezha (Albania). *Nucl. Instrum. Methods Phys. Res. Sect. B* **2009**, *267*, 2538–2544. [CrossRef]

De Raedt, I.; Janssens, K.; Veeckman, J.; Vincze, L.; Vekemans, B.; Jeffries, T.E. Trace analysis for distinguishing between Venetian and façon-de-Venise glass vessels of the 16th and 17th century. *J. Anal. At. Spectrom.* **2001**, *16*, 1012–1017. [CrossRef]

Lü, Q.-Q.; Chen, Y.-X.; Henderson, J.; Bayon, G. A large-scale Sr and Nd isotope baseline for archaeological provenance in Silk Road regions and its application in plant-ash glass. *J. Arhaeol. Sci.* **2023**, *149*, 105695. [CrossRef]

Coutinho, I.; Medici, T.; Alves, L.C.; Gratuze, B.; Vilarigues, M. Provenance studies on *façon-de-Venise* glass excavated in Portugal. *J. Archaeol. Sci. Rep.* **2017**, *13*, 185–198. [CrossRef]

Occari, V.; Freestone, I.C.; Fenwick, C. Raw materials and technology of Medieval Glass from Venice: The Basilica of SS. Maria e Donato in Murano. *J Archaeol. Sci. Rep.* **2021**, *37*, 102981. [CrossRef]

Zmaić Kralj, V.; Beltrame, C.; Miholjek, I.; Ferri, M. A Byzantine shipwreck from Cape Stoba, Mljet, Croatia: An interim report. *Int. J. Nautical. Archaeol.* **2016**, *45*, 42–58. [CrossRef]

. Bass, G.F.; Matthews, S.D.; Steffy, J.R.; van Doorninck, F.H. *Serce Limani. An Eleventh-Century Shipwreck. Volume I and II*; Texas A&M University Press: College Station, TX, USA, 2004.

. Gasparri, S.; Gelichi, S. (Eds.) *The Age of Affirmation. Venice, the Adriatic and the Hinterland Between the 9th and 10th Centuries*; Brepols: Turnhout, Belgium, 2018.

. Miholjek, I.; Zmaić, V.; Ferri, M. The Byzantine Shipwreck of Cape Stoba (Mljet, Croatia). In *Adriatico Altomedievale (VI–XI Secolo) Scambi, Porti, Produzioni*; Gelichi, S., Negrelli, C., Eds.; Studi e Ricerche 4: Venezia, Italy, 2017; pp. 227–246.

. Gorin Rosen, Y. The Hospitaller Compound: The Glass Finds. In *Akko III: The 1991–1998 Excavations: The Late Periods. Part I, The Hospitaller Compound*; Stern, E., Syon, D., Eds.; Israel Antiquities Authority Reports: Jerusalem, Israel, 2023; pp. 181–239.

Chinni, T. Le bottiglie kropfflasche: Testimonianze dal monastero di San Severo di Classe (Ravenna). *Archeol. Mediev.* **2017**, *44*, 297–304.

. Winter, T. *Lucid Transformations. The Byzantine–Islamic Transition as Reflected in Glass Assemblages from Jerusalem and Its Environs, 450–800 CE*; BAR International Series: Oxford, UK, 2019; pp. 136–139.

. Gliozzo, E.; Ferri, M.; Giannetti, F.; Turchiano, M. Glass trade through the Adriatic Sea: Preliminary report of an ongoing project. *J. Archaeol. Sci. Rep.* **2023**, *51*, 104180. [CrossRef]

. Foster, H.; Jackson, C. The composition of 'naturally colored' late Roman vessel glass from Britain and the implications for models of glass production and supply. *J. Archaeol. Sci.* **2009**, *36*, 189–204. [CrossRef]

. Pactat, I.; Čaušević-Bully, M.; Bully, S.; Perović, Š.; Starac, R.; Gratuze, B.; Schibille, N. Origines et usages du verre issu de quelques sites ecclésiaux et monastiques tardo-antiques et haut médiévaux du littoral nord Croate. In *Il Vetro in Transizione (IV–XII Secolo) Produzione e Commercio in Italia Meridionale e nell'Adriatico*; Coscarella, A., Noyé, G., Neri, E., Eds.; Edipuglia: Bari, Italy, 2021; pp. 289–302.

. Milavec, T.; Šmit, Ž. Glass from late antique hilltop site Korinjski hrib above Veliki Korinj/Steklo s poznoantične višinske postojanke Korinjski hrib nad Velikim Korinjem. *Arheol. Vestn.* **2020**, *71*, 271–282.

. Natan, E.; Gorin-Rosen, Y.; Benzonelli, A.; Cvikel, D. Maritime trade in early Islamic-period glass: New evidence from the Ma'agan Mikhael B shipwreck. *J. Archaeol. Sci. Rep.* **2021**, *37*, 102903. [CrossRef]

. Schibille, N.; Colangeli, F. Transformations of the Mediterranean glass supply in medieval Mazara del Vallo (Sicily). In *Mazara/Māzar: Nel Ventre Della Città Medievale (Secoli VII–XV). Edizione Critica Degli Scavi (1997) in via Tenente Gaspare Romano*; Molinari, A., Meo, A., Eds.; All'Insegna del Giglio: Sesto Fiorentini, Italy, 2021; pp. 491–505.

claimer/Publisher's Note: The statements, opinions and data contained in all publications are solely those of the individual hor(s) and contributor(s) and not of MDPI and/or the editor(s). MDPI and/or the editor(s) disclaim responsibility for any injury to ple or property resulting from any ideas, methods, instructions or products referred to in the content.

materials

MDPI

Article

Amber "Alchemy": Recreating and Investigating Yellow Glass Formulations

Catarina Reis Santos [1,2,3,*], Andreia Ruivo [1,2], Ana Carneiro [3], João Pedro Veiga [1,4], Teresa Palomar [2,5] and Inês Coutinho [1,2,*]

1 Department of Conservation and Restoration, NOVA School of Science & Technology, Campus Caparica, 2829-516 Caparica, Portugal; a.ruivo@fct.unl.pt (A.R.); jpv@fct.unl.pt (J.P.V.)
2 VICARTE Research Unit, NOVA School of Science & Technology, Campus Caparica, 2829-516 Caparica, Portugal; t.palomar@csic.es
3 CIUHCT Research Unit, NOVA School of Science & Technology, Campus Caparica, 2829-516 Caparica, Portugal; anacarneiro613@gmail.com
4 CENIMAT/I3N—Centro de Investigação de Materiais, NOVA School of Science & Technology, Campus Caparica, 2829-516 Caparica, Portugal
5 CSIC—Institute of Ceramic and Glass, Campus de Cantoblanco, 28049 Madrid, Spain
* Correspondence: acre.santos@campus.fct.unl.pt (C.R.S.); icoutinho@fct.unl.pt (I.C.)

Abstract: Amber glass has been produced since at least the 6th century BC. Its value derives from its ability to mimic natural amber and, later, from its ultraviolet filtering properties. Until the 19th century, amber glass was widely used for the storage of food and medicines because its protective properties had been empirically recognized. This study investigates historical methods of amber glass production by using glass recipes from four Portuguese arcana (1793–1975) and focusing on Fe-S and Fe-Mn chromophores. Five recipes were reproduced under controlled laboratory conditions, resulting in 21 experiments. Of these, only 10 produced amber glasses were with different shades. Chemical compositions were analysed by WDXRF, while DSC and dilatometry were used to assess thermal properties. Vickers hardness tests and UV–visible absorption spectroscopy provided insight into mechanical strength and chromophore presence. The study found that FeS amber glass was more difficult to produce than Fe-Mn amber glass, given the complex variables involved in the former, such as SO_3 volatility affecting the final product. Reproduction of historical recipes showed that, even without modern chemical knowledge, historical glassmakers developed practical, empirical methods for making amber glass. These findings contribute to a broader understanding of glass conservation and highlight the importance of historical glass recipes for the interpretation and conservation of glass objects.

Keywords: glass recipes; amber; Portuguese arcana; 18th–20th century

1. Introduction

Historically, amber glass has been produced and traded by humans since at least the 6th century BC [1]. Its synthesis initially served to create beads that imitated the natural and valuable amber (a fossilized resin). By the Roman era, glassmakers had mastered the production of this colour through meticulous control of their raw materials. Amber glass, a term used to describe hues ranging from golden yellow to dark brown, became popular for crafting vessels around the 4th–3rd century BCE [2,3]. Around the 19th century, it was realised that this glass had specific properties—nowadays understood as a capacity to filter the wavelength ultraviolet and visible radiations—contributing to preserve its content (e.g., food and medicines) [2]. In fact, by 2006, it was estimated that amber glass constituted 50% of all glass containers in the USA, predominantly for beer packaging [4].

In imparting colour to glass, iron is a versatile colourant in glass, producing colours from yellow (Fe^{3+}) to blue (Fe^{2+}), with green being the most common colour due to the combination of both oxidation states. However, the amber colour discussed here—a brownish yellow—differs from colours typically produced by iron alone. The amber chromophore, now understood to be a ferri–sulphide complex, was long attributed to carbon, a necessary ingredient for its production [2,4,5]. In 1959, Weyl provided a comprehensive historical review of "carbon-amber glasses", exploring earlier theories, such as those proposed by Eckert and Zschacke in 1928, who suggested that the colour was associated with iron sulphide formation, although they lacked analytical proof [6]. It is notable that this glass colour was probably among one of the most difficult to attribute to a colouring agent. It was only in 1965 that Brown and Douglas, through a series of experiments that involved the production of several variations of amber glass recipes, proposed that sulphur and iron were probably involved in the amber colouration [5]. Current understanding posits that the amber chromophore is composed of an Fe^{3+} ion with three bridging oxygens (O^{2-}) and one S^{2-} ion, as shown in Figure 1 [4].

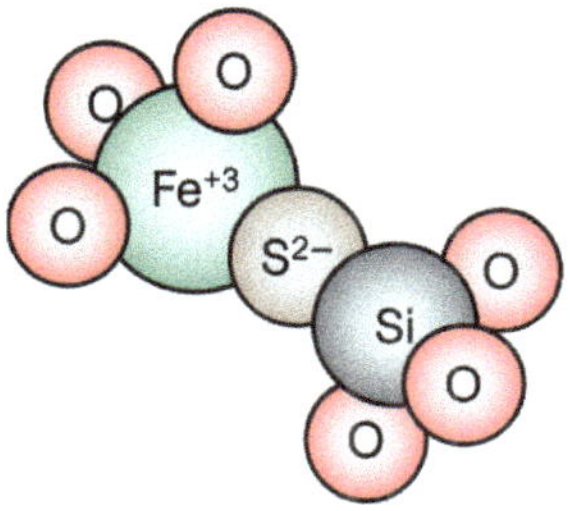

Figure 1. Illustration of the amber chromophore, adapted from Ross and Myers 2006.

Despite various advancements, producing amber glass has proven to be a challenging process. The recipe appears simple—requiring only silica (as a network former), sodium (as a flux), lime (as a stabilizer), iron (typically present in the silica source), sulphur (often from sodium sources like natron), and a reductive agent such as carbon—but it requires careful control of redox reactions. The amber colour is obtained when the co-existence of an oxidized-iron species (Fe^{3+}) and a sulphur-reduced species (S^{2-}) is achieved [2]. The success of Roman glassmakers in producing this colour was probably due to their strict control of raw materials, including a source of silica with sufficient iron oxide, a source of sodium rich in sulphur, such as natron, and a reducing agent such as charcoal [3].

The Industrial Revolution introduced more complex recipes and materials in amber glass production, as evidenced in batch books and patents. For example, the use of pyrite (FeS_2) as an iron and sulphur source became widespread [4,7]. Although amber glass can be made with inexpensive raw materials, glass factories in the mid-20th century often opted for more expensive components, such as uranium and selenium, to achieve and control the amber colour [6]. By the 1940s, researchers such as Lawton, A.J. Holland, and W.E.S. Turner had conducted extensive studies on amber glass production, showing that even with identical batch formulas, factories struggled to replicate colours consistently [6].

An alternative method for producing amber glass involves adding manganese oxide to the mix. In this case, the redox process must still be controlled, and the colour results from the balance between iron and manganese oxides (Equation (1)) and the mix of colours [2]. Achieving consistent amber colour remains to this day difficult, even when using modern techniques such as the use of real-time sensors and oxygen probes, allowing for precise adjustments to the redox environment during production.

Equation (1)—Iron and manganese redox reaction

$$Mn^{4+(brown/black)} + 2Fe^{2+(blue)} \rightleftharpoons Mn^{2+(yellow)} + 2Fe^{3+(yellow)} \tag{1}$$

This article explores how modern amber glass manufacturing can benefit from a scientific approach to historical recipes found in batch books, such as the arcana. By analyzing the technologies and knowledge of past periods, this research investigates amber glass production and its unique properties, particularly its ability to filter ultraviolet and visible radiation, while offering insights into predicting long-term glass degradation. Focusing on the industrial and technological aspects of amber glass production, this study recovers historical recipes to better understand raw materials, production methods, and the economic networks of the time. By examining amber-colored glass recipes from four Portuguese *arcana* (18th–20th centuries), this research underscores the value of continued investigation into the Portuguese glass industry, with potential benefits for modern glass production both locally and globally.

Contextualizing the Recipes: The Arcana

Batch books, often referred as arcana, are invaluable historical records that provide unique insights into the glassmaking practices of the past. The term arcana, which means "secrets" in Latin, reflects the confidential nature of the information they contain [8,9]. These notebooks are not traditional books or treatises, but practical guides used by glassmakers, especially composers, to document their recipes, techniques, and personal experiences. Serving as essential references for ensuring consistency in production, these arcana detail the proportions of raw materials, furnace conditions, and, in some cases, the origin of the ingredients. Often intended for personal use, they were usually difficult for others to interpret, functioning more as a testament to the skill and indispensability of the glassmaker who carried them from factory to factory.

The Marinha Grande Factory arcanum (MG) dates to 1793–1798 and consists of five notebooks, possibly authored by three different individuals given the handwriting variations. Like in former research, these five notebooks are to be studied as a whole [10–12]. Marinha Grande Factory, one of Portugal's most famous glass factories, operated from 1719 to 1922, producing mainly crystal and window glass [13]. The Castro and Oliveira Guerra arcanum (COG), attributed to two specific glassmakers, is dated from 1875 to 1925, and is the only arcanum in Portugal to be both published and fully contextualized [8]. Meanwhile, the Gaivotas Factory (GF) arcanum (1935–1975) likely belonged to Francisco António Rodrigues, a glassmaker whose family later donated the book for study. This factory, founded in 1811 in Lisbon, was renowned for its Art Deco glassware and laboratory bottles [13]. The most recent discovery is the arcanum from the Northern Portuguese Glass Centre (NPGC), dating to 1955, recovered from Oliveira de Azeméis, a town where the first large-scale Portuguese glass factory was established in the 16th century [13].

Across these four arcana, a total of 69 yellow and amber glass recipes have been identified. The majority (67%) rely on the iron and sulphur complex as the chromophore, while 22% depend on a balance of iron and manganese. Uranium-based recipes account for 9%, and cadmium-based recipes make up 3%. Given this distribution, this study focuses on the amber-coloured glasses derived from the iron–sulphur complex and mixture of iron and manganese, excluding uranium and cadmium-based recipes.

Since achieving a proper balance between iron and sulphur represents a significant technological challenge, four recipes using this chromophore were selected for further investigation, along with one iron–manganese recipe for comparison. The study will assess the resulting colour and thermal and mechanical properties, such as glass transition temperature (Tg) and hardness. An additional goal is to determine if the glass produced aligns with the historical recipe descriptions, such as whether a "light amber" recipe indeed results in light amber glass.

This work seeks to validate the arcanum as a key historical source, not in theory, but also potentially tied to the creation of objects currently found in Portuguese museums. By examining the intricate processes of amber glass production in the Portuguese glass industry, this research contributes to a deeper understanding of the nation's industrial heritage and its place in the global glassmaking history of the period.

2. Materials and Methods

2.1. Reproduction Methodology

Before reproducing the glass recipes in the laboratory, it is essential to establish a replicable method for interpreting and selecting relevant recipes for reproduction (Figure 2). Initially, all glass recipes of a specific colour must be identified; in this case, the focus is on amber-coloured glass recipes. Subsequently, a thorough assessment should be conducted to ensure the legibility of these recipes, discarding any that are unreadable. Once only legible recipes are considered, the next step is to select those that appear to be reproducible.

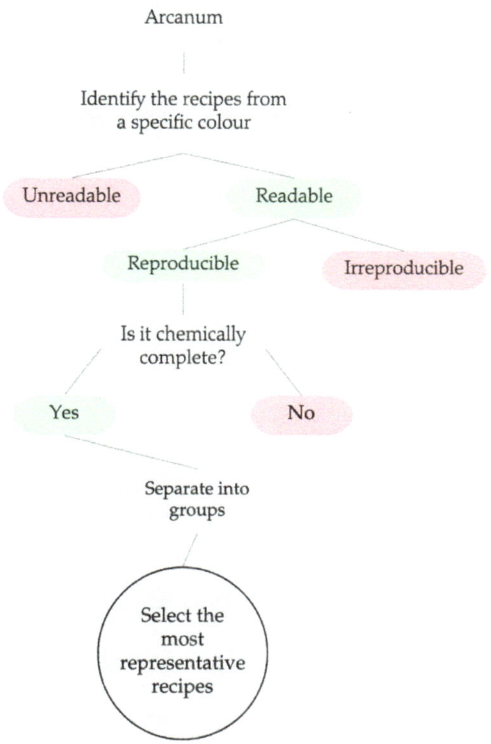

Figure 2. Decision tree of the process of interpreting and selecting the recipes to reproduce.

In terms of reproducibility, it is crucial to ascertain the meaning of each ingredient, as the legibility of a recipe does not necessarily guarantee an understanding of its components. Furthermore, it is important to determine whether a historically accurate interpretation is feasible. For instance, if a recipe from a specific factory mentions sand as an ingredient, it is imperative to ascertain whether the sand has been previously analysed. Since iron is obtained from the sand source, it is vital to understand whether its iron content is known and whether there are historical objects from that factory that can be characterized to determine their iron content.

Finally, a check for chemical completeness should be conducted. Following the composition of any glass, these recipes should include a network former, a flux, and a stabilizer. Since this is a coloured glass, they should also incorporate the respective chromophore. Given that the formation of the FeS complex implies a reductive atmosphere during melting, it was also necessary to identify the organic compounds in the recipes.

In this work, the recipes from three of the four arcana presented above were studied, and the most representative of each has been selected. The most representative is the one that is similar to the majority of recipes of that colour from each arcana. It is important to note that while a selection of recipes was made, they are quite similar and share a key trait: the amount of iron, which is crucial for producing amber glass but is never specified, as it naturally occurs in the sand. Therefore, when a recipe failed to produce amber glass, variations were developed instead of altering the original formula. Table 1 shows the complete transcript of the selected recipes as they are written in the arcana (translated into English). Recipe NP3 belongs to the NPGC arcanum and is the recipe containing Fe-Mn. Three recipes were selected from the Gaivotas arcanum—G92, G130, and G192—each with different compositions and reducing agents. Finally, from the COG arcanum, only one recipe was selected, C98, featuring a simple composition similar to most of the recipes found some of the batch books in Corning's Rakow Library

Table 1. Transcription of the recipes selected for reproduction in kilograms. Northern Portuguese Glass Centre (NP), Castro and Oliveira Guerra glassmakers (C) and Gaivotas Factory (G).

NP3		**C98**		**G92**	
Pebbles	100	SiO^3 S	200–100	Sand	30
Sand	-	$NaOCO^2$	100–44	Lime	3.60
Feldspar	60	$CaOCO^2$	72–30	Soda	15
Borax	26	Sulphur	2.8–1.4	Flour	1.5
Soda	12			Sulphur	0.200
Lime	2				
Dolomite	8	**G192**			
Boric acid	6	Sand	287.500		
Barite	-	Soda	113.410	**G130**	
Zinc oxide	4	Lime	34.000	Sand	150
Nitrate	4	Borax	1.138	Soda	66
Manganese	11	Sulphur	0.275	Lime	18
Iron oxide	13	Flour	1.900	Sulphur	0.200
Fluor-spar	1	Boric acid	3.980	Charcoal	0.950
Arsenic	0.300	NaCl	3.300	Boric acid	2

For the batch preparation, laboratory-grade reagents were used. As various studies have shown, the use of natural raw materials introduces many variables, complicating a historically accurate reproduction of glasses [14,15]. Factors such as the time of year when raw materials were collected (relevant in the case of vegetable sources), the preparation process before melting, and the initial characterization of these materials can pose challenges. In cases in which certain ingredients are known to be inherently impure, these impurities are simulated. For example, using sand as a silica source is always associated with iron oxide contamination. Reproducing a recipe from a specific batch book historically linked to a particular glass factory, using only pure silica dioxide, may not yield a historically accurate reproduction. To address this issue, and because iron oxide is essential to obtaining the amber-yellow colour, different approaches were employed to determine the amount of this oxide that should be added to the initial recipe for each arcanum. For the Castro and Oliveira Guerra arcanum recipes, the amount of Fe_2O_3 added was based on analyses conducted on the sands used by the Marinha Grande factory, as this arcanum is thought to have been used in that factory [16,17]. For the Gaivotas recipes, since there is no analysis or knowledge of the sands used, several historical colourless glasses produced at the factory were analysed to quantify the minimum amount of iron oxide required. For the NPGC, the approach was the same and the amount of iron oxide was also determined by analysing historical objects. The choice to use colourless objects to determine the percentage of iron in the sand came from the fact that they were the only historical objects we had associated with each of the factories. Considering that the sand would probably be the same for all glass compositions, the minimum iron value was determined.

Table 2 shows the amount of iron oxide used in each recipe, along with its variations. The reproduction process always starts with the original recipe, to verify its effectiveness without modifications. Next, several variables were individually modified, such as the amount of iron oxide, the addition of charcoal and the sulphur source. Given the complexity of the glass, recipe C98 was chosen to test all these variables due to its simpler composition, which facilitates the interpretation of results. In the Gaivotas arcanum, only the original interpretation of each selected recipe was tested.

Table 2. Selected recipes and variations.

ID Recipe	Version	Main Recipe	Fe/Mn and Reducing Compounds (wt.%)	Reference
NP3	a	Original recipe	Original amount Fe-Mn	-
	b	Original recipe	Half amount of Fe-Mn	-
G92		Original recipe	0.057% Fe	Historical objects
G130		Original recipe	0.057% Fe	Historical objects
G192		Original recipe	0.057% Fe	Historical objects
C98	a	Original recipe	0.008% Fe	[17]
	b	Original recipe	0.2% Fe	[3]
	c	Original recipe	0.2% Fe + 10% charcoal	[3]
	d1	Natron as sulphur source	0.1% Fe + 20% charcoal	[3]
	d2	=	=	[3]
	d3	=	=	[3]
	e1	Natron as sulphur source	0.2% Fe + 20% charcoal	[3]
	e2	=	=	[3]
	e3	=	=	[3]
	f	=	0.2% Fe + 30% charcoal	[3]
	g	Double amount of e1	0.2% Fe + 20% charcoal	[3]
	h	Natron as sulphur source	0.2% Fe + 10% charcoal	[3]
	i	Natron as sulphur source	0.1% Fe + 10% charcoal	[3]
	j	Natron as sulphur source	0.05% Fe + 20% charcoal	[3]
	k	Original recipe	Fe-Mn from NP3a	-
	l	Natron as sulphur source	0.15% Fe + 20% charcoal	[3]

The total amount of raw materials weighed for each recipe ranged from 50 to 60 g (Table 3), and each recipe mixture was homogenized by mixing it in a powder mixer (turbula) for at least 30 min. The mixture was then placed in the furnace at room temperature in an alumina crucible. This is one of the proofs that a large-scale reproduction is not exactly the same as a small-scale reproduction, and at this is one of the major problems in the study and reproduction of historical recipes.

Regarding the glass melting process, no recipes indicate information on the furnace or the melting temperature. Therefore, a temperature of 1450 °C has been chosen for melting, with a two hours' dwell. This temperature may potentially exceed historical practices; however, it helps to optimise the time required to obtain an proper glass homogenization and, therefore, for each reproduction, making this study less time-consuming and more sustainable. When necessary, adjustments to the temperature and, consequently, the soaking time will be made, as the latter plays a crucial role in the success of glass production. It is important to note that the atmosphere inside the furnace was not controlled or measured. Previous studies have demonstrated that the atmosphere within the glass batch itself has more significant impact on the final result than the furnace atmosphere [18]. As for the procedure following the melting of the glass, some researchers left it in the furnace to cool, while others poured it onto a metal plate [14,15,19,20]. In previous work [9], the glass was left in the furnace to cool down to room temperature prior to testing; however, this caused difficulties in obtaining samples for analysis without cracks or fractures. Therefore, in this study, it was decided to pour the glass onto a metal plate and then subject the sample to an annealing process to remove any internal stresses, at approximately 470 °C for one hour, and to leave it to cool naturally in the furnace.

Table 3. Weighted quantities in grams.

	$CaCO_3$	$CaMg(CO_3)_2$	Charcoal	Fe_2O_3	Flour	H_3BO_3	MnO_2	$Na_2[B_4O_5(OH)_4]\cdot 8H_2O$	Na_2CO_3	Na_2SO_4	$NaAlSi_3O_8$	NaCl	$NaHCO_3$	$NaNO_3$	SiO_2	S	ZnO
NP3 a	0.403	1.623		2.641		1.218	2.237	5.284	2.436		12.197			0.816	20.327		0.811
NP3 b	0.41	1.625		1.1329		1.225	1.112	5.287	2.443		12.193			0.816	20.329		0.815
G92	3.578			0.022	1.495				14.909						29.834	0.199	
G130	3.794		0.202	0.017		0.421			13.935						31.604	0.043	
G192	3.815			0.023	0.213	0.449		0.132	12.726			0.374			32.252	0.033	
C98 a	9.615			0.002					13.348						26.678	0.374	
C98 b	9.603			0.101					13.343						26.679	0.374	
C98 c	9.605		5.014	0.1					13.344						26.684	0.376	
C98 d1	10.348		10	0.05						4.765		4.765	6.354		23.767		
C98 d2	8.874		10.002	0.052						4.943		4.942	6.59		24.654		
C98d3	8.878		10.026	0.051						4.938		4.939	6.587		24.65		
C98 e1	10.348		10	0.1						4.765		4.765	6.354		23.767		
C98 e2	8.875		10.005	0.102						4.941		4.941	6.592		24.651		
C98 e3	8.876		10.008	0.1						4.943		4.941	6.59		24.663		
C98 f	10.348		15	0.1						4.765		4.765	6.354		23.767		
C98 g	20.696		20	0.2						9.53		9.53	12.708		47.534		
C98 h1																	
C98 h2	8.875		5.003	0.105						4.943		4.945	6.593		24.657		
C98 i	8.874		5.001	0.049						4.941		4.943	6.592		24.653		
C98 j	8.873		10.006	0.027						4.945		4.945	6.593		24.656		
C98 k	9.605			3.468			2.934		13.339						26.673		
C98 l	8.877		10.004	0.076						4.944		4.943	6.591		24.653		

Letters: identification of the different versions of the original recipe. Numbers: identification of the several reproductions of the same recipe.

2.2. Characterisation of the Reproduced Samples

The obtained glass samples were characterised using the following techniques: wavelength dispersive X-ray fluorescence spectrometry (WDXRF), UV–vis absorbance spectroscopy, differential scanning calorimetry (DSC), dilatometry, the Vickers hardness test, colorimetry measurements and image capture with the stereomicroscope (SMZ Microscope with a colour camera Nikon DS-Fi3, Tokyo, Japan). The WDXRF was used to analyse the elemental chemical composition of the samples; UV–Vis spectroscopy to confirm the colourants imparting the amber colour; colorimetry was used to determine the colour coordinates and to compare tonalities between the FeS amber with the Mn-Fe amber recipes; and DSC and dilatometry were used to assess the workability of the glasses. Finally, the Vickers hardness test was used to test the possibility of cutting and engraving, or the hardness of the glass, in general.

The UV–Vis absorbance spectroscopy analyses were performed with Avantes AvaSpec-2048 fibber optic spectrometer (Avantes, Eebeek, The Netherlands), which operates at 200–1100 nm with a resolution of 2.4 nm. The emitted light was measured using a 200 μm reflection probe (Avantes FCR 7-UV-200), which consists of a central reading fibber, surrounded by six lighting fibers, each having a diameter of 200 μm. The spectra were obtained in absorbance mode, between 350 and 1050 nm, with an integration time of 13–18 ms and 20 scans.

The thermal analyses were performed using a Netzsch Pegasus® DSC 404 F3 (Selb, Germany), which is equipped with a furnace that can reach temperatures up to 1550 °C. The experiments were conducted using a platinum crucible at a heating rate of 20 $°C \cdot min^{-1}$ in a nitrogen atmosphere. This technique provides the glass-transition temperature (Tg), and three analyses were carried out on each sample. In dilatometry, the glass transition temperature (Tg) was determined by dilatometry, employing a Netzsch Gerätebau dilatometer (model 402 PC) with a heating rate of 5 $°C \cdot min^{-1}$, and the equipment was operated by Cristina Ruiz Santa-Quiteria (Institute of Ceramic and Glass, Madrid, Spain).

The Vickers hardness of the glass samples was measured using Zwick-Roell Indentec test equipment (West Midlands, UK). Ten indentations were performed on each sample, with a measurement time of 10 s and a load of 0.5 kg.

Elemental analysis by WDXRF was conducted using a PANalytical XRF-WDS 4 kW AXIOS sequential spectrometer (PANalytical B.V., Almelo, The Netherlands) equipped with a Rh X-ray tube and four analyser crystals. The equipment was operated by Fernanda Carvalho (NOVA School of Science & Technology, Caparica, Portugal). The measurements were carried out under a helium flow in scanning mode to detect the maximum number of elements in the sample ($Z > 8$). Spectral deconvolution was performed using the iterative least squares method, and elemental quantification was based on the fundamental parameter approach, utilizing 15 certified secondary standards. This was achieved using the SuperQ software package and PANalytical's standardless analytical program IQplus ((v.5.3A, PANalytical B.V., Almelo, The Netherlands). This methodology enabled a multi-element analysis of samples with varying quantities and complex chemical compositions in a relatively fast and straightforward manner. Due to the high resolution of the technique, spectral interferences, such as overlaps of characteristic X-rays from different elements, were minimal, allowing detection of elements even at trace levels (ppm).

Three colorimetric measurements of each sample were conducted with a Lovibond® TR520 handheld spectrophotometer, (Dortmund, Germany) configured with an optical geometry of 8° viewing angle, diffused illumination, and a measurement aperture of 8 mm. Colour coordinates were calculated using CIE Illuminant D65, and are presented in accordance with the CIE L*a*b* system.

3. Results and Discussion

3.1. Recipe Identification

It should be noted that all four arcana are handwritten. In some cases, the calligraphy prevents a correct interpretation or identification of the written ingredients. However, the discussion around the reproducibility of a recipe is far more complex than the one around the calligraphy. The reproducibility is indelibly associated with the secrecy surrounding the arcanum (Figure 2). Sometimes these notebooks were written with code words and hidden information, but it also turns out that this non-reproducibility can be a consequence of the fact that the authors are so familiar with the subject that writing something so obvious leads them to often resort to abbreviations or expressions, which were then current or were part of the factory jargon, but no one knows their meaning nowadays. It is also important to note that most recipes are entitled "yellow glass". However, various colouring agents can be included in this designation, such as uranium, cadmium, iron–sulphur and iron–manganese. All recipes whose composition includes Fe-S or Fe-Mn, regardless of their denomination, are considered to be amber glass recipes. The others, with uranium and cadmium, are called yellow glass. The following section is dedicated to identifying and positioning the amber glasses within each arcanum.

3.1.1. Marinha Grande Arcanum (MG)

Given the type of objects the factory produced, the MG arcanum is the one with the fewest amber glass recipes, a glass normally associated with utilitarian objects. Out of a total of 136 glass recipes, only 12 are of yellow/amber glass, 9% of the total. Considering that this is an 18th-century arcanum, most recipes are made of lead glass, as is the case for the other coloured recipes in the arcanum, and the chromophore is FeS, which, although not specified as an ingredient, was assumed to be the colouring agent in most recipes due to the presence of iron as an impurity in the sand, and the presence of sulphate compounds from the other ingredients. Some of the recipes even specify the origin of the sand (sometimes different from the other sand sources in other recipes), reinforcing the view that the iron comes from the sand and that it was a raw material known to the glassmakers who knew the characteristics of the sands from different locations and which one was suitable for producing this colour.

Some recipes' names mention using this glass to produce beads. Unfortunately, with the information and context we have about the MG Factory and the arcanum today, almost all recipes have been defined as irreproducible, because although the basic recipe for producing amber glass is relatively simple, the most important information is the ratio and origin of the raw materials. Given that the industry has been studied emphasising political and social issues rather than technical ones, the lack of information on the raw materials used in Marinha Grande during this period makes it impossible to interpret most of the recipes.

3.1.2. Castro and Oliveira Guerra Arcanum (COG)

The COG arcanum that belonged to two glassmakers, Castro and Oliveira Guerra, respectively, has a total of 161 glass recipes, 23 of which (14%) refer to yellow/amber glass. Most recipes have FeS or Fe-Mn as chromophore agents, and, like in the other arcana, the base composition is soda–lime silicate glass. The remaining recipes are associated with the manufacture of glass containing uranium, either alone or mixed with iron or copper. As we have seen, the glass made with uranium, according to these recipes, always contains potassium, alone or together with sodium. Again, there is no use of the word amber in these recipes, but instead the use of terms such as composition for French- or English-style bottles, gold or olive-oil coloured glass and topaz yellow in the Fe-Mn or FeS recipes and sulphur or canary colour in the recipes using uranium. Although all the recipes were considered legible, only eight were classified as reproducible. Like in other arcana, this is due to the use of cullet mixtures, or very specific cullet compositions, as for example, thick cullet, iron cullet, and green cullet. together with yellow cullet, among others. The basic composition is simple, but there are some variations in the type of reducing agent used (fir bark, corn flour, sulphur and coal were the most frequently mentioned), which may be explained by the fact that this arcanum accompanied the professional careers of both glassmakers who worked in various factories throughout time.

3.1.3. Gaivotas Factory Arcanum (GF)

The arcanum with the greatest number of recipes is the GF arcanum, with 201 glass recipes, 15% of which are for producing amber/yellow glass. Most recipes use soda–lime glass as the base glass and FeS as the chromophore. Although the composition makes it easy to distinguish which recipes are for yellow glass or for amber glass, only one of the recipes is named as being for amber glass. The titles of all the other recipes mention yellow, champagne yellow, yellow for containers and China yellow (in the recipes with uranium). The recipes with cadmium or uranium, on the other hand, always have potassium in their vitreous matrix. Of the 31 recipes, only one is unreadable, but 15 are reproducible because most recipes include a cullet in the composition, indicating the use of dark and light cullet—a possible colour control technique—which makes it impossible to correctly interpret the recipe. The composition of most recipes is relatively simple, mentioning only silica, sodium, calcium, sulphur and flour. Recipes that include cullet also include flour and sulphur. It is worth noting that, except for two recipes that use charcoal, the most used reducing agent at GF was flour.

3.1.4. Northern-Portuguese Glass Centre Arcanum (NPGC)

Regarding the arcanum belonging to the NPGC, it has a total of twenty-one glass recipes, and three are for amber glass, which corresponds to 14% of the recipes in total. All the recipes were classified as being Fe-Mn amber, although only two indicate both ingredients in their composition. The third recipe only mentions the use of manganese, but, as mentioned in the literature, manganese can only produce the amber colour in glass together with iron [6]. The iron in this recipe could, therefore, be an impurity derived from a raw material, namely from the sand. However, this recipe, like most of the recipes in this arcanum, mentions the use of quartz pebbles, which, according to the literature, have a low iron composition [21]. It is worth mentioning that this third recipe uses corn flour and sulphur as ingredients, which reinforces the view that iron is present in the composition and perhaps the manganese was just a routine addition, as it is an ingredient often used in glass production. The same recipe has a high percentage of cryolite (Na_3AlF_6), which could indicate that it is a yellow opal glass. All the recipes contain arsenic and zinc oxide, as a fining agent and a stabilizer, respectively. The names of the recipes are very basic and make no reference to what kind of objects could be produced with a particular glass composition.

Although the four arcana belong to different chronologies and are different in length and number of recipes, the percentage of amber glass recipes is consistent and similar—9–15% (Figure 3). Most amber glass recipes are made from soda–lime glass, with iron or iron and manganese identified as chromophores (Figure 4). As mentioned earlier, iron alone does not produce the desired yellow/amber colour in the glass, but rather the chromophore FeS, which complexes in the glass matrix during the glass melting process. It should also be noted that most recipes are entitled as "recipe for yellow glass", while just one uses clearly the word "amber" (in a recipe entitled "black glass (bottle amber)"). To better understand the interests of each factory, all recipes for yellow glass were considered, whether producing amber glass or not.

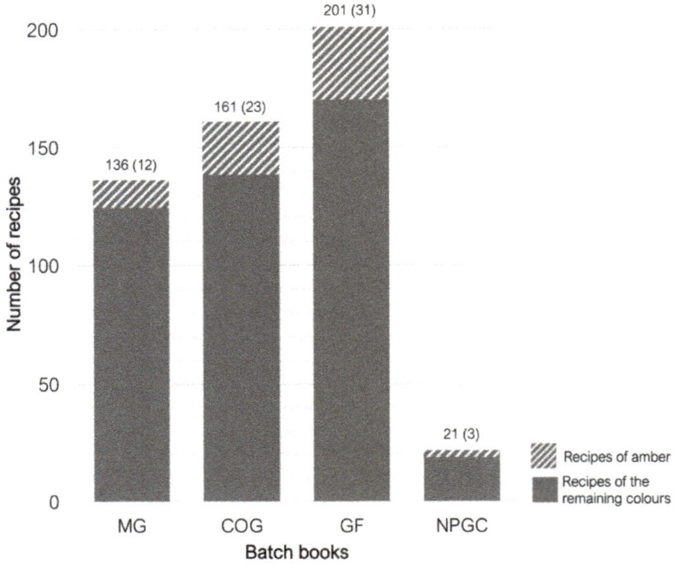

Figure 3. The total number of recipes in each arcanum. Diagonals indicate the number of yellow/amber recipes in each arcanum.

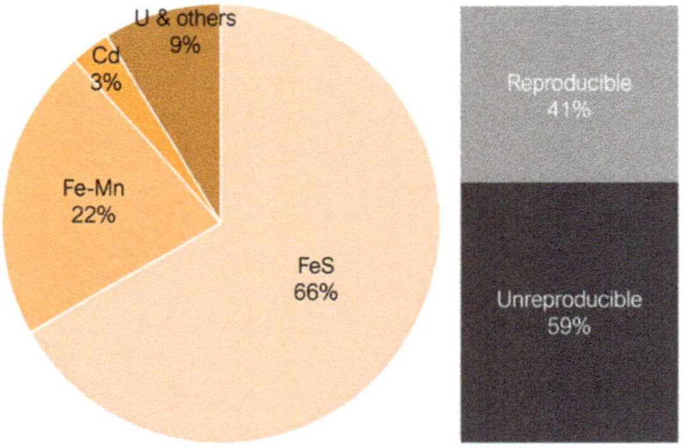

Figure 4. Different chromophores of glass recipes in the arcana and percentage of reproducible recipes for all arcana.

3.2. Recipe Reproduction

From the five selected recipes, 21 attempts to reproduce amber glass were made and only 10 had a result between light and dark amber, as shown in Figure 5. It is worth noting that only two of the unsuccessful samples (because the result was either colourless or bluish colourless glass) have been considered in Figure 5 as examples, to make the data easier to read. Two versions of the NP3 recipe were reproduced, one of each of the GF recipes (G92, G130 and G192). The C98 appeared to be the simplest, with only four ingredients; for this reason, it was chosen to be explored in more depth, with more variables, and 15 versions were reproduced.

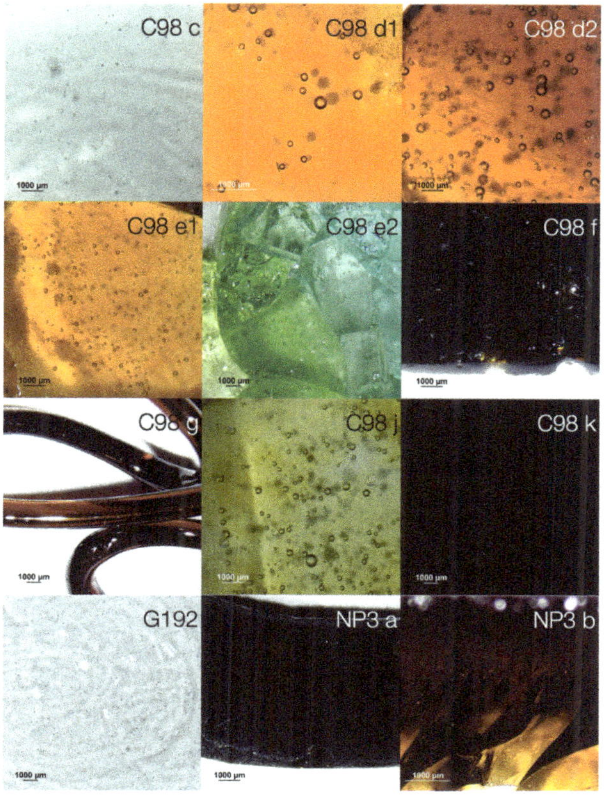

Figure 5. Pictures taken with the stereomicroscope of the samples produced.

The two reproductions of the NP3 (recipe with Fe-Mn) resulted in a dark amber. The sample from the original recipe, **NP3 a**, produced a very dark, almost black glass. **NP3 b** had half the amount of Fe-Mn, which slightly altered the colour of the sample, but the difference is only visible in the finer areas.

Sample **G92** had elemental sulphur in its composition and flour as a reducing agent, but the result was a completely colourless glass. **G130** had charcoal as a reducing agent, but the same result was obtained. **G192**, despite having a more complex composition and more ingredients, resulted in the same as before, a colourless glass. In all three recipes, the problem is believed to be the amount of iron oxide added (0.057 wt.%), so for future work on amber glass from this factory, more information on its context is needed to ascertain the raw materials that were used.

Regarding recipe C98, it was chosen to be explored in more depth and with more variables. Sample **C98 a** used the amount of iron oxide from the analyses of the sand used by the Marinha Grande Factory (Barros 1969). As expected, due to the lack of a reducing agent, this recipe resulted in a colourless glass. In sample **C98 b**, more iron oxide was added, according to existing studies about the reproduction of amber glass [3]. Since there was still no reducing agent, the result was again a colourless glass. Therefore, in sample **C98 c,** charcoal was added to the composition. The result was a bluish colourless glass, as can be seen in Figure 5. According to other studies about natron composition, the problem seems to be the way sulphur is added [22]. The following results will therefore be presented together with the chemical characterisation of the glasses using WDXRF, to understand what may or may not be affecting the formation of the ferric sulphide complex.

3.3. Chemical Characterization

Although the original interpretation of the recipe did not produce the expected result, it provided an opportunity to explore and refine various factors. By successively adjusting several variables, we aimed to gain valuable insights into the role of each ingredient and to improve our understanding of the characteristics of the recipe. Table 4 shows the quantification of all the samples discussed in this section.

Table 4. Wavelength-dispersive X-ray fluorescence spectroscopy (WDXRF) results in weight percentage of oxides.

SAMPLE	Sum Before Normalization	Na_2O	MgO	Al_2O_3	SiO_2	SO_3	Cl	K_2O	CaO	MnO	Fe_2O_3	Zn
NP3 a	80.1	8.07	0.95	11.0	64.5	-	-	0.20	2.11	4.27	6.85	1.8
NP3 b	74.3	8.24	0.89	13.0	67	-	-	0.21	2.34	2.25	3.92	2
G92	73.5	18.4	-	0.8	73.9	0.91	-	0.05	5.83	-	0.07	0.0
G130	69.7	16.5	-	0.2	76.5	0.18	-	0.06	6.44	-	0.06	0.0
G192	74.4	16.3	-	0.9	76.1	0.1	0.48	0.05	6.05	-	0.06	-
C98 a	81.2	16.8	-	0.6	66.4	0.85	-	0.03	15.2	-	-	-
C98 b	81.1	16.9	0.27	1.27	65.3	0.50	-	0.02	15.4	-	0.32	-
C98 c	78.2	15.3	-	1.4	66.1	0.76	-	0.13	16	-	0.27	-
C98 d1	84.4	11.1	0.19	2.24	66.6	0.08	0.29	0.12	19.1	-	0.18	0.0
C98 d2	80.9	9.96	0.14	6.45	66.4	0.08	0.18	0.09	16.5	-	0.18	0.0
C98 d3	80.4	8.54	0.22	0.84	70	0.23	0.51	0.11	19.3	-	0.19	-
C98 e1	58.4	9.91	0.2	1.88	66.3	0.05	0.32	0.13	20.8	-	0.41	0.0
C98 e2	76.3	11.4	0.18	2.16	68.2	-	0.22	0.13	17.3	-	0.36	0.0
C98 e3	74.3	9.99	-	0.63	69.9	0.24	0.56	0.12	18.1	-	0.37	-
C98 f	68.2	10.1	0.18	2.75	64.6	1.07	0.63	0.19	20.1	-	0.37	0.0
C98 g	82.5	8.62	0.19	5.17	67.5	0.37	0.94	0.44	16	-	0.64	0.0
C98 h	86	11.8	-	1.28	69.9	0.31	0.57	0.05	15.7	-	0.34	-
C98 i	71.5	9.81	-	3.24	67.6	0.17	0.4	0.09	18.5	-	0.19	0.0
C98 j	82.7	10.4	0.14	4.55	67.8	-	0.18	0.12	16.6	-	0.11	0.0
C98 k	81.1	13.1	-	3.57	54.4	-	-	0.08	13.5	5.8	9.47	0.0
C98 l	83.3	10	-	1.71	69.1	0.3	0.36	0.12	18.1	-	0.22	0.0

Letters: identification of the different versions of the original recipe. Numbers: identification of the several reproductions of the same recipe.

The authors are aware that some samples—C98 d2, C98 g and C98 j—present a higher content of alumina than expected. This is due to the alumina crucible contamination. Since the focus of the paper is on the colour control and variation, we believe that alumina does not play a role in this. However, it will have an impact on the mechanical properties of the glasses, which will be reflected in the Tg and Vickers Hardness values. When compared to historical samples, these will not have this influence from the crucible, as the extent of the contamination area of the crucibles used in the past (which were much bigger than the ones used in the current work) has already been studied [23].

Glass production in the 19th century already used chemical reagents produced in factories and laboratories, but their purity was not as rigorous and controlled as it is nowadays. For this reason, although the original recipe mentions the use of sodium carbonate, with the information we have today it is not possible to determine what impurities would be associated with it or even whether a specific mixture of sodium sources was used to produce amber. Therefore, based on these facts and the literature, the basic recipe was modified by changing the sodium source to a mixture of sodium bicarbonate, sodium sulphate and sodium chloride, in a ratio of 40:30:30 wt.% [3].

Sample **C98 d1** was the first successfully reproduced amber glass, with the following differences from the original: an increase in the amount of iron, the addition of sulphur with sodium and the addition of charcoal. From this recipe, two more samples were reproduced, **C98 d2** and **C98 d3**, based on the exact same recipe, with the aim to test the reproducibility. In the last one, the result was a colourless glass, despite having tried to reproduce it under the same conditions. This difference in result may have been due to an electrical fault during the melting process, or even differences in the percentage of water in the raw materials or batch moisture [4]. Looking at the characterization of the amber (**C98 d1** and **C98 d2**) and colourless (**C98 d3**) samples, the main difference is the SO_3 value, which in the case of **C98 d3** is much higher than in the other samples. Samples **G92, G130** and **G192** also have a higher SO_3 amount than samples **C98 d1** and **C98 d2**. According to the literature, the lower the amount of SO_3, the more reduced the composition of the glass [4,6]. This shows that for the samples **G92, G130** and **G192,** the proportion, melting conditions or atmosphere in the glass matrix were not the most suitable for reducing the batch [4]. The differences in the percentage of SO_3 in recipes of the same composition (for example, **C98 d1** and **C98 d3**) may indicate a greater volatilisation of SO_3, which influences the redox conditions of the batch. Again, this may be due to batch moisture conditions.

As, according to the literature, the ideal proportion of iron oxide in amber glass is between 0 and 0.5 wt.%, the subsequent samples were made increasing this oxide from 0.05 wt.% to 0.2%. Again, three attempts were made with the same composition, and only **C98 e1** produced the amber colour. The **C98 e2** sample, although not completely colourless, has a much more subtle and heterogeneous colour. It is possible to see some areas which are more yellowish and others more bluish, possibly due to the higher concentration of Fe^{3+} and Fe^{2+} ions. The significant difference in characterization is again the SO_3 amount, which in the case of sample **C98 e2** is below the detection limit, indicating that it has perhaps volatilised more than in the other samples. The difference in colour may have been due to a lack of homogenisation of the composition. Sample **C98 f** (30% charcoal) has the same amount of iron as **C98 e1** (20% charcoal) and different charcoal amounts, so a slight difference in colour was to be expected, but not too pronounced, as the amount of colourant is the same. However, the result for **C98 f** was an almost black glass, with the amber colour only visible in very thin areas, clearly due to the increased amount of reducing agent. The amount of reducing agent plays a fundamental role, sometimes even more important than the different amounts of iron. Samples **C98 d1** and **C98 e1**, with 0.1 and 0.2 wt.% iron oxide, respectively, and both with 20 wt.% charcoal, have a very similar visual result. However, samples **C98 e1** and **C98 f**, with the same amount of iron oxide (0.2 wt.%) and 20 and 30 wt.% charcoal, respectively, produced two completely different colours on the glass (amber, and a very dark amber resembling black, respectively).

Sample **C98 g** was made using the exact same recipe as sample **C98 e1**, but testing with the double of the total amount of raw materials. Instead of 60 g in total, 120 g was produced, and the result obtained was quite different for the two samples, with sample **C98 g** being noticeably darker. This shows that the result depends on several variables, including the batch quantity.

According to the literature, the ideal percentage of charcoal is between 10 and 20% [3]. Samples **C98 h** and **C98 i** contain 10% of charcoal and 10% and 20% of iron respectively. Both resulted in colourless glass, suggesting that although 10% of charcoal is sufficient to make amber glass according to the literature, it may not be enough to work with the proportions in this recipe.

Returning to 20% of charcoal, two more variations were tested. Sample **C98 j** has 0.05% iron oxide, an amount very similar to the G samples. The result was a very light, greenish amber, but it proved that it is possible to produce amber with extremely low iron content. Sample **C98 l** has 0.15% iron and although this is theoretically an amount that should produce an amber colour, since the samples with 0.10% and 0.20% (**C98 d1** and **C98 e1**) worked perfectly, the result was not satisfactory. Table 4 shows a higher SO_3 value compared to the amber samples.

Finally, sample **C98 k** was an attempt to understand the influence of the base composition on Fe-Mn amber, starting from the C98 recipe, but with the same Fe-Si and Fe-Mn ratio as the **NP3 a** recipe. Despite the different base composition, the result was a glass that is nearly black, and indistinguishable from the **NP3 a** sample.

3.4. Thermal Properties

Thermal characterisation was carried out using DSC and dilatometry techniques (Table 5), which allowed for the measurement of the glass transition temperature (Tg), the dilatometric softening point (Td) and the thermal expansion coefficient (αL) of seven of the samples.

Table 5. Results of the thermal analyses.

	C98 d1	C98 d2	C98 e1	C98 f	C98 k	NP3 a	NP3 b
$T_{g\,(onset)}$ DSC (°C)	608.9	617.4	607.3	-	-	582.9	583.0
$T_{g\,(onset)}$ DIL (°C)	608.7	623.3	611.7	605.1	545.9	580.7	593.4
Td (°C)	642.9	663.6	649.3	650	583.6	643.9	659
αL (50–500 °C) × 10^{-6} (°C^{-1})	9.7	9.6	9.5	10.6	12.2	7.1	7.0

The glass transition temperatures of the samples measured by dilatometry range from 545.9 °C to 623.3 °C, where the samples coloured with FeS have a higher Tg, of around 600 °C, and the samples coloured with Fe-Mn have a lower Tg, of around 550 °C. It is worth noting that sample **C98 k** has the lowest Tg compared to **NP3 a** and **NP3 d**. This could be caused by the fact that the first sample has a higher sodium content, which has a direct impact on Tg. The lower Tg of NP3 a and NP3 b glasses have the advantage of requiring lower energy/temperature to melt the glass, making the process faster and less expensive.

As for the softening temperature, it gives us valuable information, and can even justify some choices about the uses of amber glass, mainly used for containers rather than for elaborate pieces. The Td of the samples varies between 583.6 °C and 663.6 °C. These temperatures are considered quite high, especially when compared to other compositions known to have been used for blown glass [9]. This characteristic possibly conditioned the type of object that could be produced with these compositions, because it needs a higher temperature for work that requires being constantly reheated to allow it to be blown.

With regard to αL, its values are between 7.0 and 12.2, which indicates that, despite the different compositions, these values remain similar. Samples with a lower αL have a higher percentage of Al_2O_3, since the presence of metal oxides like this means that there are fewer non-bridging oxygens available in the composition, making these glasses more resistant to expansion differences [24].

3.5. Vickers Hardness

The hardness tests were carried out to understand the properties of the produced historical glasses. By carrying out hardness analyses on samples produced from historical recipes and creating a database of their characterisation, when a reproduction is attributed to a historical object it is possible to hypothesize the thermal and mechanical properties of that object, without damaging it, in a kind of reverse engineering.

Nine samples were then measured to obtain the Vickers hardness, as shown in Table 6.

Table 6. Vickers hardness of the amber samples.

	C98 d1	C98 d2	C98 e1	C98 e2	C98 j	C98 j	C98 k	NP3 a	NP3 b
HV 0.5	565	574	590	574	627	567	605	568	655
STDEV	19.8	8.8	15.2	4.4	15.4	8.1	22.7	14.5	16.2
Mean percentage	3.51%	1.53%	2.63%	0.77%	2.46%	1.41%	3.75%	2.55%	2.48%

The hardness of most of the samples ranges between 565 HV0.5 and 655 HV0.5, values that are significantly high compared to other studies based on historical recipes [9], and even similar to the hardness of borosilicate glass [25].

The samples with the highest hardness are **C98 j**, **C98 k** and **NP3 b**, and the samples with the lowest hardness are **C98 d1**, **C98 e2** and **C98 f**. As can be seen in Figure 5, hardness does not seem to be connected to colour or colourants at all. However, by comparing their chemical characterisation in Table 4, we can see that the samples with the highest percentage of CaO and the lowest percentage of Al_2O_3 are the ones with the lowest hardness, coinciding with what is described in the literature [26].

3.6. Colour Characterisation

3.6.1. Colorimetry

Since iron can give several colours to glass, such as blue, green and yellow, colour measurement is essential to understand the small differences in the colours of the various samples and which ones resemble each other despite the differences in composition. The scatter plot in Figure 6 represents the a* and b* coordinates of the CIE L*a*b* system.

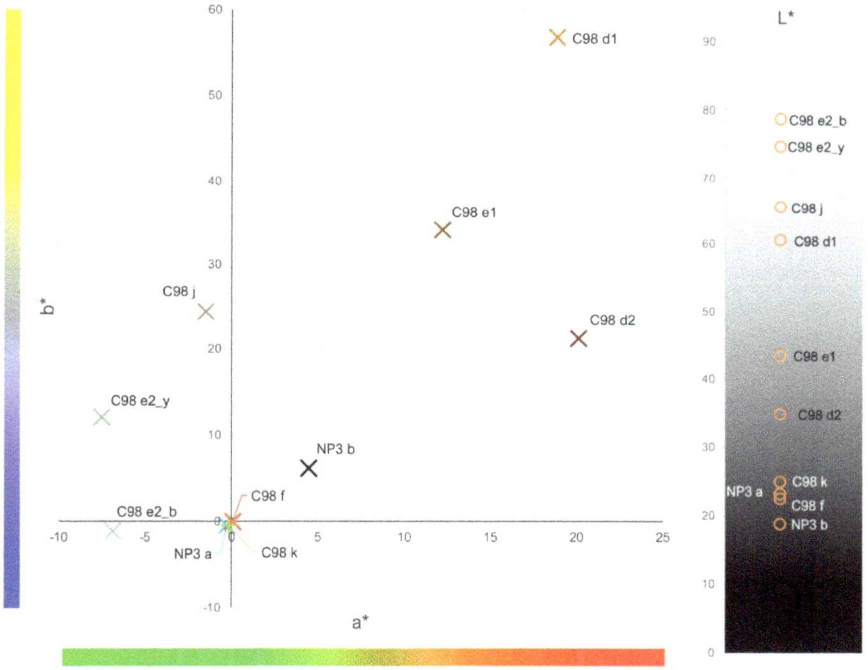

Figure 6. Scatter plot of the L*a*b* coordinates of the colorimetry measurements. Negative L* values closer to black and positive values closer to white; negative a* values closer to green and positive values closer to red; negative b* values closer to blue and positive values closer to yellow.

Figure 6 shows that sample **C98 d1** is the closest to the yellow and red colours and sample **C98 d2** is much closer to the blue, despite their same composition. According to the literature, the most similar sample to the amber colour, in terms of a* and b* values, is sample **C98 e1** [27]. In the **C98 e2_b** and **C98 e2_y** samples (measurements taken in the bluest and yellowest areas of the sample C98 e2) it is noticeable that, although the colour change is not so significant, the analysis indicated that **C98 e2_y** has a higher b* value and is more on the yellow side of the plot. Point **C98 e3_b,** on the other hand, has negative values, and is closer to the blue area of the plot. In relation to sample **NP3 b,** it is visually quite dark, but the measurement clearly shows that its colour is much lighter than that of sample **NP3 a**. Samples **NP3 a**, **C98 f** and **C98 k**, as seen from Figure 6, are the darkest, almost black, so the b* and a* values are very close to zero. The L* colourimetric measurement confirms this, as samples C98 k, C98 f, NP3 a and NP3 b are the ones with the highest colour intensity.

3.6.2. UV–Visible Absorbance Spectroscopy

The chromophores were characterised using UV–Vis absorbance spectroscopy in six of the samples shown in Figure 7. The rest of the amber samples were not characterised by this technique because of their dark colour, which saturates the signal.

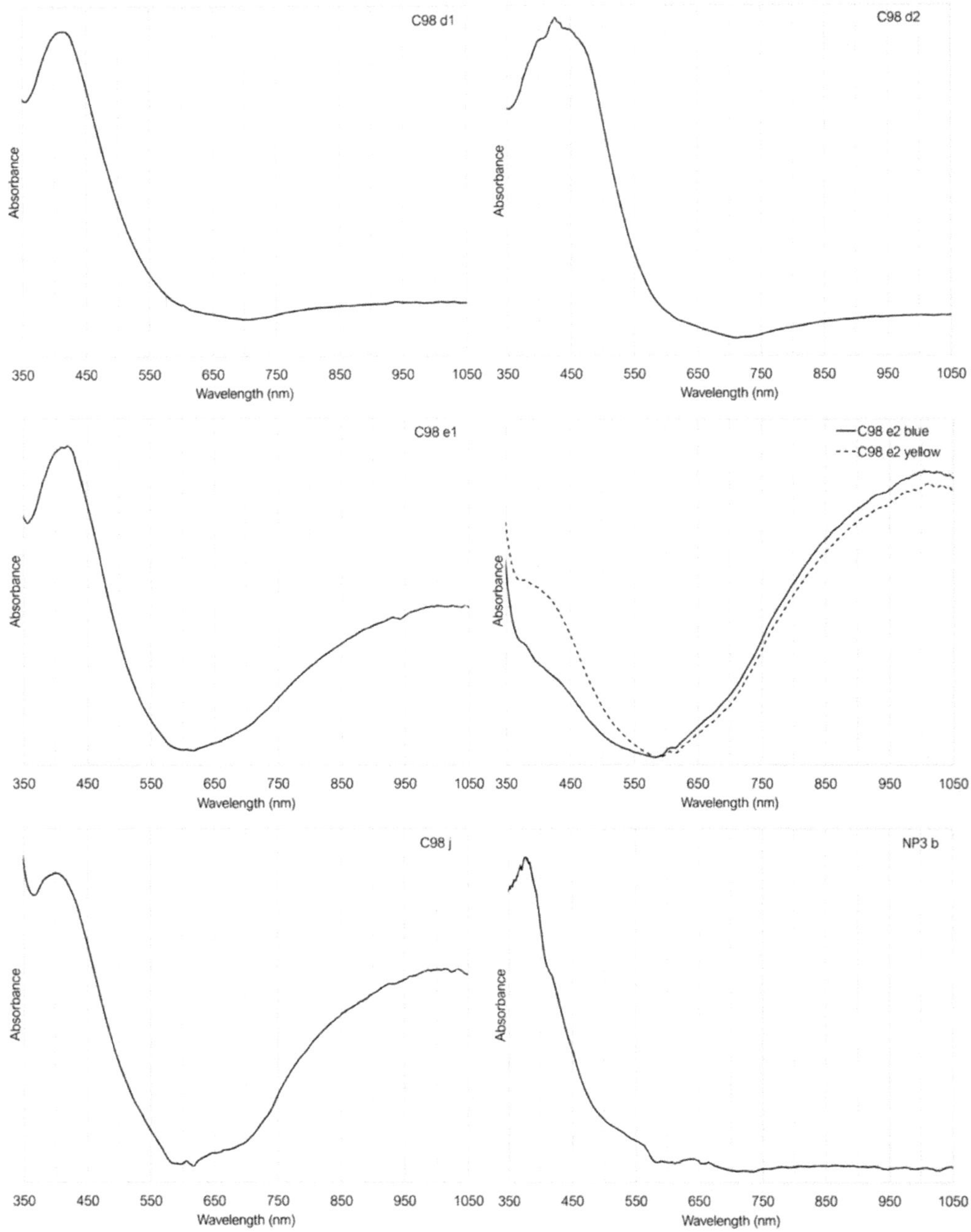

Figure 7. Absortion spectra of the amber glasses.

Samples **C98 d1** and **C98 d2,** as expected, show an absorption band around 425 nm, characteristic of the FeO_3S sulphide ion. Additionally, a slight increase in absorption from 650 nm onwards is observed, indicating the possible presence of Fe^{2+} ions. Similarly, samples **C98 e1** and **C98 j** also exhibit a band at 425 nm; however, their spectra appear more absorbent beyond 650 nm, further suggesting again the presence of Fe^{2+} ions. Their characteristic bands are found at 1100 and 2100 nm, beyond the detection limit of the equipment used, but the broad absorption band presented between 650 and 1050 nm indicates its presence. The existence of Fe^{2+} ions (responsible for the blue colour) and Fe^{3+} ions (responsible for the yellow colour) produces the green colour in glass, one of the most common and easiest colours to produce, due to the natural presence of both iron oxidation states in glass [2]. Looking at Figure 6, it is possible to see that the colorimetry of the samples **C98 e1** and **C98 j** is shifted towards the green side of the graph, which may help to confirm the view that both oxidation states are present in these samples, although they were visually classified as amber glasses. In sample **NP3 b**, the Fe-Mn amber, the absence of Fe2+ ions are noticeable, indicating complete equilibrium between iron and manganese, as indicated by the equation (Equation (1)).

As mentioned above, the **C98 e2** sample is not homogeneous, so the analysis was carried out in two different areas, to see if this change was noticeable (Figure 7). It became apparent that the **C98 e2 yellow** point has a much more intense band at 425 nm from the Fe^{3+}, while the Fe^{2+} region is slightly less intense. The **C98 e2 blue** point is exactly the opposite, as it is the bluest sample. It has less absorbance in the Fe^{3+} region and slightly more in the Fe^{2+} region.

4. Conclusions

In conclusion, this paper has explored the complex amber-glass colour technology from the 18th to 20th centuries, using historical recipes as primary sources. The research began by decoding and interpreting amber glass recipes from four Portuguese arcana, focusing on those with FeS or Fe-Mn as chromophores. These recipes were then reproduced in the laboratory and characterized chemically, thermally, and mechanically. The study revealed that producing FeS amber glass is highly intricate, with numerous variables that must be carefully controlled throughout the process. One recipe was selected for deeper analysis, and underwent significant adjustments as regards its interpretation. This was necessary because when converting historical recipes into modern laboratory protocols, it can be challenging to determine whether a recipe is ineffective or simply misinterpreted. By modifying the method of adding certain ingredients—while keeping the same ratio of elements as the original—we were able to explore whether specific raw materials were used in alternative forms that had not been initially considered.

This study introduced the critical observation that deeper knowledge of glassmaking processes does not always guarantee better results. Despite the widespread use of amber glass throughout history, it remains a challenge to reproduce it today with precise control over raw materials. Our results show that successful experimental results do not necessarily equate to a full understanding of all the variables involved, leading to inconsistencies in reproducing results. This adds a layer of complexity to the study of historical glass recipes and underlines the importance of not only interpreting historical methods, but also recognising the limitations that remain, even with modern advances.

As a result of this investigation, the creation of systematic guidelines for reproducing historical glass is proposed. These guidelines can help future studies on glass from different centuries, ensuring that new knowledge can continuously refine the process. By comparing reproduced glass with historical objects and sources, we can assess whether the information aligns or contradicts existing interpretations, which is essential to understanding the technological capabilities of the Portuguese glass industry during the period under consideration.

In addition to this impact, it is important to highlight information derived from the experiments that were carried out:

- Producing FeS amber glass proved far more complex than Fe-Mn amber. While Fe-Mn chromophores were easier to work with, they resulted in very dark colours, making it difficult to achieve softer, lighter amber colours.
- The percentage of SO_3 and its volatility or complexation is critical. WDXRF analysis showed that samples with higher SO_3 content tended to yield colourless glass. The method of sulphur addition also proved significant, with sulphur acting as expected, and producing the amber colour only when combined with sodium.
- Samples coloured with Fe-Mn (NP3 a and b) exhibited the lowest Tg, though all samples had high softening points.
- Vickers hardness tests showed that the samples were quite hard, with values approaching those of borosilicate glass. However, samples with higher calcium oxide and lower alumina oxide content were slightly less hard.
- Colorimetry results indicated that most amber samples skewed toward yellow and red, which is consistent with desired amber hues. While FeO_3S is responsible for amber coloration, some UV–Vis spectra suggested the presence of Fe^{2+}.

By uncovering the secrets of Portuguese glass factories through the study of batch books and recipes, this research bridges the past and present, offering critical insights into the craft behind the historic objects in our museums. Understanding these recipes not only improves the conservation and interpretation of these objects, but also contributes to the wider field of glass conservation. In this way, we ensure that the knowledge and techniques of past glassmakers are not lost to time, but continue to inform and enrich the future of glassmaking and conservation.

Author Contributions: Conceptualization, C.R.S. and I.C.; methodology, C.R.S. and I.C.; validation, C.R.S., A.R. and I.C.; formal analysis, C.R.S., A.R., J.P.V. and T.P.; investigation, C.R.S. and I.C.; resources, C.R.S., A.R., J.P.V., T.P. and I.C.; data curation, C.R.S.; writing—original draft preparation, C.R.S. and I.C.; writing—review and editing, C.R.S., A.C., A.R., J.P.V., T.P. and I.C.; visualization, C.R.S.; supervision, A.C. and I.C.; project administration, C.R.S. and I.C.; funding acquisition, C.R.S., I.C., A.C. and A.R. All authors have read and agreed to the published version of the manuscript.

Funding: This research was funded by the FCT-Portuguese Foundation for Science and Technology, grant number UIDB/00729/2020, UIDP/00729/2020, LA/P/0140/2020, 2023.05135.RESTART, UIDP/50025/2020, UIDB/50025/2020, CEECIND/02249/2021 (T. Palomar), 2020.00252.CEECIND (A. Ruivo) and 2020.09269.BD (C. Reis Santos).

Institutional Review Board Statement: Not applicable.

Informed Consent Statement: Not applicable.

Data Availability Statement: The original contributions presented in the study are included in the article, further inquiries can be directed to the corresponding authors.

Acknowledgments: The authors would like to thank Fernanda Carvalho, of the NOVA School of Science & Technology (Portugal) for the WDXRF analyses and Cristina Ruiz Santa-Quiteria, of the Institute of Ceramic and Glass (Spain) for the dilatometry.

Conflicts of Interest: The authors declare no conflicts of interest. The funders had no role in the design of the study; in the collection, analyses, or interpretation of data; in the writing of the manuscript; or in the decision to publish the results.

References

1. Costa, M.; Barrulas, P.; Arruda, A.M.; Dias, L.; Barbosa, R.; Vandenabeele, P.; Mirão, J. An insight into the provenance of Phoenician-Punic glass beads of the necropolis of Vinha das Caliças (Beja, Portugal). *Archaeol. Anthr. Sci.* **2021**, *13*, 149. [Cross
2. Navarro, J.M.F. *El Vidrio*, 3rd ed.; CSIC: Madrid, Spain, 2003.
3. Paynter, S.; Jackson, C.M. Mellow yellow: An experiment in amber. *J. Archaeol. Sci. Rep.* **2017**, *22*, 568–576. [CrossRef]
4. Ross, C.P.; Myers, D.D. Amber Glass—40 years of lessons learned. In *66th Conference on Glass Problems—Ceramic Engineering Science Proceedings*; Kriven, W.M., Ed.; John Wiley & Sons, Inc.: Hoboken, NJ, USA, 2006; pp. 129–137.
5. Brown, D.; Douglas, R.W. Carbon-sulphur' amber glass. *Glass Technol.* **1965**, *6*, 190–196.
6. Weyl, W.A. *Coloured Glasses*; Dawson's of Pall Mall: London, UK, 1959.

Rakow Research Library CMOG. Batch Book (1964–1981) from Beaumont Glass Company. Available online: https://cmog.primo.exlibrisgroup.com/permalink/01CORNING_INST/1531shl/alma991452843504126 (accessed on 14 October 2024).
Barosa, J.P. *O 'Arcanum' de João Augusto de Castro e Augusto de Oliveira Guerra*; Santos Barosa Museum: Marinha Grande, Portugal, 2000.
Santos, C.R.; Vilarigues, M.; Dabas, P.; Coutinho, I.; Palomar, T. Reproducing crystal glass from three 18th-20th centuries Portuguese glass *arcana*. *Int. J. Appl. Glas. Sci.* **2020**, *11*, 743–755. [CrossRef]
Lopes, F.; Lima, A.; de Matos, A.P.; Custódio, J.; Cagno, S.; Schalm, O.; Janssens, K. Characterization of 18th century Portuguese glass from Real Fábrica de Vidros de Coina. *J. Archaeol. Sci. Rep.* **2017**, *14*, 137–145. [CrossRef]
De Matos, A.P.; Queiroz, C.; Lopes, F.; Ruivo, A.; Lima, A.; Vilarigues, M. Glass Colours at Marinha Grande by the Last Quarter of the 18th Century. In Proceedings of the Annual Meeting of the SGT, Washington, DC, USA, 4–8 October 2009; Available online: www.stainedglassconservation.co.uk (accessed on 14 October 2024).
De Matos, A.P.; Lopes, F.; Vilarigues, M.; Lima, A.; Coutinho, I. A journey of the glass colours through three portuguese arcana. In Proceedings of the SGT Centenary Conference & ESG2016, Sheffield, UK, 4–8 September 2016.
Valente, V. *O Vidro em Portugal*; Editora Portucalense: Porto, Portugal, 1950.
Dussubieux, L.; Fenn, T.R.; Abraham, S.A.; Kanungo, A.K. Tracking ancient glass production in India: Elemental and isotopic analysis of raw materials. *Archaeol. Anthr. Sci.* **2022**, *14*, 226. [CrossRef]
Govantes-Edwards, D.J.; Duckworth, C.N.; Córdoba, R. Recipes and experimentation? The transmission of glassmaking techniques in Medieval Iberia. *J. Medieval Iber. Stud.* **2016**, *8*, 176–195. [CrossRef]
Santos, C.R. A Indústria dos Segredos: Os Arcanos de Vidro Portugueses dos Séculos XVIII-XX. Master's Thesis, FCT NOVA, Caparica, Portugal, 2018.
da, S. *Barros, C.V. Real Fábrica de Vidros da Marinha Grande: II Centenário 1769–1969*; Edições Magno: Leiria, Portugal, 1969.
Vilarigues, M.; Ruivo, A.; Hagendijk, T.; Bandiera, M.; Coutinho, M.; Alves, L.; Dupré, S. Red glass in Kunckels' *Ars Vitraria Experimentalis*: The importance of temperature. *Int. J. Appl. Glas. Sci.* **2023**, *14*, 201–215. [CrossRef]
Gherardi, F.; Hole, C.; Campbell, E.; Cotte, M.; Tyson, R.; Paynter, S. Unravelling the role of iron and manganese oxides in colouring Late Antique glass by micro-XANES and micro-XRF spectroscopies. *J. Phys. Photonics* **2024**, *6*, 025001. [CrossRef]
Bandiera, M. Technological Study and Chemical-Archaeometric Characterization of Roman Opaque Red Glass from Opus Sectile Decoration in the Lucius Verus Villa, 2nd Century AD. Ph.D. Thesis, NOVA School of Science and Technology, Caparica, Portugal, 2021.
Coutinho, I.; Medici, T.; Gratuze, B.; Ruivo, A.; Dinis, P.; Lima, A.; Vilarigues, M. Sand and Pebbles: The Study of Portuguese Raw Materials for Provenance Archaeological Glass. *Minerals* **2022**, *12*, 193. [CrossRef]
Jackson, C.M.; Paynter, S.; Nenna, M.-D.; Degryse, P. Glassmaking using natron from el-Barnugi (Egypt); Pliny and the Roman glass industry. *Archaeol. Anthr. Sci.* **2018**, *10*, 1179–1191. [CrossRef]
Dungworth, D. Glass-ceramic reactions in some post-medieval crucibles: An instrumental analysis study of archaeological samples. *Glass Technol. Eur. J. Glass Sci. Technol. Part A* **2008**, *49*, 157–167.
Shelby, J.E. *Introduction to Glass Science and Technology*, 2nd ed.; The Royal Society of Chemistry: London, UK, 2005.
Barlet, M.; Delaye, J.-M.; Charpentier, T.; Gennisson, M.; Bonamy, D.; Rouxel, T.; Rountree, C.L. Hardness and toughness of sodium borosilicate glasses via Vickers's indentations. *J. Non-Cryst. Solids* **2015**, *417–418*, 66–79. [CrossRef]
Hand, R.J.; Tadjiev, D.R. Mechanical properties of silicate glasses as a function of composition. *J. Non-Cryst. Solids* **2010**, *356*, 2417–2423. [CrossRef]
ABabini, A.; Green, P.; George, S.; Hardeberg, J.Y. Comparison of Hyperspectral Imaging and Fiber-Optic Reflectance Spectroscopy for Reflectance and Transmittance Measurements of Colored Glass. *Heritage* **2022**, *5*, 1401–1418. [CrossRef]

claimer/Publisher's Note: The statements, opinions and data contained in all publications are solely those of the individual hor(s) and contributor(s) and not of MDPI and/or the editor(s). MDPI and/or the editor(s) disclaim responsibility for any injury to ple or property resulting from any ideas, methods, instructions or products referred to in the content.

materials

Article

Non-Destructive Ellipsometric Analysis of the Refractive Inde of Historical Enamels

Teresa Palomar [1,2,*], Trinitat Pradell [3] and Jadra Mosa [2]

1 VICARTE Research Unit, NOVA School of Science & Technology, Campus Caparica, 2829-516 Caparica, Portugal
2 Institute of Ceramic and Glass (ICV-CSIC), c/ Kelsen 5, Campus de Cantoblanco, 28049 Madrid, Spain; jmosa@icv.csic.es
3 Departament de Física, Centre de Recerca en Ciència i Enginyeria Multiescala de Barcelona, Universitat Politècnica de Catalunya·BarcelonaTech (UPC), Campus Diagonal Besòs, Av. Eduard Maristany, 10-14, 08019 Barcelona, Spain; trinitat.pradell@upc.edu
* Correspondence: t.palomar@csic.es

Abstract: The refractive index is an important parameter for the restoration of historical cultural heritage and for non-destructive optical techniques. In this study, different mathematical models for lead glasses were assessed in order to analyze their feasibility to calculate the theoretical refractive index of the historical enamels in stained-glass windows. The models selected were those specifically developed for lead glasses: the Appen method (1949), the Fanderlik and Skrivan model (1972), and the Bonetti and Salvagno method (1983). The results of the mathematical methods were compared with the real values analyzed via ellipsometry. The historical enamels were determined on non-prepared samples, taking into account the Cauchy model in order to avoid damaging the historical pieces. We show that the measured refractive indices of the historical enamels (1.59–1.66) are higher than the values of the lead glasses in the literature (1.55–1.57). The PbO and B_2O_3 were the compounds that most influenced the value of the refractive index; nevertheless, the presence of metallic elements increased their value compared to pure lead glasses. In addition, the presence of a thin layer of grisaille on the historical enamels and the formation of alteration layers could also modify the real value of the refractive index. As far as theoretical calculations are concerned, the mathematical model of Bonetti and Salvagno (1983) seems to be the most accurate model for this material, with errors < 0.04 units. None of the three models work for glasses with >60% PbO, which is not common in cultural heritage.

Keywords: refractive index; ellipsometry; historical enamels; glass

1. Introduction

The refractive index (n) is one of the optical properties of glasses, along with transparency and coloration. It is manifested when a ray of light enters in a glass from the air, where it possesses the maximum velocity, and its velocity is reduced as a result of the interaction of the light with the ions in the glass. In the case of the vertical incidence of light, the path of the light ray is not altered, but in the case of oblique incidence, a deviation occurs [1].

The refractive index is related to the electronic density in the glass and the polarizability of its constituent elements. The bonding oxygens in the glass matrix are strongly polarized and poorly deformable, so the refractive index is low. The addition of modifier oxides, such as alkaline and alkaline-earth elements, increases the number of non-bonding oxygens and, hence, the polarizability of the oxygen and the refractive index of the glass. The polarization of cations, especially those with small ionic radii, is very low compared to anions. Voluminous cations with partially filled outer shells significantly increase the refractive index [2].

The refractive index has also been used for new materials with optical properties [3], solar concentrators [4,5], smart windows [6,7], or to imitate gemstones [8]. In forensic investigations, the chemical composition and refractive index of glass micro-fragments are also analyzed to determine their origin (i.e. automobile windows, headlamps, side mirrors, beverage container glasses. . .) [9].

Few works have analyzed the refractive index of historical glasses, probably because it is necessary to prepare a sample with a moderate size, which should not be allowed in cultural heritage objects. Tennent and Townsend [10] compiled 53 analyses of historical glasses from different museums and collections. They were clustered in four groups. The Roman and Egyptian glasses showed $1.490 < n < 1.540$, the post-Medieval vessel glasses had $1.505 < n < 1.530$, the Medieval stained glasses presented $1.540 < n < 1.585$, and the lead glasses were $1.550 < n < 1.570$. Vassas [11] also analyzed Medieval stained glasses, whose values ranged between $1.5102 < n < 1.5331$ for soda-lime silicate glasses, $1.5379 < n < 1.5514$ for potash-lime silicate glasses, and $1.5546 < n < 1.5670$ for high-lime low-alkali glasses, agreeing with the Tennent and Townsend groups. Saminpanya et al. [12] analyzed ancient Indo-Pacific beads made of soda-alumina glasses, whose values ranged from 1.51 to 1.60, with this variation related to their coloration and, therefore, to their chemical composition. Further, Drozdov et al. [13] analyzed some opaque potassium lead silicate glasses for mosaics and bijouterie produced by Mikhail Lomonosov in the Ust-Ruditsa factory in the 18th century. Their refractive indices ranged from 1.493 to 1.825. Glasses with high lead content have higher refractive indices. The refractive index is an important parameter in glass restorations [10,14,15]. The adhesive should have a similar refractive index to conceal the fissures' reparation. The closer the refractive index of the adhesive is to that of the glass, the less noticeable the repair will be.

It is also important in optical techniques used for the non-destructive analysis of cultural heritage. Optical coherence tomography (OCT) uses the reflection and scattering of light to identify the multilayer structure and to measure the thickness of the different layers. The tomograms are corrected for the refractive index of the materials penetrated by the probing beam, assuming that the refractive index remains relatively constant in space and time [16,17]. Another example is nonlinear optical microscopy (NLOM), which provides compositional and structural information based on the detection of the emission fluorescence of fluorophores through multiphoton excitation fluorescence microscopy (MPEF) and local differences in refractive indices via third-harmonic generation microscopy (THG) [18,19]. Three-dimensional Confocal Microscopy also allows measurements to be made, considering that the refractive index mismatch between the immersion fluid and sample can affect the focus position and, therefore, making axial distances appear more elongated [20]. Terahertz time-domain spectroscopy (THz-TDS) uses THz pulses to epi-detect the signal reflected from the crossed interfaces between materials characterized by different refractive indices, allowing cross-sectional images of the object's stratigraphy to be reconstructed [18].

These techniques have been widely applied to cultural heritage objects. OCT has been applied to characterize the surface alteration in glass objects [21–25]; NOLM has been used to determine the thickness of alteration layers in medieval-like glasses [26] and historical glass grisailles [27,28]; and Confocal Microscopy has allowed for the measurement of the thickness of laminated altered layers in historical glasses [29]. The refractive index of glasses and their decoration is, therefore, a key parameter for their characterization and restoration.

Several techniques have been developed to measure the refractive index. The most common methods are the deflection of a beam of light in a prism with an angle of about 60°; the refractometer, in which a glass polished on one side is pressed onto the refractometer prism with a liquid of a high index of refraction; the immersion method, in which small glass pieces are immersed in liquids of known refractive indices until the glass "disappears"; or with an interferometer that detects the interference produced by the glass on one of the two light beams emitted by the instrument [1,2]. Most of these techniques require sample preparation and sufficient size (usually a few centimeters) and are, therefore, not suitable for historical glasses. Another method of measuring the refractive index is ellipsometry, which is based on the measurement of the elliptical polarization experienced by a beam of polarized light when reflected from the surface of a glass [1,2]. This technique permits the non-destructive analysis of surface layers, even those only a few nanometers thick. For this reason, this technique should be suitable for cultural heritage material. In this study, ellipsometry was employed to determine the refractive indices of historical enamel samples. Ellipsometric measurements are based on the changes in the polarization of light reflected from the sample surface, from which information about the optical properties of the sample is extracted. However, obtaining reliable refractive index values requires careful analysis of the ellipsometric data through advanced modeling techniques. Specialized software tools were used to fit the experimental ellipsometric data to models describing the optical behavior of the enamel films. These models take into account various parameters, such as layer thickness, roughness, and the refractive index itself, among others. In this regard, the refractive index of surface enamels from stained-glass windows is very complicated to measure because they are highly fusible colored glasses that are applied as surface paintings [30], and lead and/or borax ($Na_2B_4O_7{\cdot}10H_2O$) are usually used to reduce the melting temperature [31]. Therefore, ellipsometry is proposed as a good technique to carry out these measurements.

The main objective of this study is to determine the refractive index of the enamel layers (high-lead glasses) on historical glasses. For this purpose, different mathematical models from the bibliography for silicate glasses [32,33] and lead glasses [34,35] were evaluated for glasses with chemical compositions similar to historical enamels. The feasibility of applying non-destructive ellipsometry to historical enamels was also assessed, and, finally, the agreement between mathematical and the experimental results was evaluated.

2. Materials and Methods

2.1. Samples

Three sets of samples were analyzed. The first set (Figure 1a), produced in VICARTE laboratories, consisted of blue enamels prepared according to the recipes of the historical treatise The Handmaid to the Arts (1758) by Robert Dossie [31]. The second set (Figure 1b), produced by J.M. Bonet Vitralls S.L., was prepared using historical enamels from the Rigalt, Granell & cia workshop [36] on modern glass. And the third set (Figure 1c) consisted of historical pieces of Catalan Modernist stained-glass windows from the beginning of the 20th century [37]. The samples were analyzed without any preparation in order to preserve the historical material. To minimize measurement errors, each enamel was measured 3 times in each area, and each sample was analyzed in at least 3 areas ($n > 9$).

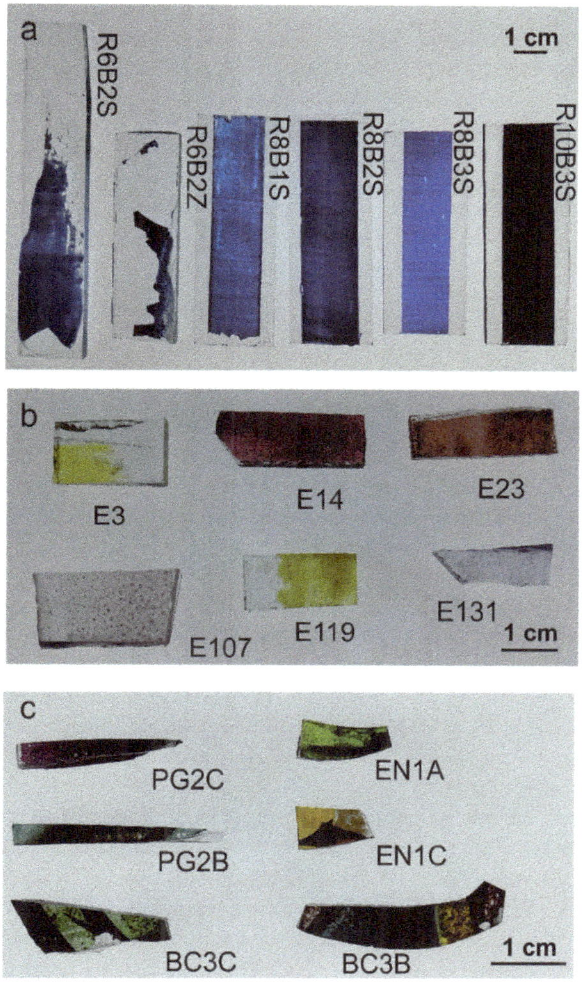

Figure 1. Sample sets: (**a**) Set 1: produced enamels; (**b**) set 2: historical enamels painted; (**c**) set 3: historical enamels from stained-glass windows.

2.2. Theoretical Calculations

Enamels are glasses that melt at a lower temperature than the glass to which they are applied [30]. Lead and/or borax ($Na_2B_4O_7{\cdot}10H_2O$) are usually used to lower the melting temperature [31]. As the refractive index is directly influenced by the ions in the glass matrix, 33 lead borosilicate glasses (Bansal and Doremus, 1986) [38], 34 lead silicate glasses (Bonetti and Salvagno, 1983) [35] and 16 lead silicate glasses (Fanderlik and Skrivan, 1972) [34] were considered for the assessment of the feasibility of the proposed methods.

Three mathematical methods were assessed. The Appen method was proposed for silicate glasses. In it, the refractive index is calculated from the portions of the individual elements by using Equation (1) and the factors contained in (Appen, 1949) [32,33]. p_i is the molar percentage (mol %) of each oxide and n_i is their specific factor.

$$n = \frac{\sum n_i \cdot p_i}{100} \quad (1)$$

To simplify the calculations, Fanderlik and Skrivan [34] formulated an equation to calculate the refractive index of the "sonoro superiore" lead glasses, in which the percentage of PbO is between 0 and 5 wt.% (Equation (2)). Some years later, Bonetti and Salvagno [35] developed another equation for glasses with a PbO content between 21 and 27 wt.% (Equation (3)). In these two models, the p_i is the weight percentage (wt.%).

$$n_D = 1.46037 + 0.00079p_{Na_2O} + 0.00290p_{CaO} + 0.00167p_{BaO} + 0.00228p_{MgO} + 0.00327p_{ZnO} + 0.00076p_{Al_2O_3} + 0.00141p_{B_2O_3} + 0.00337p_{PbO} + 0.000053p_{PbO} \cdot p_{CaO} + 0.000068p_{PbO} \cdot p_{BaO}$$

$$n_D = 1.46221 + 0.00151p_{Na_2O} + 0.0011p_{K_2O} + 0.00316p_{CaO} + 0.0023p_{BaO} + 0.00198p_{ZnO} + 0.00258p_{PbO} + 0.00215p_{B_2O_3}$$

2.3. Analytical Methods

Enamels were measured via ellipsometry to determine their refractive index. Spectral ellipsometric measurements were performed using a Variable Angle Spectroscopic Ellipsometer (WVASE32, M-2000UTM, J.A. Co., Woollam, Lincoln, NE, USA) to characterize the refractive index of glazes deposited onto glass slides. Spectra were recorded in the visible range between 250 and 900 nm at variable angles of incidence of 65°, 70° and 75° and photon energies in the range of 0.7–4.0 eV (1770–310 nm wavelength). The illuminated area of the sample at these angles is approximately 3×7 mm^2. The data were fitted using the WVASE32 software and considering the Cauchy model [39]. WVASE32 employs sophisticated fitting algorithms to minimize the difference between the measured and modelled data, allowing for accurate extraction of optical constants. The software can handle measurements taken at multiple angles of incidence, which is critical for accurate modelling of optical properties. Different optical models are available, including single-layer, multi-layer, and complex models, depending on the sample structure. Among these, the Cauchy model describes the refractive index of a transparent material as a function of wavelength (λ) and is particularly suitable for materials that are transparent in the visible to near-infrared range, making it ideal for glass and enamel. The model is relatively simple and requires fewer parameters than more complex models, making it easier to fit and interpret.

The data collected from the samples were first fitted with a transparent Cauchy layer to provide an initial value of n and e. Using this thickness, the data were then fitted by varying the refractive index and extinction coefficient (k) of the layer; in all cases, only very small values of k (weak absorption) were obtained. These best-fit values were then parameterized with a general oscillator model to ensure Kramers–Kronig consistency of the optical model [40]. Specifically, the model can include multiple layers to account for glass and the enamel coating. Each layer presents its own refractive index and thickness. Calibration was performed using standard silicon as reference material with known optical properties to ensure accurate measurements and then aligning the optical components and verifying the zero point before each measurement. Measurements were performed at a constant humidity level (typically around 40%) to prevent moisture from affecting the measurements and at a temperature (around 25 °C) to minimize thermal effects on the measurements. Samples were thoroughly cleaned with lint-free wipes to remove any contaminants and then handled carefully to avoid scratches or other damage. WVASE32 was used for data acquisition and analysis, allowing the ellipsometry data to be fitted to a selected appropriate optical model. Experimental data were fitted to an optical model, and thickness and refractive index were adjusted to achieve the best fit. Historical enamels can exhibit significant heterogeneity in their chemical composition due to variations in raw materials and manufacturing processes. This can affect their optical properties and may result in the sample being analyzed in different zones. The mean and standard deviation of the measured ellipsometry angles (Ψ and Δ) over multiple measurements were performed to analyze the central tendency and variability in the data. A linear regression was then performed. This can be used to model the relationship between Ψ and Δ and other variables, such as thickness or refractive index, helping to understand how these parameters interact. A fit test was carried out to assess how well the experimental data fit the chosen optical model, thereby evaluating the validity of the model parameters derived from Ψ and Δ. The error must be less than 10%, and the number of measurements per sample varies between 6 and 10.

To improve the fit to the experimental data, addition of surface roughness, modelled as a thin surface layer consisting of 50% underlying material and 50% air, was performed by evaluating the Mean Squared Error (MSE) between the fit and experiment. Surface roughness can cause light scattering and alter the measured ellipsometry angles (Ψ and Δ), leading to an inaccurate determination of the refractive index. A common approach is to model the surface roughness as a layer with optical properties different from the bulk material. This layer is often described using a "graded" or "effective medium" approximation, where the refractive index gradually transitions from the air to the bulk value of the material. This helps to correct the measured values. One way to correct for these surface layers is to incorporate them into the ellipsometry model as additional thin layers above the enamel material. By including the thickness and refractive index of this layer, the correct refractive index of the enamel can be determined. In this case, however, the heterogeneity of enamels does not allow for the layer to be identified. If the enamel is applied to a substrate, the optical properties of the substrate may affect the ellipsometric measurement. The refractive index of the substrate must be included in the ellipsometry model by considering the substrate as a background layer and fitting the data to separate the contributions of the enamel and the substrate. For complex systems, where there are multiple layers (e.g., enamel, oxide layer, substrate), optical interference between the layers can complicate the ellipsometry measurement. Therefore, a Bruggeman effective medium approximation was used to account for surface roughness in the data fitting process.

The experimental phases considered in this study are as follows: theoretical calculations from literature models, determination of refractive index via ellipsometry, and comparison of data. The comparison between the analyzed results and the mathematical methods was made by directly comparing the theoretical value with its real value. The methodology followed in this study is shown in Figure 2.

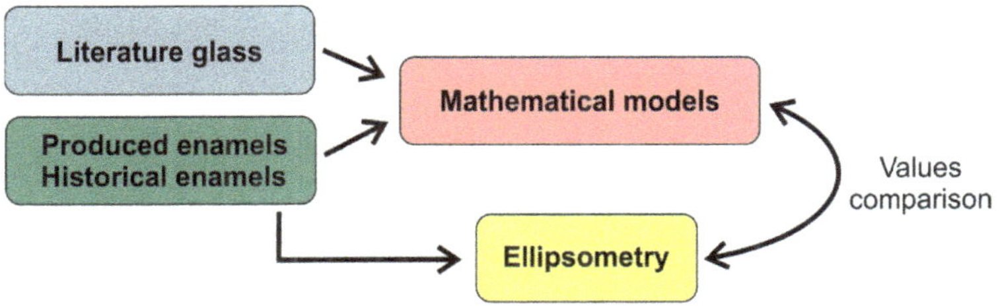

Figure 2. Schematic representation of the methodology used in this study.

3. Results and Discussion

3.1. Theoretical Calculations

In Figure 3a, the refractive index of the glasses is represented as a function of the concentration of the lead oxide, as it is the most polarizable ion in the glass matrix. It is observed that the concentration of lead oxide in the glass is directly related to its refractive index. A higher concentration of PbO means a higher refractive index, especially for [PbO] > 70 wt.%.

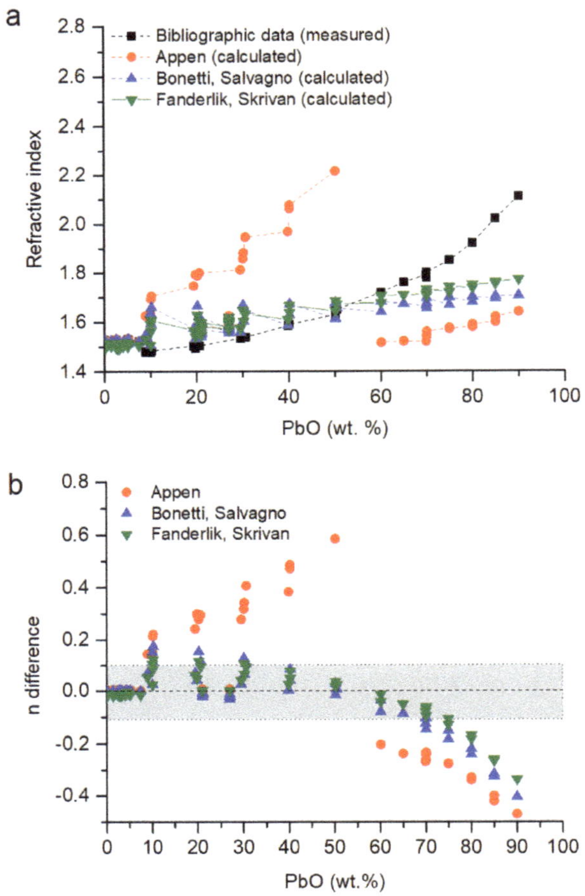

Figure 3. (**a**) Refractive index of lead borosilicate glasses. Data from [34,35,38,41] compared with data calculated following the Appen model [1], the Bonetti and Salvagno model [35] and the Fanderlik and Skrivan model [34]. (**b**) Difference in the values of the refractive index following $n_{difference} = n_{model} - n_{measured}$. In grey, the values with $n \pm 0.1$.

From the three models (Section 2.3), the Appen model overestimated the refractive index by up to 0.6 units for concentrations with 20–50 wt.% PbO, and for concentrations > 60 wt.%, it was underestimated by up to 0.4 units (Figure 3b). The models proposed by Fanderlik and Skrivan and Bonetti and Salvagno are similar; they have a very good agreement for [PbO] < 10 wt.%, with errors between them of ±0.1 for glasses with chemical compositions with [PbO] < 70 wt.%. For glasses with [PbO] > 70 wt.%, the refractive index was progressively underestimated (Figure 3b). There was no good model for [PbO] > 60 wt.% because the proposed ones are focused on glasses with [PbO] < 30 wt.%, and the high polarizability of lead ions significantly increases the refractive index [2].

3.2. Experimental Results

The enamels prepared in the Ateliers, sets 1 and 2, were measured without problems using spectral ellipsometry, and the measurements were fitted using the Cauchy model [39]. Figure 4 shows the adjustment in the fitting of the measurement of sample R8B3S, which is representative of all the samples analyzed. The historical stained glasses were more difficult to measure. Some of the historical samples presented a very thin layer of grisaille, which is formed by metallic particles embedded in a lead glass matrix that can scatter the light, compromising the analysis [42]. Additionally, in general, they show an altered surface. The irregularities in the surface can also affect the measurement [43].

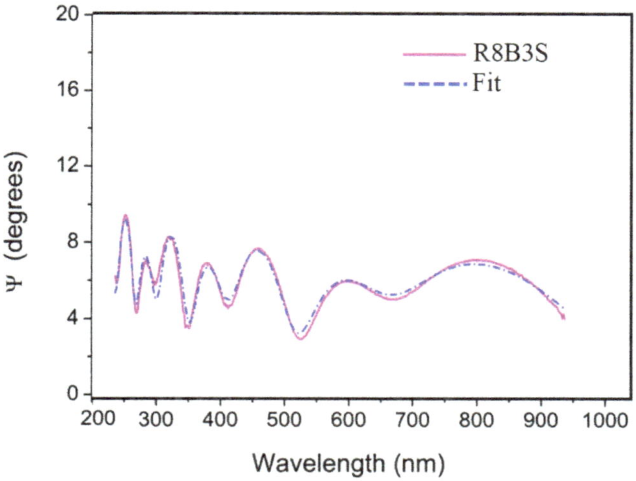

Figure 4. Adjustment in the ellipsometry measurement of the sample R8B3S. It is representative of all the samples analyzed.

The refractive index of the different enamels was measured at between 1.5 and 1.7 (Table 1). Due to the large variability in the enamel matrixes, their refractive index was not as linear as the Geller and Bunting data (Figure 3a).

Table 1. Refractive index and thickness of surface enamels measured via ellipsometry.

Set	Sample	n Calculated			Results Ellipsometry Refractive Index	Main Elements of the Chemical Composition (wt.%)										
		Appen	Bonetti, Salvagno	Fanderlik, Skrivan		B_2O_3	Na_2O	MgO	Al_2O_3	SiO_2	K_2O	CaO	CuO	ZnO	BaO	PbO
et 1. Produced enamels [31]	R6B2Z	1.65	1.62	1.62	1.616 ± 0.004	10.9	4.7	-	-	22.4	15.8	-	-	-	-	40.8
	R6B2S	1.67	1.62	1.62	1.610 ± 0.011	11	4.7	-	-	22.6	16.3	-	-	-	-	40.9
	R8B1S	1.45	1.59	1.59	1.604 ± 0.006	-	3.7	-	0.2	34.3	17.9	-	-	-	-	38.2
	R8B2S	1.6	1.59	1.59	1.597 ± 0.011	5.1	2.2	-	0.2	34.1	17.2	-	-	-	-	35.8
	R8B3S	1.36	1.49	1.47	1.489 ± 0.009	1.3	4.3	-	0.2	71.2	17.3	-	-	-	-	-
	R10B3S	1.18	1.49	1.46	1.554 ± 0.063	1.1	3.8	-	0.1	59.4	14.6	-	16.3	-	-	-
et 2. Replica enamels [36]	E3	1.68	1.63	1.67	1.613 ± 0.003	7	5.1	0.1	0.3	30.4	0.1	2.6	-	-	-	53
	E14	1.73	1.67	1.71	1.638 ± 0.004	21.1	2.1	0.1	0.3	8.6	0.2	0.9	-	11.2	-	53.2
	E23	1.63	1.63	1.67	1.652 ± 0.007	16.6	0.8	-	0.3	9.3	0.1	0.5	-	14.0	-	40.8
	E107	1.73	1.67	1.72	1.645 ± 0.003	19.8	0.9	-	0.2	8.9	0.3	0.6	-	13.8	-	52.6
	E119	1.74	1.67	1.72	1.652 ± 0.003	20.4	1.1	-	0.2	8.2	0.3	0.5	-	13.9	-	53.4
	E131	1.73	1.67	1.71	1.684 ± 0.002	21.3	1.5	0.1	0.5	9.3	0.1	0.7	-	12.9	-	52.0
et 3. Historical 20th cent. stained-glass windows [37]	BC3B blue	1.59	1.61	1.65	1.591 ± 0.008	5.2	4.9	0.1	3.6	28.3	0.1	1.2	-	8.1	0.2	42.2
	BC3C green	1.58	1.62	1.66	1.645 ± 0.003	7.5	6.3	0.1	1.9	26.5	0.2	7.1	1.6	2.2	0.1	40.5
	PG2C bluish	1.55	1.6	1.64	1.644 ± 0.006	4.5	5.5	0.1	2.1	32.9	0.1	7.2	1.9	3.7	0.4	35.9
	PG2C purple	1.46	1.57	1.59	1.593 ± 0.005	3.5	6.7	0.2	0.7	41.6	0.3	5.8	-	0.2	0.4	27.9
	EN1A green	1.71	1.66	1.7	1.641 ± 0.015	15.7	1.4	-	0.8	15.8	0.1	0.8	1.6	1.9	-	61.3
	EN1C yellow	1.75	1.66	1.71	1.693 ± 0.029	14.5	1.5	-	0.5	15	0.1	1.0	-	7.5	-	58.4

In the produced enamels (set 1), the refractive index was between 1.489 and 1.616 (Table 1, Figure 5a). Samples R6B2Z and R6B2S were prepared with six parts of base glass and one part of coloring agent [31]. These proportions of raw materials produced the highest content of PbO and, therefore, the highest refractive index. Samples R8B1S, R8B2S, and R8B3S were made with four parts of base glass and one part of coloring agent, but each sample was made with different base glasses. Sample R8B1S was made with a lead glass without borax, sample R8B2S with a lead glass with borax, and sample R8B3S with common flint glass and borax [31]. The addition of borax diminished the refractive index of the enamels, decreasing up to 0.07 units (Table 1). Finally, sample R10B3S was prepared with five parts of base glass, one part of coloring agent, and one part of copper [31]. It had a relatively high refractive index value because the enamel contains 16.3 wt.% of CuO, which compensates for the content of lead.

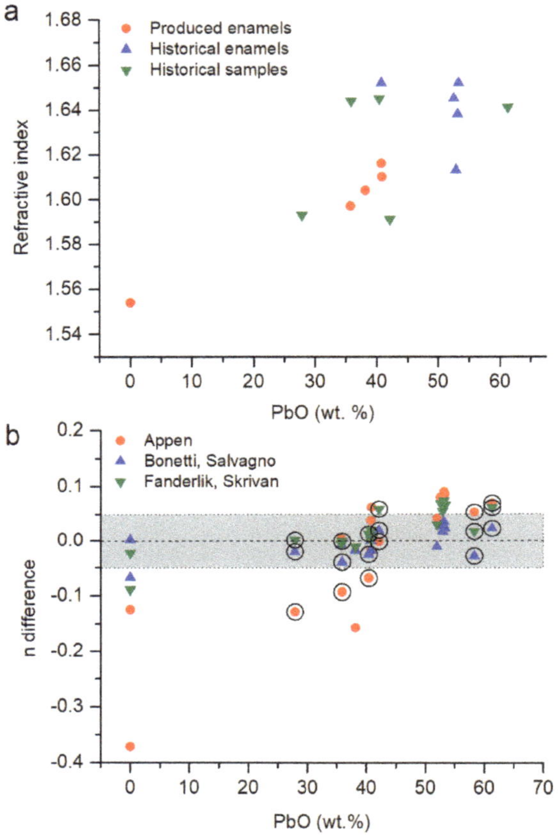

Figure 5. (**a**) Representation of the refractive index in function of the PbO content on the samples from sets 1, 2 and 3. (**b**) Difference between the refractive index value ($n_{difference} = n_{model} - n_{measured}$) calculated following the Appen model [1], the Bonetti and Salvagno model [35] and the Fanderlik and Skrivan model [34] model and measured via ellipsometry (Table 1). The circles indicate the altered historical enamels. In grey, the values with $n \pm 0.05$.

The three mathematical models showed average variations of 0.4 units. The model of Bonetti and Salvagno showed the most similar result to the measured data (Figure 5b). Samples R8B3S and R10B3S have the highest error because they did not have lead oxide in their composition (Table 1). In addition, none of the models include CuO in their calculations (Equations (1)–(3)), increasing the error in sample R10B3S.

The historical enamels, from set 2, show a good correlation between the PbO content and the measured refractive index (Figure 5a). Samples E3, E14, E107, E119, and E131 are chemically similar; however, their refractive indices are slightly different (Table 1). Sample E23 had less PbO, but it was compensated by a lower content of B_2O_3. In the case of sample E3, the high content of SiO_2 decreased the refractive index of the enamel, even its high content of PbO. Samples EN1A and EN1C showed the highest errors in the average results of ellipsometry (Table 1), because a thick layer of grisaille was applied over the enamel, and the PbO content was low [37]. Regarding the accuracies of the models, they were overestimated by up to 0.1 unit (Figure 5b), with the largest errors in the Appen model. The model of Bonetti and Salvagno was the most accurate, with an error < 0.04 units.

Finally, six historical enamels from stained-glass windows were measured (set 3). They have chemical compositions similar to some of the enamels produced nowadays (sets 1 and 2); however, they have slightly higher refractive indices (Table 1, Figure 5a). This behavior could be due to two reasons: the presence of a thin layer of grisaille on the enamel or the formation of an alteration layer.

The presence of metallic particles from the grisaille over the enamel layer can alter the measured data. The refractive index of the most common grisaille's particles is Fe_2O_3: 2.91, Fe_3O_4: 2.42, MnO_2: 2.13, SiO_2: 1.544 [44,45]. Most of these particles have refractive indices higher than simple enamels (Table 1, sets 1 and 2), so they could increase their value.

Another reason could be the alteration in the surface due to its interaction with the environment. Analysis of the alteration products in the surface revealed compounds enriched in lead [37] that could be lixiviated from the glassy matrix. This loss would decrease the refractive index of the enamel due to a reduction in the electronic density in the glass, which contradicts the experimental results.

Previous studies have investigated the change in the refractive index due to the formation of an alteration layer. Lind and Hartman [46] assessed different soda-lime silicate glasses exposed to a semi-arid environment for more than 30 years. In these glasses, the refractive index was diminished due to the formation of alteration layers of 148–228 nm. Similarly, Casparis-Hauser and Guenther [47] also observed a decrease in the refractive index due to the formation of an alteration layer on the surface of a glass enriched in BaO and ZnO (N-BAK4). Kaspar et al. [40] characterized the effect of induced corrosion on a borosilicate glass (ISG glass), and they also observed that the refractive index of the alteration layer was lower than the glass substrate. Only in a study by Portal and Sempere [48] was the formation of a two-layer system observed in a soda silicate glass with a higher refractive index, which was attributed to the densification of the leached layer.

Therefore, the increase in the refractive index of the enamels could be due to the application of the thin layer of grisaille on the surface of the enamels.

For these samples, the mathematical models from Fanderlik and Skrivan and Bonetti and Salvagno agreed with the experimental results (Figure 5b). The error in the Fanderlik and Skrivan model was up to 0.06 units, but the error in the model of Bonetti and Salvagno was <0.04 units. The latter model was developed for "sonoro superiore" lead glasses with 21–27 wt.% PbO. Nevertheless, it also works for glasses with a higher content of lead.

Very few studies have analyzed the refractive index of historical glasses, and none of them characterized the glazed surface paintings. The historical lead glasses analyzed by Tennent and Townsend [10] had refractive indices between 1.550 and 1.570; however, the chemical composition of these pieces has not been published. Vassas [11] did not analyze lead glasses, and Saminpanya et al. [12] did not publish the value of each glass bead, so it is not possible to compare the n value with the chemical composition. Finally, Drozdov et al. [13] analyzed some opaque potash-lead silicate glasses with n values between 1.493 and 1.825. For a similar lead content (sample 1), the n was 1.825, a higher value than that obtained in the present study, but its chemical composition did not present boron, unlike the historical enamels. This light element reduces the value of the refractive index [2].

4. Conclusions

In general, enamels and grisaille paints have a higher refractive index (1.59–1.66) than historical glasses due to their high lead oxide content. When compared with historical lead glass, they have higher values than measured in the literature (1.55–1.57), probably because the surface paints contain metallic elements that increase the electronic density of the glass and, therefore, the n value.

This study evaluated the accuracy of three mathematical models. The Appen model was found to be less accurate than the models proposed by Bonetti and Salvagno and by Fanderlik and Skrivan. The latter is similar, with the Bonetti and Salvagno model being slightly better because it was developed for glasses with a higher PbO content. Nevertheless, the three models do not work for glasses with >60% PbO, which are not common in cultural heritage.

This study also showed that ellipsometry is a suitable technique for analyzing the refractive index of glass and surface paints without damaging the stained-glass window's fragments. A limitation of this technique is that historical samples are not homogeneous, so a large number of measurements are needed to reduce the error. In addition, the presence of very thin surface layers such as grisaille or alteration layers inherent to the cultural object can also modify the refractive index value, requiring more measurements to obtain an average value. Polishing or damaging the object is not allowed under any circumstances.

Future lines of research include the application of ellipsometric analyses to samples of different chronologies, locations, and chemical compositions to provide a database useful for restorers and heritage scientists.

Author Contributions: T.P. (Teresa Palomar) designed the study; T.P. (Teresa Palomar) and J.M. carried out the tests and analyses; T.P. (Teresa Palomar), J.M. and T.P. (Trinitat Pradell) discussed the results; T.P. (Teresa Palomar) and J.M. prepared the original draft. All authors have read and agreed to the published version of the manuscript.

Funding: This research has been funded by the Fundação de Ciência e Tecnologia from Portugal (project ref. UIDB/EAT/00729/2020, UIDP/00729/2020, 2023.05135.RESTART, and researcher grant CEECIND/02249/2021), the Ministerio de Ciencia e Innovación from Spain (project ref. PID2022-137783OB-I00 and Ramón y Cajal contract (RYC2023-045699-I), the Generalitat de Catalunya (Consolidated Group 2021 SGR 00343),and the Regional Government of Madrid through the project TECHeritage-CM (TEC-2024/TEC-39). The authors wish also to acknowledge the professional support of the Interdisciplinary Thematic Platform from CSIC Open Heritage: Research and Society (PTI-PAIS).

Institutional Review Board Statement: Not applicable.

Informed Consent Statement: Not applicable.

Data Availability Statement: The data that support the findings of this study are available from the corresponding author upon reasonable request.

Acknowledgments: The authors wish to thank VICARTE for providing one of the sets of enamels and Jordi Bonet from J.M. Bonet Vitralls S.L. who kindly supplied all the materials from the Rigalt, Granell & cia workshop as well as historic enameled glasses.

Conflicts of Interest: The authors declare no conflicts of interest.

Abbreviations

k	extinction coefficient
MPEF	Multiphoton excitation fluorescence microscopy
MSE	Mean Squared Error
n	Refractive index
NLOM	Nonlinear optical microscopy
OCT	Optical coherence tomography
THG	Third-harmonic generation microscopy
THz-TDS	Terahertz time-domain spectroscopy
wt.%	Weight percent

ferences

Scholze, H. *Glass: Nature, Structure, and Properties*; Springer: New York, NY, USA, 1991.
Fernández Navarro, J.M. *El Vidrio*, 3rd ed.; Consejo Superior de Investigaciones Científicas, Sociedad Española de Cerámica y Vidrio: Madrid, Spain, 2003; ISBN 8400081587.
Sonal; Sharma, A. Energetic Argon Beam Stimulated Growth of Plasmonic Silver Nanoparticles in Ag+-Exchanged Soda Glass: A Study on the Structural, Optical, Photoluminescence and Electrical Behavior. *Mater. Sci. Eng. B* **2021**, *263*, 114860. [CrossRef]
Yuan, Y.; Chen, Y.; Chen, W.L.; Hong, R.J. Preparation, Durability and Thermostability of Hydrophobic Antireflective Coatings for Solar Glass Covers. *Sol. Energy* **2015**, *118*, 222–231. [CrossRef]
Liu, G.; Mazzaro, R.; Sun, C.; Zhang, Y.; Wang, Y.; Zhao, H.; Han, G.; Vomiero, A. Role of Refractive Index in Highly Efficient Laminated Luminescent Solar Concentrators. *Nano Energy* **2020**, *70*, 104470. [CrossRef]
Chowdhary, A.K.; Sikdar, D. Design of Electrotunable All-Weather Smart Windows. *Sol. Energy Mater. Sol. Cells* **2021**, *222*, 110921. [CrossRef]
Liu, X.; Wu, Y. Monte-Carlo Optical Model Coupled with Inverse Adding-Doubling for Building Integrated Photovoltaic Smart Window Design and Characterisation. *Sol. Energy Mater. Sol. Cells* **2021**, *223*, 110972. [CrossRef]
Zhang, S.; Li, K.; Pu, J.; Ni, W. Preparation and Basic Properties of Praseodymium-Neodymium-Chromium Containing Imitation Gemstone Glass. *Materials* **2022**, *15*, 7341. [CrossRef]
Palásti, D.J.; Kopniczky, J.; Vörös, T.; Metzinger, A.; Galbács, G. Qualitative Analysis of Glass Microfragments Using the Combination of Laser-Induced Breakdown Spectroscopy and Refractive Index Data. *Sensors* **2022**, *22*, 3045. [CrossRef]
Tennent, N.H.; Townsend, J.H. The Significance of the Refractive Index of Adhesives for Glass Repair. *Stud. Conserv.* **1984**, *29*, 205–212. [CrossRef]
Vassas, C.D. Étude Chimique, Thermographique et Physique de Verres de Vitraux Du Moyen-Age. In Proceedings of the Ninth International Congress on Glass, Versailles, France, 27 September–2 October 1971; pp. 241–266.
Saminpanya, S.; Bavornyospiwat, N.; Homklin, S.; Danyutthapolchai, S.; Bupparenoo, P. Physical and Chemical Properties of the Ancient Glass Beads from the Highland Log-Coffin Culture and the Lowland Areas, Thailand: Considerations on Their Colors and Technology. *J. Archaeol. Sci. Rep.* **2016**, *8*, 366–380. [CrossRef]
Drozdov, A.; Andreev, M.; Belousov, Y.; Lobanov, A.N.; Kovalenko, K.V.; Pettinari, C. The Coloration of Smalt Produced at Ust-Ruditsa Glass Factory from 1753 to 1768. *Inorganica Chim. Acta* **2022**, *542*, 121104. [CrossRef]
Mills, A.A. The Conservation and Restoration of Mediaeval Stained Glass Windows by "Gel-Plating". *Stud. Conserv.* **1987**, *32*, 122. [CrossRef]
II, J.M.M.; Lansbury, P.T. Controlling the Refractive Index of Epoxy Adhesives. *J. Am. Inst. Conserv.* **1989**, *28*, 127. [CrossRef]
Everall, N.J. Modeling and Measuring the Effect of Refraction on the Depth Resolution of Confocal Raman Microscopy. *Appl. Spectrosc.* **2000**, *54*, 773–782. [CrossRef]
Targowski, P.; Kowalska, M.; Sylwestrzak, M.; Iwanicka, M. OCT for Examination of Cultural Heritage Objects. In *Optical Coherence Tomography and Its Non-Medical Applications*; Wang, M., Ed.; IntechOpen: London, UK, 2020; pp. 147–164.
Dal Fovo, A.; Castillejo, M.; Fontana, R. Nonlinear Optical Microscopy for Artworks Physics. *Riv. Nuovo Cimento* **2021**, *44*, 453–498. [CrossRef]
Filippidis, G.; Tserevelakis, G.J.; Selimis, A.; Fotakis, C. Nonlinear Imaging Techniques as Non-Destructive, High-Resolution Diagnostic Tools for Cultural Heritage Studies. *Appl. Phys. A* **2015**, *118*, 417–423. [CrossRef]
Besseling, T.H.; Jose, J.; Blaaderen, A. Van Methods to Calibrate and Scale Axial Distances in Confocal Microscopy as a Function of Refractive Index. *J. Microsc.* **2015**, *257*, 142–150. [CrossRef]
Palomar, T.; Iwanicka, M.; Pombo Cardoso, I.; Vilarigues, M.; Targowski, P. Assessing the Decorative Techniques of Two Art Nouveau Glass Windows by Optical Coherence Tomography (OCT). *Herit. Sci.* **2023**, *11*, 203. [CrossRef]
Kunicki-Goldfinger, J.; Targowski, P.; Góra, M.; Karaszkiewicz, P.; Dzierzanowski, P. Characterization of Glass Surface Morphology by Optical Coherence Tomography. *Stud. Conserv.* **2009**, *54*, 117–128. [CrossRef]
Brostoff, L.B.; Ward-Bamford, C.L.; Zaleski, S.; Villafana, T.; Buechele, A.C.; Muller, I.S.; France, F.; Loew, M. Glass at Risk: A New Approach for the Study of 19th Century Vessel Glass. *J. Cult. Herit.* **2022**, *54*, 155–166. [CrossRef]
Read, M.; Cheung, C.S.; Ling, D.; Korenberg, C.; Meek, A.; Kogou, S.; Liang, H. A Non-Invasive Investigation of Limoges Enamels Using Both Optical Coherence Tomography (OCT) and Spectral Imaging: A Pilot Study. In Proceedings of the Optics for Arts, Architecture, and Archaeology VII, Munich, Germany, 24–27 June 2019; Targowski, P., Groves, R., Liang, H., Eds.; SPIE: St Bellingham, WA, USA, 2019; Volume 11058, p. 2.
Brostoff, L.; Zaleski, S.; Ward-Bamford, C.L.; Montagnino, E.; Muller, I.; Buechele, A.; Loew, M.; France, F. Nineteenth Century Glass Manufacture and Its Effect on Photographic Glass Stability. *J. Inst. Conserv.* **2020**, *43*, 125–141. [CrossRef]
Oujja, M.; Palomar, T.; Martínez-Weinbaum, M.; Martínez-Ramírez, S.; Castillejo, M. Characterization of Medieval-like Glass Alteration Layers by Laser Spectroscopy and Nonlinear Optical Microscopy. *Eur. Phys. J. Plus* **2021**, *136*, 859. [CrossRef]

27. Oujja, M.; Agua, F.; Sanz, M.; Morales-Martin, D.; García-Heras, M.; Villegas, M.A.; Castillejo, M. Multiphoton Excitat Fluorescence Microscopy and Spectroscopic Multianalytical Approach for Characterization of Historical Glass Grisailles. *Tala* **2021**, *230*, 122314. [CrossRef]
28. Machado, C.; Oujja, M.; Cerqueira Alves, L.; Martínez-Weinbaum, M.; Maestro-Guijarro, L.; Carmona-Quiroga, P.M.; Castil M.; Vilarigues, M.; Palomar, T. Laser-Based Techniques for the Non-Invasive Characterisation of Grisaille Paints on Stained-G Windows. *Herit. Sci.* **2023**, *11*, 85. [CrossRef]
29. Schalm, O.; Anaf, W. Laminated Altered Layers in Historical Glass: Density Variations of Silica Nanoparticle Random Packi as Explanation for the Observed Lamellae. *J. Non Cryst. Solids* **2016**, *442*, 1–16. [CrossRef]
30. Schalm, O.; Van der Linden, V.; Frederickx, P.; Luyten, S.; Van der Snickt, G.; Caen, J.; Schryvers, D.; Janssens, K.; Cornelis Van Dyck, D.; et al. Enamels in Stained Glass Windows: Preparation, Chemical Composition, Microstructure and Cause Deterioration. *Spectrochim. Acta Part B At. Spectrosc.* **2009**, *64*, 812–820. [CrossRef]
31. Machado, A.; Vilarigues, M. Cobalt Blue—Reproduction and Characterisation of Blue Enamel Recipes from The Handmaid to Arts by Robert Dossie. *Glass Technol. Eur. J. Glass Sci. Technol. Part A* **2016**, *57*, 131–140. [CrossRef]
32. Appen, A.A. Calcul des Propriétés Optiques, de la Masse Volumique et du Coefficient de Dilatation des Verres de Silicates Fonction de Leur Composition. *Dokl. Akad. Nauk.* **1949**, *69*, 841–844.
33. Scholze, H. Calculation Based on Composition. In *Glass: Nature, Structure and Properties*; Springer: New York, NY, USA, 1 pp. 223–227.
34. Fanderlik, I.; Skrivan, M. Utilisation de la Méthode Mathématico-Statistique des Expériences Planifiées Pour Suivre la Varia des Propriétés Physiques des Verres de Cristal Au Plomb En Fonction de Leur Composition. *Verres Refract.* **1972**, *26*, 19–23.
35. Bonetti, G.; Salvagno, L. Utilizzazione di un Metodo Matematico-Statistico per il Calcolo Dell'indice di Refrazione dei V "Sonoro Superiore" in Finzione della Loro Composizione Chimica. *Riv. Stn. Sper. Vetro* **1983**, *13*, 99–108.
36. Beltrán, M.; Schibille, N.; Brock, F.; Gratuze, B.; Vallcorba, O.; Pradell, T. Modernist Enamels: Composition, Microstructure Stability. *J. Eur. Ceram. Soc.* **2020**, *40*, 1753–1766. [CrossRef]
37. Beltran, M.; Schibille, N.; Gratuze, B.; Vallcorba, O.; Bonet, J.; Pradell, T. Composition, Microstructure and Corrosion Mechani of Catalan Modernist Enamelled Glass. *J. Eur. Ceram. Soc.* **2021**, *41*, 1707–1719. [CrossRef]
38. Bansal, N.P.; Doremus, R.H. *Handbook of Glass Properties*; Academic Press: Orlando, FL, USA, 1986.
39. Boissiere, C.; Grosso, D.; Lepoutre, S.; Nicole, L.; Bruneau, A.B.; Sanchez, C. Porosity and Mechanical Properties of Mesopor Thin Films Assessed by Environmental Ellipsometric Porosimetry. *Langmuir* **2005**, *21*, 12362–12371. [CrossRef]
40. Kaspar, T.C.; Reiser, J.T.; Ryan, J.V.; Wall, N.A. Non-Destructive Characterization of Corroded Glass Surfaces by Spectrosc Ellipsometry. *J. Non Cryst. Solids* **2018**, *481*, 260–266. [CrossRef]
41. Geller, R.F.; Bunting, E.N. The System PbO-B_2O_3-SiO_2. *J. Res. Natl. Bur. Stand.* **1939**, *23*, 275–283. [CrossRef]
42. Machado, C.; Vilarigues, M.; Palomar, T. Historical Grisailles Characterisation: A Literature Review. *J. Cult. Herit.* **2021** 239–249. [CrossRef]
43. Todorov, R.; Tasseva, J.; Lozanova, V.; Lalova, A.; Iliev, T.; Paneva, A. Ellipsometric Characterization of Thin Films fr Multicomponent Chalcogenide Glasses for Application in Modern Optical Devices. *Adv. Condens. Matter Phys.* **2013**, *2013*, 308 [CrossRef]
44. Lide, D.R. *CRC Handbook of Chemistry and Physics*; CRC Press: Boca Raton, FL, USA, 2003.
45. Hendi, A.H.Y.; Al-Kuhaili, M.F.; Durrani, S.M.A. Chemical and Optical Properties of MnO_2 Thin Films Prepared by Reac Evaporation of Manganese. *Int. J. Res. Eng. Technol.* **2016**, *5*, 320–328.
46. Lind, M.A.; Hartman, J.S. Natural Aging of Soda-Lime-Silicate Glass in a Semi-Arid Environment. *Sol. Energy Mater.* **198** 81–95. [CrossRef]
47. Casparis-Hauser, E.; Guenther, K.H.; Tiefenthaler, K. Spectrophotometric and Ellipsometric Study of Leached Layers Forr on Optical Glass by a Diffusion Process. In Proceedings of the 1983 International Technical Conference—SPIE 0401, Gen Switzerland, 28 November 1983; Thin Film Technologies I. Jacobsson, J.R., Ed.; SPIE: St Bellingham, WA, USA, 1983; pp. 211–
48. Portal, S.; Sempere, R. Study of Alkali Silicate Glass Corrosion Using Spectroscopic Ellipsometry and Secondary Ion M Spectrometry. *Phys. Chem. Glas.* **2003**, *44*, 303–307.

Disclaimer/Publisher's Note: The statements, opinions and data contained in all publications are solely those of the indivic author(s) and contributor(s) and not of MDPI and/or the editor(s). MDPI and/or the editor(s) disclaim responsibility for any injur people or property resulting from any ideas, methods, instructions or products referred to in the content.

materials

MDPI

ticle

haracterization and Making Techniques of Calcareous onstruction Materials for Phaya Thon Zu Temple in Bagan listorical Area, Myanmar

e Ri Yang [1,2], Gyu Hye Lee [1], Dong Min Kim [3] and Chan Hee Lee [1,*]

1 Department of Cultural Heritage Conservation Sciences, Kongju National University, Gongju 32588, Republic of Korea; 1212heri@naver.com (H.R.Y.)
2 Conservation Science Division, National Research Institute of Cultural Heritage, Daejeon 34122, Republic of Korea
3 International Cooperation Center, Korea Heritage Agency, Seoul 06153, Republic of Korea
* Correspondence: chanlee@kongju.ac.kr

Abstract: The calcareous materials used in constructing the Phaya Thon Zu temple at the Bagan historical sites in Myanmar are mortars, plasters, and stuccos. Among them, the mortars and plasters are a mixture of original and new materials used for recent conservation treatments. In this study, the making techniques were examined through analysis of calcareous materials by production period. All calcareous materials have a mineral composition similar to soil, except calcite. Stuccos have the most refined aggregates, homogeneous particle size, and the highest lime and organic contents. They were designed to improve ease of carving and weathering resistance, considering the unique characteristics of the stuccos. Because all calcareous materials were mixed with soil, the origin of the clay materials was analyzed. It was concluded that the mortars were produced by mixing clay and sandy soil, and the original mortars showed characteristics similar to soil. It is highly possible that sandy soil from around the Htillominlo temple was used to produce new plasters, and it is estimated that a mixture of clay soil was used for the original plasters and stuccos. A clear provenance interpretation of the original and raw materials used for each construction and the mixing ratio of clay materials need to be discussed through experiments, along with the estimated provenance area of the raw calcareous materials.

Keywords: Phaya Thon Zu temple; mortar; plaster; stucco; manufacturing techniques

1. Introduction

The Bagan historical area is located at a bend in the Ayeyarwady River developed in the central plains of Myanmar, and it is one of the three greatest Buddhist sites in the world designated by UNESCO, along with Angkor monuments in Cambodia and Borobudur temple in Indonesia, with the most enormous scale among them (Figure 1A). Numerous archaeological sites, such as pagodas, temples, and monasteries, as well as murals and sculptures, are distributed in the Bagan region [1]. They are a testament to the climax of the Bagan civilization in Myanmar, which flourished during the 11th to 13th century, and its dedication to Buddhism [2]. In recognition of their value, seven sectors of Bagan were registered as UNESCO World Heritage Sites in June 2019 (Figure 1B).

The subject of this study, the Phaya Thon Zu temple, with the literal meaning of three pagodas, is believed to have been constructed in the late 13th century [1,3]. It consists of three independent temples with antechambers and a main chamber covered with domed ceilings that are connected by two passages, a unique structure that differentiates it from other temples in the Bagan area (Figure 1C). The antechamber of each temple houses a primary Buddha statue, and the interior walls and ceilings are decorated with a wide range of artistic elements, including various murals such as painted representations of Sakyamuni

Buddha statues and lotus patterns, thereby endowing a significant archaeological and historical value [4].

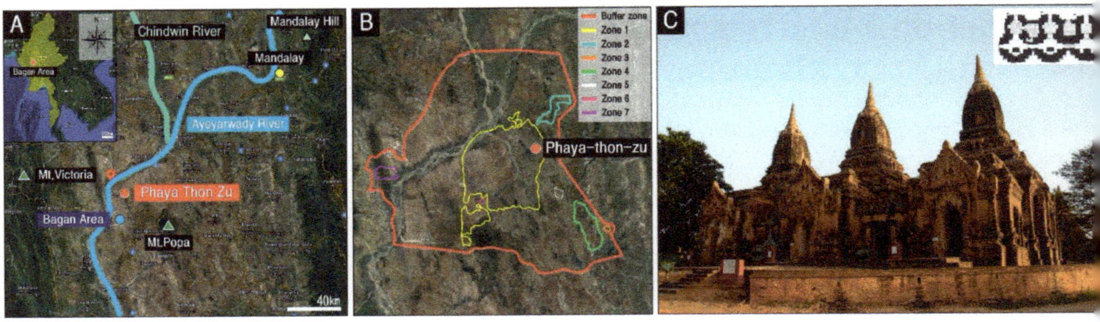

Figure 1. Location and overview of the study area. (**A**) Detailed location of the Phaya Thon Zu temple and Bagan area in Myanmar. (**B**) Map showing the Bagan historical area. (**C**) General view of the Phaya Thon Zu temple.

It is believed that approximately 4000 pagodas were constructed between the 11th and 13th centuries in the Bagan area, and there are currently about 3000 pagodas and temples distributed throughout the region. However, although many of these pagodas and temples were damaged by earthquakes with magnitudes of more than six on the Richter scale that occurred in 1975 and 2016, there is insignificant research on their repair and conservation management [5–10].

Diversified damages to the Phaya Thon Zu temple have also progressed significantly due to damage from prolonged exposure to natural weathering, disasters, and artificial deteriorations [11]. Due to the absence of technology to cope with frequently earthquakes and anthropogenic damages, it is essential to conduct scientific analysis and precision diagnosis to research on the selection and application of alternative materials that reflect the results thereof [6,12].

Therefore, based on the investigation history and records of local experts, the original and new materials used for repair works were distinguished, and various specimens of the original and new ones were obtained from multiple repair processes on the site. In this study, the same analysis method was applied to all acquired specimens to examine the differences and homogeneity between the new ones for repair work and the original materials.

Investigations on the main calcareous materials of architectural and cultural heritages have also been conducted relatively diversely in Korea [13–19]. However, research on the calcareous materials used as the functional materials of the primary structures has been only limitedly reviewed, thereby needing substantially more extensive studies in the future [6,7,20,21].

This study strove to analyze the material characteristics of mortars, plasters, and stuccos used in the construction of the Phaya Thon Zu temple and examine the construction technology system based on the results obtained. Accordingly, the data gathered in this study contributes towards the establishment of appropriate conservation measures to maintain the outstanding universal value, authenticity, and integrity of the original form of the damaged construction materials as a world heritage.

2. Materials and Methods

2.1. Conservation and Classification of Materials

Materials including brick, mortar, plaster, stucco, and rock were utilized in the construction of the Phaya Thon Zu temple (Figure 2). Bricks were made by firing clay and were the structural material used in the construction of the entire temple. Almost all have a reddish-brown color, are rectangular in shape, and are classified into small and large bricks with relatively constant dimensions of about 18 × 10 × 5 cm and 28 × 16 × 3 cm, respectively.

Figure 2. Photographs showing representative construction materials of the Phaya Thon Zu temple.

Mortar was used as the joint filler of the bricks, which is categorized into two types: red series, for which the clay is the main component, and light gray series, for which the lime is the main component. In addition, the surface is composed of brick and mortar and finished with plaster. Plaster is made up of lime and sand as the main components and was used as the finishing material for the base layer of the murals, walls, and ceilings inside the temple at the time of construction. It is also the material used most often to repair temple walls nowadays.

Although the Phaya Thon Zu temple is situated on relatively firm ground and maintains a stable structure, a closer examination reveals that damages are distinct and progressive for each type of material. Moreover, due to the repeated repairs and numerous conservation treatments, the distinction between the original materials and the new materials used for repairs is ambiguous, and it is difficult to identify trends or uniformity in the locations at which conservation treatments were rendered.

The representative types of damages in the interior and exterior of the temple can be broadly categorized into physical damages, such as cracks, exfoliation, and detachment, and chemical and biological damages, such as discoloration by contaminants and infestation. Physical damages are particularly evident in a wide area with cracks and detachment of the brick, mortar, and plaster layers that make up the walls (Figure 3A).

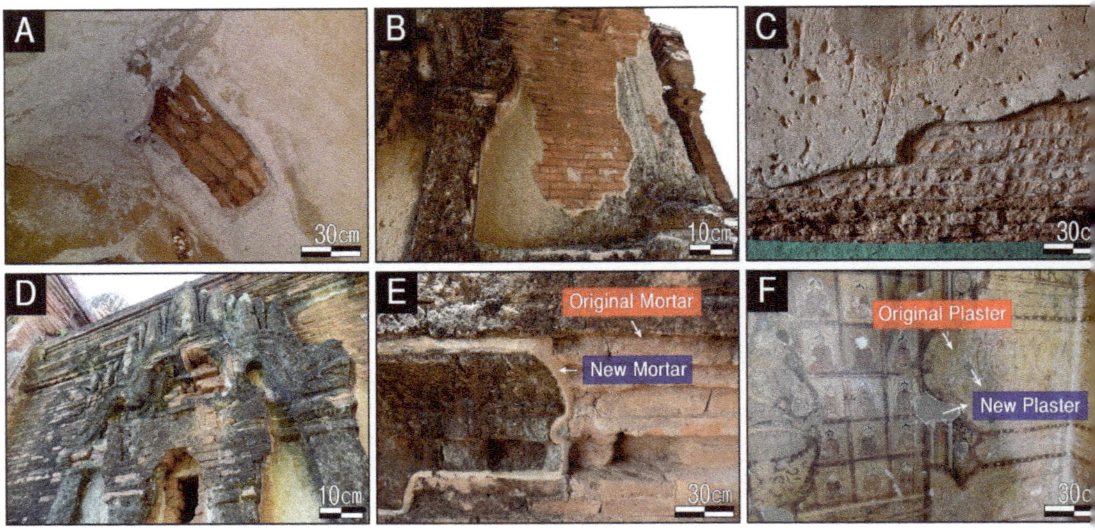

Figure 3. Photographs showing representative deteriorations and conservation treatments of the Phaya Thon Zu temple. (**A**,**B**) Physical damage. (**C**,**D**) Chemical and biological damage. (**E**,**F**) Examples of conservation and restoration interventions.

Although it is presumed that the stuccos covered the entire exterior walls of the temple at the time of construction, much of it has fallen and is lost due to natural deterioration and natural disasters, such as earthquakes. The physical damages in the remaining stuccos are also very severe, with an extensive distribution of cracks and detachment throughout the entire area (Figure 3B).

Chemical and biological damages are also observed throughout the temple. The temple's interior and exterior brickwork and calcareous auxiliary materials are interspersed with cavities throughout. These appear to have been caused by insects, implying the possibility of the addition of organic components, such as molasses or sap, at the time of brick production (Figure 3C). Among the calcareous materials, the stuccos used for exterior decoration displayed the most pronounced biological contamination with extremely severe black stains and lichen coverage, making it impossible to discern its original color (Figure 3D).

Although the conservation history of the Phaya Thon Zu temples has still not been documented or confirmed, it is presumed that conservation measures were taken intermittently, over several periods, based on the lack of consistency of locations and methods of repair implemented. It can be confirmed that the original mortar used as joint filler has been overlaid with relatively recent new mortars along the sections of cracks and detachments (Figure 3E). The sections with detached coloring layers of the mural are interspersed with traces of relatively recent repairs with new plasters (Figure 3F).

The original materials presented in the study and the new materials used for the repair are distinguished visually on site. In particular, the two parts differ in color and texture. However, in order to obtain samples for analysis, the original and new repair mortars were selected with the help of local experts who were in charge of the actual repair of the temple.

2.2. Analytical Specimens

In this study, samples of 6 mortars, 6 plasters, and 11 stuccos were used to analyze the material characteristics of the materials used at the time of the initial construction and the new materials used in repairs after that. All the stuccos are original. In the case of the samples of mortars used as joint fillers for the bricks, some are original and some are new; the first ones are brown, and the second have a milky white hue.

The original mortars have been detached from the temple's exterior walls in a powdery state and have a clay-like consistency with a high alumina content. As mentioned above, the original mortar and the mortar used for repair have differences in color and texture. On the other hand, the new mortars are relatively cohesive and have a distinct calcareous substrate (Figure 4, Table 1).

Figure 4. Photographs showing the samples analyzed. Sample numbers are the same as those of Table 1.

The sample selection for the study was quite limited. However, since the temple had already undergone several repairs, there were relatively many and diverse samples preserved on site. Fortunately, through discussions with the local administrator, we were able to secure the minimum number of samples required for analysis.

Two of the plaster samples are new materials used for the repairs. The original plaster was collected from those detached from the exterior wall in a powdery state, along with the agglutinated portion with a milky white color. The new plaster, in part, has a relatively high degree of solidification, a grayish-brown color, and a mixture of white substances, thereby displaying different characteristics.

Also, all the stucco samples are in their original form, produced at the time of the initial construction. Although they generally have a milky white color, some samples have a light brown color. Even though they display decorative features that may have adorned the temple's exterior walls, their decorative elements are either unclear or unconfirmed since some samples were collected only in minute quantity. The stucco samples are generally dominated by calcareous substrate, but phaneritic crystals were not confirmed (Figure 4).

Table 1. Sample list of analytical materials and characteristics from the Phaya Thon Zu temple in Bagan historical area, Myanmar.

Type	Group	Sample No.	Location	Use
Mortars	PMO	PMO-1	On the outside surface of Phaya Thon Zu	Original mortars
		PMO-2		
		PMO-3		
	PMN	PMN-1		New mortars
		PMN-2		
		PMN-3		
Plasters	PPO	PPO-1	On the outside surface of Phaya Thon Zu	Original plasters
		PPO-2		
		PPO-3		
		PPO-4		
	PPN	PPN-1		New plasters
		PPN-2		
Stuccos	PST	PST-1	Near the Phaya Thon Zu temple	Original stuccos
		PST-2		
		PST-3		
		PST-4		
		PST-5		
		PST-6		
		PST-7		
		PST-8		
		PST-9		
		PST-10		
		PST-11		
Soils		PS-1	Near the Sulamani temple	Clayey soil
		PS-2	Near the Htilominlo temple	Sandy soil
		PS-3	Brick factory in the Thu Htay Kan	Clayey soil
		PS-4		Sandy soil

2.3. Research Methods

To examine the material characteristics of the calcareous auxiliary materials used in the Phaya Thon Zu temple, a precision field survey of the construction materials that compose the temple was conducted first to collect calcareous samples that fell off by the time of their respective productions. Their physicochemical characteristics, such as chromaticity, magnetic susceptibility, and P-XRF analysis, were reviewed on the site to select the representative samples based on their outcomes.

The studied samples were concurrently examined under a stereomicroscope (Nikon Eclipse LV 100N POL, Nikon Instruments Inc., Melville, NY, USA) to investigate the status of their yield, voids, mineral composition, and weathering. Thin sections were also prepared and examined under a polarizing microscope (Nikon Eclipses E 600W, Nikon Instruments Inc., Melville, NY, USA), and scanning electron microscopy (Tescan, MIRA-3, Brno, Czech Republic) and attached energy-dispersive elemental spectroscopy (Bruker Quantax 200, Bruker, Bremen, Germany) were used for microstructure and qualitative analysis. The samples were coated with platinum (Pt) to increase their electrical conductivity and minimize the influence of the composition ratio.

To precisely identify the mineral composition of all samples, they were pulverized into sizes less than 20 μm and subjected to X-ray diffraction analysis (Rigaku MiniFlex 600, Rigaku, Tokyo, Japan). The X-ray irradiation source was CuKα, and the anode accelerating voltage and filament current were 15 mA at 40 kV, respectively. The measurement range and speed were 3° to 60° and 1°/min, respectively.

Meanwhile, a differential thermal analyzer and thermogravimetric analyzer (TA Instruments, SDTQ600, Hüllhorst, Germany) were used to track the thermal history and phase transitions of the constituent minerals during the reheating process of the representative samples. The α-Al_2O_3 was used as a standard sample, and the thermal history change was measured in the room temperature range to 1000 °C at a rate of 10 °C/min.

Geochemical analysis was performed to quantify each sample's major, some trace, and rare earth elements. Analyses were performed at ACT LAB in Canada using inductively coupled plasma-atomic emission spectroscopy (ICP-AES), mass spectrometry (ICP-MS), and an instrumental neutron activation analyzer (INAA). In addition, all analytical data were validated for their quantitation by utilizing blank, duplicate, and standard samples.

3. Results and Interpretation

3.1. Textural Characteristics

To examine the microtexture of the calcareous materials to be investigated, including the substrate, particle size of minerals, voids, and organic matter distribution, stereoscopic and polarizing microscopy observations and SEM-EDS analysis were conducted. As a result, although the substrate of all the mortars utilized as joint sealers was cryptocrystalline regardless of their production time, a relatively large quantity of clay was observed in the substrates of the raw materials (Figure 5). The constituent minerals are mainly quartz, with a size of approximately 1 mm and a roundness ranging from subcircular to subangular. Such characteristics were also visible under SEM, and EDS analysis of some raw materials displayed the highest content of Si (Figure 6, Table 2).

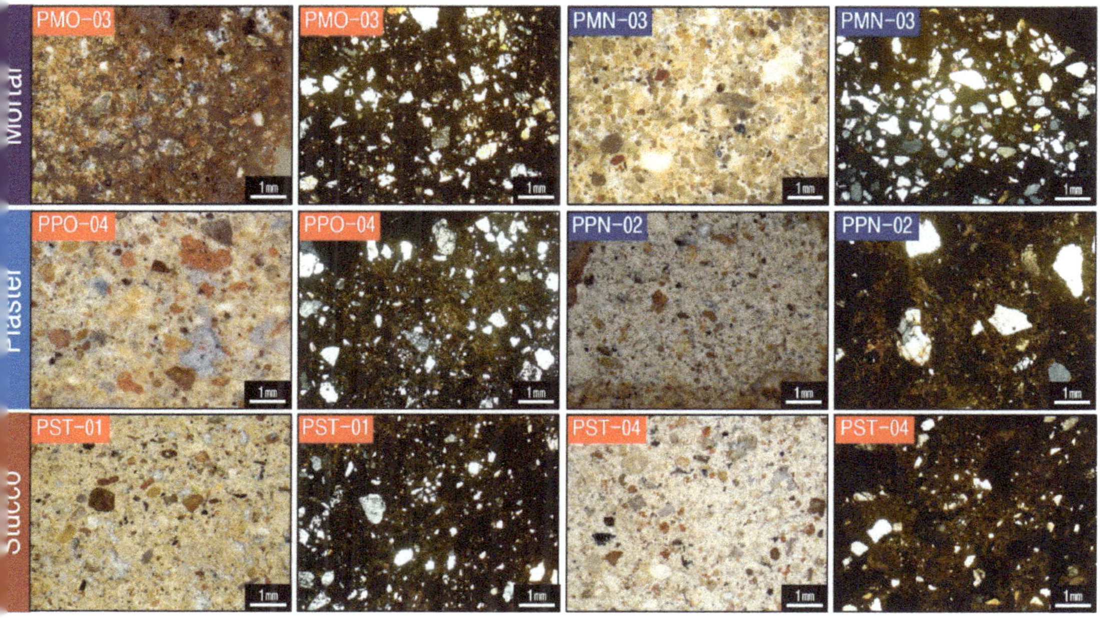

Figure 5. Photographs showing representative stereoscopic and polarizing microscopic images of calcareous materials. Sample numbers are the same as those of Table 1.

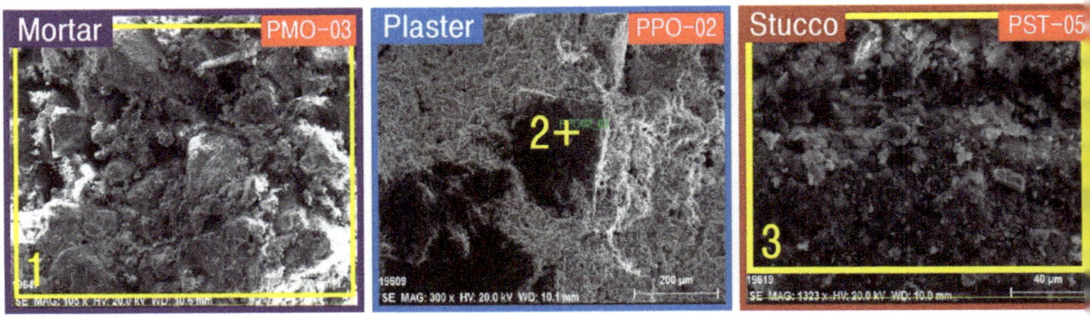

Figure 6. Microphotographs showing the SEM images and analytical area for SEM-EDS of calcareous materials. Sample numbers are the same as those of Table 2.

Table 2. Chemical compositions (wt.%) by SEM-EDS analysis of calcareous materials. Sample numbers are the same as those of Figure 6.

Sample	Si	Al	Fe	Ca	Mg	Na	K	Ti	O
Mortar 1	20.67	9.11	2.48	1.50	1.84	0.93	1.45	-	62.02
Plaster 2	16.57	2.48	-	16.42	3.52	-	1.39	-	0.43
Stucco 3	10.74	4.77	2.27	20.94	3.87	-	1.13	-	56.28

Plaster appeared as a covering layer on bricks and mortar, forming the base layer of murals. According to the microtextural characteristics of the different production periods, the original plaster was a yellowish cryptocrystalline matrix with the detection of many subangular quartz and opaque minerals with sizes of about 1 mm. The new application had a milky white microcrystalline substrate with observation of subangular to angular quartz with dimensions of more than 1 mm, thereby illustrating some differences in composition minerals and grain sizes depending on the production time (Figure 5). Under the scanning electron microscopy, although massive substrate was dominant in both the original and new plasters, calcite of idiomorph crystals was observed in the original plaster (Figure 6, Table 2).

The stucco displayed milky white to the white microcrystalline substrate without significant differences, although some samples (PST-1, PST-4) had high chrominance values and others (PST-5) had low chrominance values. They generally contained a significant quantity of poorly apportioned subangular to angular prismatic quartz with sizes of around 1 mm, along with observation of a small amount of clayey material and opaque minerals (Figure 5). Numerous organic matters along with the peculiar substrate were confirmed in SEM-EDS analysis with the detection of Ca with the highest content at 20.94 wt.% and Si and Fe subsidiary component elements (Figure 6, Table 2).

3.2. Mineralogical Compositions

X-ray diffraction analysis was performed to investigate the mineral composition of the calcareous materials. As a result, quartz, alkali feldspar, plagioclase, mica, and amphibole were detected in the original mortar. In contrast, calcite appeared in the new mortar and amphibole was not seen in the new mortar. Thus, mortars used as joint fillers displayed differences in their mineral compositions depending on the time of production and utilization (Figure 7).

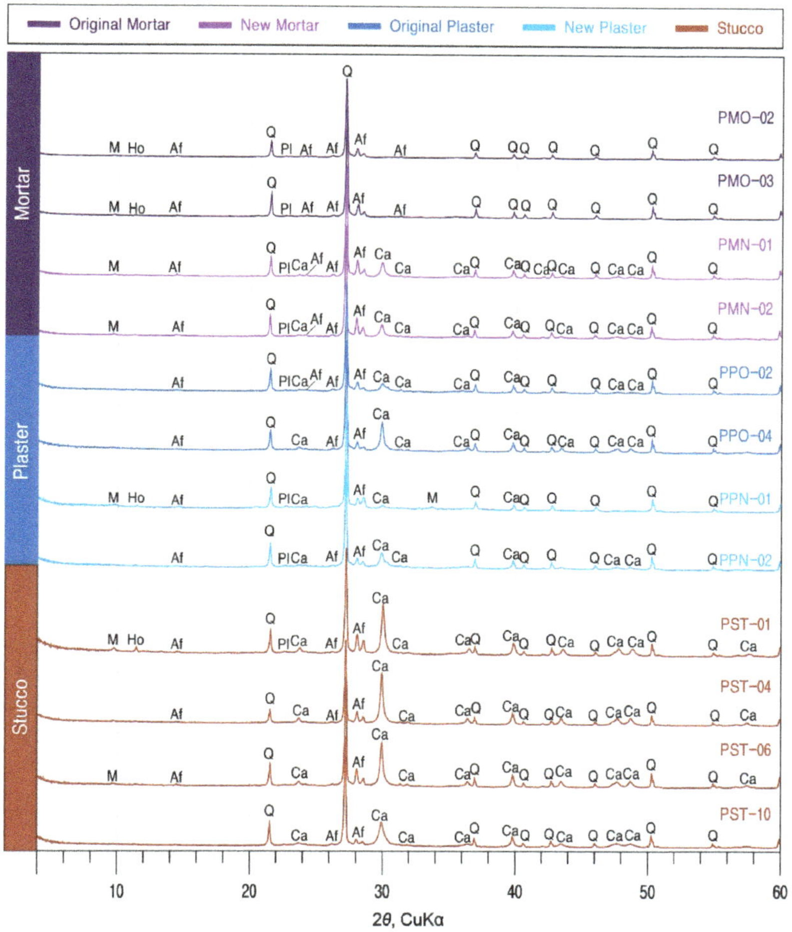

Figure 7. X-ray diffraction analysis of calcareous materials. M; mica, Ho; hornblende, Af; alkali feldspar, Q; quartz, Pl; plagioclase, Ca; calcite. Sample numbers are the same as those of Table 1.

Quartz was the most dominantly detected mineral in the plaster regardless of the production, and calcite and alkali feldspar are commonly confirmed as subordinate minerals. The mineral composition of some of the new plasters used in the repair displayed some variations with the presence of mica and amphibole. However, secondary minerals that can occur from the hydration reaction of calcareous materials did not appear, so more quantitative analysis was required to detect them.

Although the minerals comprising the stucco were mainly quartz, SEM-EDS analysis displayed a high content of Ca, implying that the microcrystalline substrate contains numerous minerals with Ca as the main element. As a result of X-ray diffraction analysis of the stucco, a mineral composition mainly of calcite and quartz was found with alkali feldspar, mica, and traces of amphibole (Figure 7).

3.3. Thermal Analyzes

The analyzed calcareous materials were composed by mixing raw materials with different composition ratios, and although their mineral composition could be determined by X-ray diffraction analysis, it was not easy to discern the changes in the mineral phase. Therefore, differential thermal analysis (DTA) and thermogravimetric analysis (TGA) were performed concurrently to track the changes in the mineral phases and weight loss due to temperature change.

The mineral composition of the mortar varied depending on the production time, and the thermal characteristics were also different. As a result of thermal analysis, the original mortar had an average weight loss of 4.40 wt.%, while the modern repair material had a very high weight loss of 14.61 wt.% (Figure 8, Table 3). In addition, only a 573 °C endothermic peak by the phase transition of quartz, along with an endothermic peak in the low temperature domain, was confirmed in the raw material.

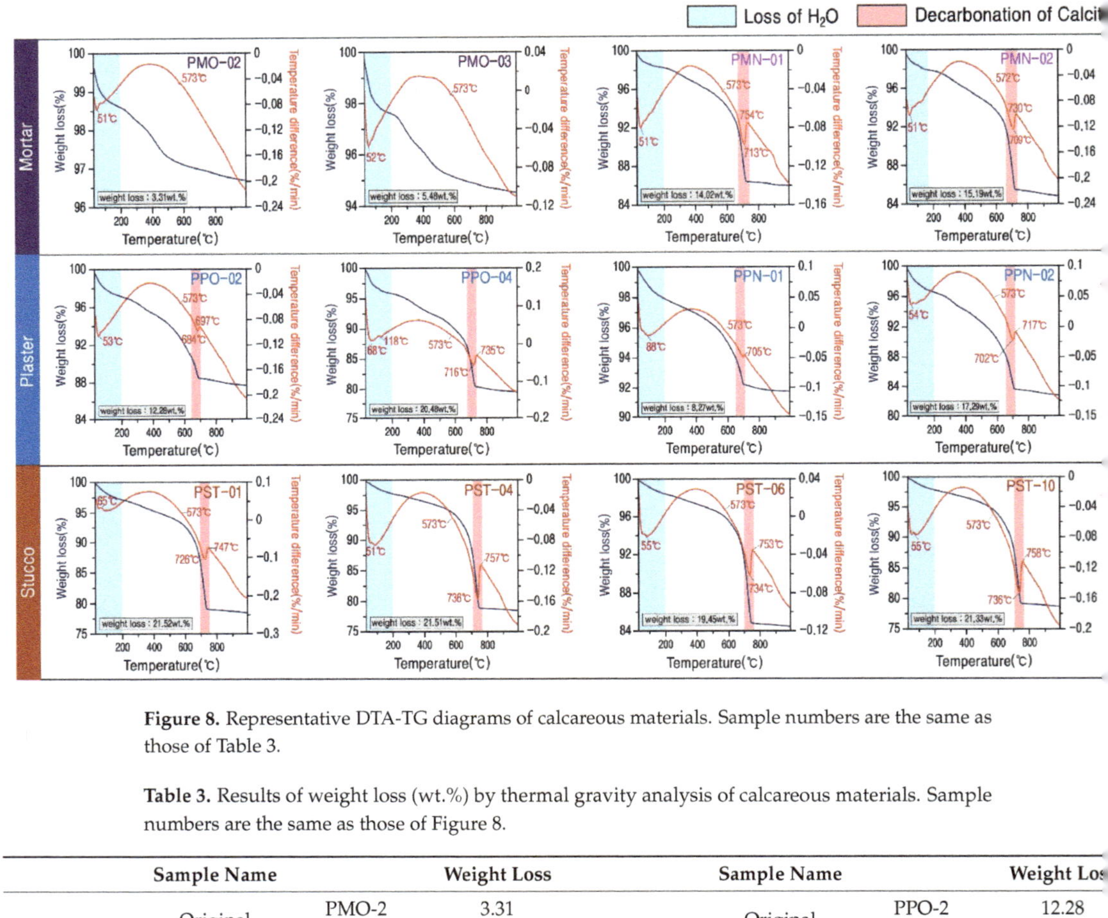

Figure 8. Representative DTA-TG diagrams of calcareous materials. Sample numbers are the same as those of Table 3.

Table 3. Results of weight loss (wt.%) by thermal gravity analysis of calcareous materials. Sample numbers are the same as those of Figure 8.

	Sample Name		Weight Loss		Sample Name		Weight Los
Mortar	Original	PMO-2	3.31	Plaster	Original	PPO-2	12.28
		PMO-3	5.48			PPO-4	20.48
	New	PMN-1	14.02		New	PPN-1	8.27
		PMN-2	15.19			PPN-2	17.29
Stucco	Original	PST-1	21.52	Stucco	Original	PST-6	19.45
		PST-4	21.51			PST-10	21.33

However, an endothermic reaction at about 750 °C is confirmed simultaneously, along with a large amount of weight loss at around 750 °C in the new mortar (Figure 8, Table 3). This peak is interpreted as a characteristic of the decarbonization reaction of calcite. Therefore, proving that the analyzed mortar has different mineral compositions depending on the production time was possible.

The plaster displayed an endothermic peak at 573 °C and an endothermic response at around 700 °C due to the phase transition of quartz in all samples (Figure 8). Moreover, there was a sharp weight loss along with an endothermic reaction at around 700 °C due to the decarbonization reaction of calcite. The weight loss rate of the whole plaster was relatively high, ranging from 8.27 to 20.48 (average of 14.58) wt.%, and the reduction rate of the original plaster was high (Table 3).

Stuccos displayed the highest weight loss rate, ranging from 19.45 to 21.52 (average 20.95) wt.% (Table 3). They had a common trend of weight loss that decreases slightly up to about 680 °C and then sharp decreases after that. Along with the weight loss, an extensive range of endothermic peak at about 750 °C caused by the decarbonization reaction of calcite was detected in all samples (Figure 8).

3.4. Geochemical Analyzes

The geochemical characteristics of the studied calcareous samples were analyzed to quantitatively review their material characteristics and homogeneity (Table 4). Major and some trace elements of all calcareous materials were standardized based on the average compositions of rocks presented by [22,23], respectively. The initial ratio of meteorite presented by [24] was applied to rare earth elements and the primitive mantle compositions presented by [25] to compatible and incompatible elements.

The mortar had the highest SiO_2 content with an average of 68.84 wt.%. It contained CaO (average 7.82 wt.%) and Al_2O_3 (average 7.01 wt.%) as subsidiary components (Figure 9). The contents of all the elements are differentiated according to the time of their production, and, in particular, the CaO content is high in the new mortars from which calcite was detected. The original mortar with low CaO content has high Al_2O_3, K_2O, Fe_2O_3, MnO, and TiO_2 contents, while the MgO content was similar regardless of the time of production but varied depending on the content of SiO_2 (Table 4). Although the compatible and incompatible elements had similar overall behavior, they differed slightly in their levels of enrichment and deficiency, while P_2O_5 content varied by the samples.

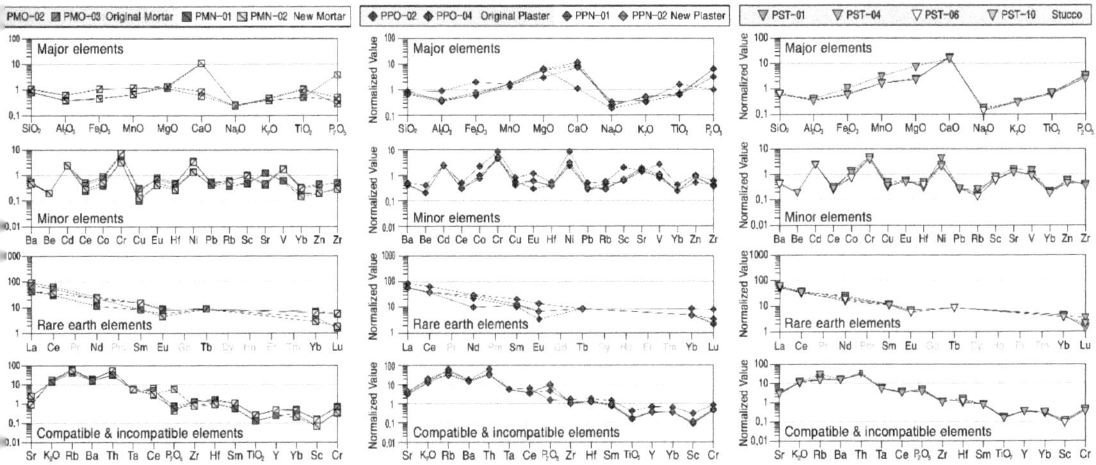

Figure 9. Normalized pattern diagrams showing the major, minor, rare earth, compatible and incompatible elements of calcareous materials. Sample numbers are the same as those of Table 1.

Table 4. Chemical compositions of major elements (wt.%), some minor and rare earth elements (ppm) for calcareous materials. Sample numbers are the same as those of Table 1.

No.	Mortar				Plaster				Stucco			
	PMO-2	**PMO-3**	**PMN-1**	**PMN-2**	**PPO-2**	**PPO-4**	**PPN-1**	**PPN-2**	**PST-1**	**PST-4**	**PST-6**	**PST-**
SiO_2	78.66	76.64	60.63	59.42	61.62	48.58	59.27	51.78	43.98	44.83	49.17	45.8
Al_2O_3	8.72	8.60	5.47	5.24	5.4	4.88	12.33	5.54	5.64	4.71	5.01	4.7
Fe_2O_3	2.69	2.73	1.18	1.12	1.52	1.57	5.08	1.92	2.66	1.51	1.62	1.4
MnO	0.07	0.07	0.04	0.04	0.08	0.10	0.10	0.10	0.19	0.10	0.10	0.1
MgO	0.59	0.72	0.66	0.74	2.96	3.46	1.60	3.29	3.99	1.23	1.33	1.2
CaO	0.78	1.12	14.59	14.78	11.92	17.99	10.05	15.76	20.53	24.18	21.62	23.7
Na_2O	0.71	0.70	0.80	0.79	0.73	0.57	1.02	0.79	0.54	0.41	0.43	0.4
K_2O	2.58	2.46	2.08	2.12	2.97	1.82	2.01	2.72	1.59	1.74	1.94	1.6
TiO_2	0.39	0.37	0.19	0.18	0.23	0.23	0.60	0.26	0.27	0.23	0.25	0.2
P_2O_5	0.05	0.09	0.07	0.71	1.18	0.58	0.18	1.13	0.59	0.46	0.63	0.5
LOI	3.26	5.35	14.60	14.98	12.11	20.14	8.13	17.14	20.66	21.31	18.67	20.8
Total	98.51	98.84	100.30	100.10	100.70	99.91	100.40	100.40	100.60	100.70	100.70	100.
Ba	390	362	326	322	327	295	351	287	272	317	317	33
Be	1	1	1	1	1	1	2	1	1	1	1	1
Cd	0.5	0.5	0.5	0.5	0.5	0.5	0.5	0.5	0.5	0.5	0.5	0.5
Co	7	6	3	4	6	8	18	6	11	9	6	6
Cr	177	156	87	83	105	109	204	125	115	93	116	97
Cu	7	9	3	4	18	13	25	17	15	11	11	10
Hf	4	3.4	2.6	2.1	3.2	3	4.3	3.2	2.4	3.9	3.1	2.5
Ni	38	41	17	16	26	27	92	36	48	24	27	23
Pb	18	21	15	15	12	11	19	13	11	10	11	11
Rb	130	130	80	110	100	70	120	60	40	60	30	30
Sc	5.7	6.0	2.8	2.7	3.5	3.7	12.1	4.3	4.9	3.6	3.7	3.5
Sr	104	119	305	316	446	352	395	502	420	321	344	34
V	57	64	21	20	27	33	88	36	49	34	33	31
Zn	28	32	13	14	59	37	70	60	44	31	38	32
Zr	125	113	79	67	90	103	148	99	91	105	103	101
La	34.3	26.8	16.0	15.7	20.6	21.0	31.0	21.5	19.0	20.2	21.5	22.
Ce	62	51	29	34	37	35	62	37	33	34	31	37
Nd	18	15	8	16.0	16	7	22	20	18	11	13	15
Sm	3.5	3.5	1.9	1.9	2.4	2.7	4.7	3.0	2.7	2.5	2.6	2.7
Eu	0.8	0.7	0.5	0.4	0.3	0.6	1.2	0.6	0.6	0.5	0.5	0.6
Tb	0.5	0.5	0.5	0.5	0.5	0.5	0.5	0.5	0.5	0.5	0.5	0.5
Yb	1.5	1.7	1.0	0.7	1.2	1.2	2.1	1.2	1.0	1.1	1.0	0.9
Lu	0.21	0.21	0.06	0.07	0.13	0.09	0.30	0.08	0.13	0.05	0.06	0.0

In particular, although the original mortar had a high alumina content with no detection of calcite, the new mortar displayed the properties of a lime mixture. Even though quartz and feldspar are also detected in the new mortar obviously, their physical and mineralogical characteristics were significantly different from those of the original mortar. However, despite the differences in material characteristics by the time of their projection, their geochemical behavior tended to be similar (Figure 9, Table 4).

SiO_2 content was the highest, with an average of 55.31 wt.% in all plasters, followed by CaO and Al_2O_3, with average contents of 13.93 wt.% and 7.04 wt.%, respectively. When the correlation of elements with SiO_2 was examined, CaO content was found to be higher, and MgO content was also relatively higher in the original plaster than in the new plaster. However, the contents of K_2O and Na_2O were relatively low (Figure 9). Although the contents of trace and rare earth elements also varied somewhat depending on the extent of enrichment and deficiency of the main component element, the overall trend was very similar.

The behavior of the main element of stuccos is highly homogeneous, with SiO_2 being the highest at an average of 45.96 wt.%, followed by CaO and LOI at averages of 22.51 wt.%

and 20.37 wt.%, respectively (Figure 9). However, level enrichment of CaO, MnO, MgO, and P_2O_5 with the most significant level of enrichment for CaO was displayed. Trace elements had a homogeneous tendency to be enriched for Cd, Cr, Ni, Sr, and V while deficient for other elements.

In particular, some rare earth elements in stuccos displayed the tendency of a gradual decrease in the level of enrichment from light to heavy rare earths, and all the compatible and incompatible elements also behaved by displaying the same enrichment and deficiency. Therefore, since the stuccos had very homogeneous geochemical characteristics, it can be interpreted that raw materials with the same cause of their genesis were utilized (Figure 9, Table 4).

4. Discussion

4.1. Material Considerations

This study examined the material scientific diversity of the calcareous materials utilized in the construction of the Phaya Thon Zu temple by categorizing them into original and new materials. In particular, plaster and stuccos are distinguished by whether they are used as decorative elements on the exterior walls of the temple or not, and, upon summarization of their material analysis, they displayed very similar textural, mineral, and geochemical characteristics.

Therefore, to examine the differences in the characteristics of the lime mixtures based on CaO, the distribution of the main component elements was reviewed by illustrating them on the SiO_2-(CaO + MgO)-Al_2O_3 triangle (Figure 10A) and preparing a basic oxide (RO + R_2O) correlation chart by the acidic oxide (RO_2) (Figure 10B). As a result, the contents of CaO and basic oxides were relatively the highest in stuccos.

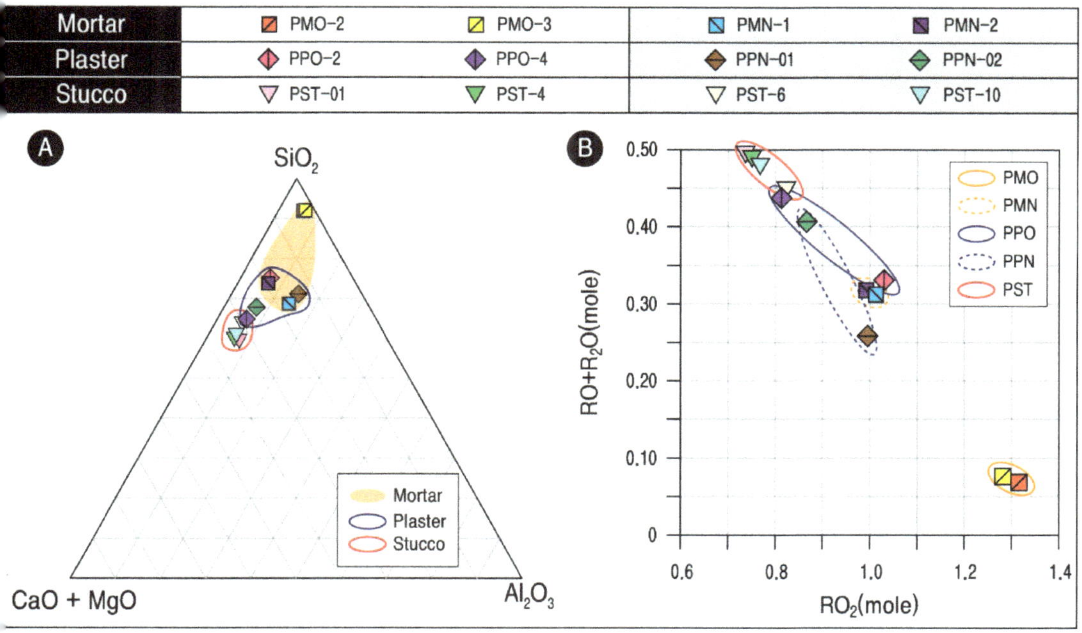

Figure 10. Plotted diagrams showing the comparison results of calcareous materials. (**A**) Ternary diagram showing the relationship between SiO_2-(CaO + MgO)-Al_2O_3. (**B**) Correlation of (RO_2-RO) + R_2O. Sample numbers are the same as those of Table 4.

The stuccos had a relatively good distribution compared to the plaster and mortar, with high roundness and small particle size. While the other materials displayed different appearances with low roundness and uneven distribution (Figure 5). As illustrated, mixing a higher proportion of lime in the selected aggregates with a homogeneous particle size would have been advantageous for the ease of carving and detailed expressions. It seems that the lime used to make stucco would have been more readily available based on the fact that the diversity of basic oxides acts to lower the melting point of the subject material for firing.

The function of stucco is to protect walls exposed to the external environment and prevent water leakage by sealing the joints between the individual masonry pieces and the joint filler as finishing materials. According to the archaeological evidence such as inscriptions and murals, it can be discerned that organic materials such as buffalo hide glue, Aegle Marmelos sap, fishtail palm fruit, and neem tree resin were used as additives in the production of stucco in construction works during the Bagan Dynasty [7,26–28].

Accordingly, given the unique nature of the stuccos being continuously exposed to the outside environment by decorating the outermost aspects of the structures, it is deemed that an increase in the binding force and enhancement of the durability were sought by increasing the contents of organic additives in their production. This may explain why the stuccos have slightly higher lime and organic component content than the finishing materials and joint filler inside the temple.

However, the effectiveness of organic additives has yet to undergo a stage of scientific verification because it is not possible to quantitatively evaluate the organic components only by means of the LOI. In addition, it would be necessary to investigate the exact composition and mixing ratio of bonding material, aggregates, and organic additives used in the lime mixture, including the finishing material and joint filler. As such, for a clear material distinction between lime mortar, plaster, and stuccos, the characteristics of organic components used as additives, the mixing ratio of raw materials, and the strength characteristics of lime mixtures must be researched.

Although the original mortars displayed characteristics close to ordinary soil, such as shallow CaO content and almost no detectable calcite, the new mortar was characterized by a lime mixture similar to stuccos and plasters. This CaO content was also the same in the level of RO_2-($RO + R_2O$) correlation, which can be interpreted as an adjustment of the material differences based on CaO by the production times (Figure 10).

All the plaster samples had been illustrated in the domain close to the SiO_2-CaO line, thereby displaying the tendency to be distributed in different locations depending on the SiO_2 content (Figure 10A). The level of RO_2-($RO + R_2O$) correlation also displayed a broader range of acidic oxides of the original plasters (Figure 10B), thereby implying that there was a slight difference in its composition.

4.2. Provenance Considerations

It is interpreted that all types of soils had been mixed for the calcareous materials since the Ca content is high and Si, Al, and K were detected except the original mortar. In particular, the original mortar was primarily produced using soil, while the new mortar was made by mixing some soil as the aggregate with the primary lime material.

Therefore, clayey and sandy soils were collected from the area around the Phaya Thon Zu temple to analyze their microstructure, mineral composition, and geochemical behavioral characteristics. This analysis enabled the examination of the clay used as a raw material for the calcareous materials, its origin, and the production technology system. Soil samples were collected from clay pits within a 3.5 km radius of the Phaya Thon Zu temple, which are currently used to produce bricks for various purposes, including repairing the temple site (Figure 11).

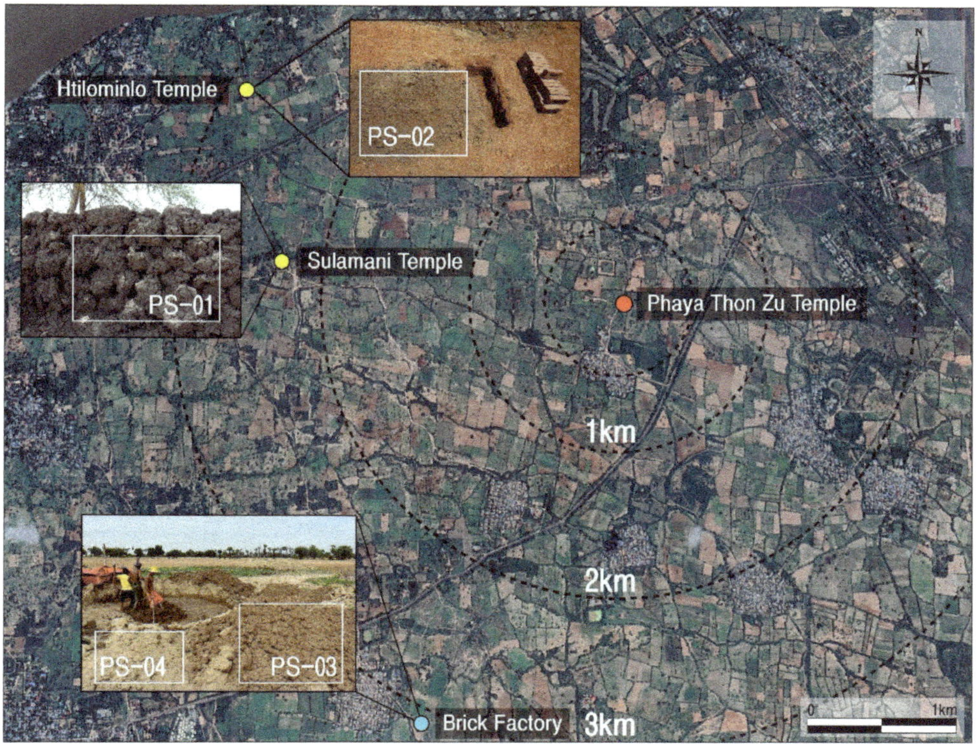

Figure 11. Locations showing the soil samples of origin for clay materials used to produce calcareous materials around the Phaya Thon Zu temple. Sample numbers are the same as those of Table 1.

Sample PS-1 is viscous alluvium recovered from the marshy deposit in front of the Sulamani temple in its dried and peculiar state and produced in a lumpy form. PS-2 is a sandy subsoil distributed in the flat area in the vicinity of the Htillominlo temple (Figure 12). Samples PS-3 and PS-4 were collected during a visit to a brick factory in Thu Htay Kan village, situated approximately 3.5 km southwest of the Phaya Thon Zu temple.

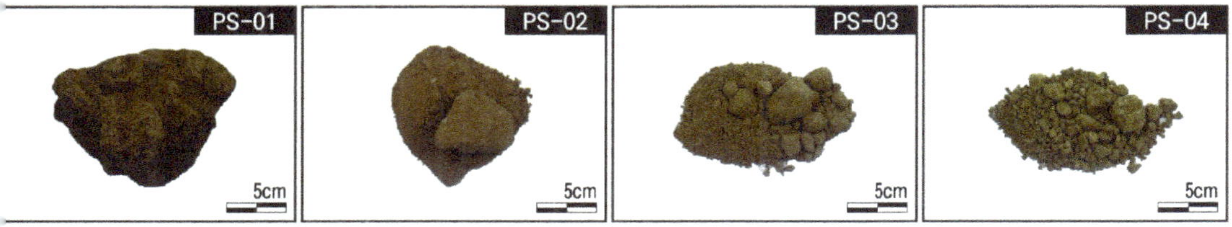

Figure 12. Occurrences of soil samples around the Phaya Thon Zu temple. Location of the collection point shown in Figure 11. Sample numbers are the same as those of Figure 11.

These are the clay for the bricks that are used locally for the construction and repair of the temples as mentioned above (Figure 12). These marshy clays were collected at the suggestion of local experts. According to them, although there is no record in the literature, locals testified that in the past and present, the same materials were obtained from the same place in the same way and used to repair the monuments. Therefore, this study also respects their methods and reflects them in the study.

In the Bagan region of Myanmar, tropical grassland (Savana) soil has developed due to the dry climate with low rainfall. Therefore, the soil layer is characterized by porosity and reddish color due to the large quantity of iron concentrated by leaching action due to rapid drainage. Accordingly, when the soil samples were observed under a polarizing microscope, they all displayed a reddish color and cryptocrystalline substrate, and quartz was regarded as the main mineral, while having slight differences in their particle sizes and contents (Figure 13).

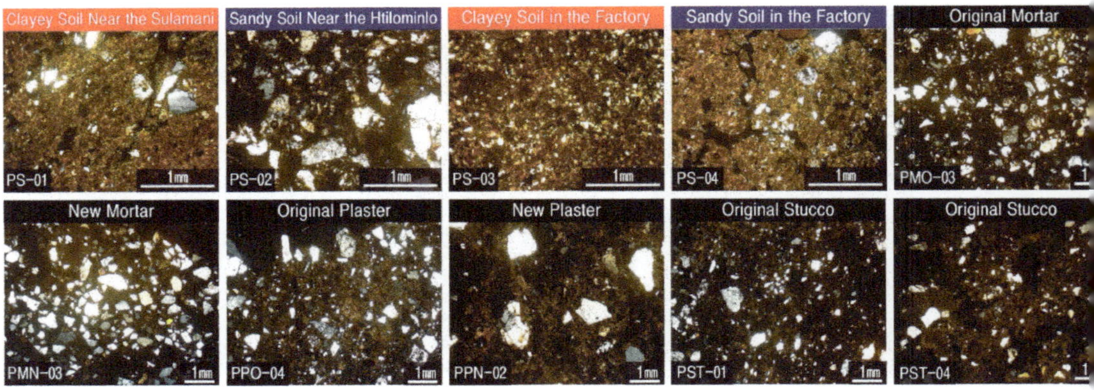

Figure 13. Microphotographs showing the polarizing microscopic images of soil and building materials around the Phaya Thon Zu temple. Sample numbers are the same as those of Figure 11.

Regardless of the production time, all of the studied mortars had relatively small particles of the constituent minerals compared to sandy soil. Still, the content of sand was observed to be high. While the stuccos and original plasters were characterized by a mineral distribution similar to that of clayey soil, the new plaster had characteristics that were most similar to those of sandy soil near Htillominlo temple, with mineral particle sizes of more than 1 mm (Figure 13).

When the homogeneity of these soil samples with calcareous materials was examined through X-ray diffraction analysis, it was found that they all contained quartz and alkali feldspar as major components, with the inclusion of trace amounts of mica and plagioclase (Figure 14). In addition, montmorillonite was detected in soils (PS-3 and PS-4) recovered from a brick factory in Thu Htay Kan. Thus, the calcareous materials used in the Phaya Thon Zu temple were found to have a very similar mineral composition to the soil except calcite. The results on calcareous substances obtained in the study are summarized in Table 5.

When the elemental behavior of the plasters and stuccos of the Phaya Thon Zu temple was compared with that of the soil, it was found that they had considerable geochemical homogeneity, thereby suggesting that they all utilized soil distributed in the general region of the temple (Figure 15). Therefore, the calcareous materials used in the temple were categorized according to the time of their production, and their behavior was examined by calculating the average values for the constituent elements of each material (Figure 16).

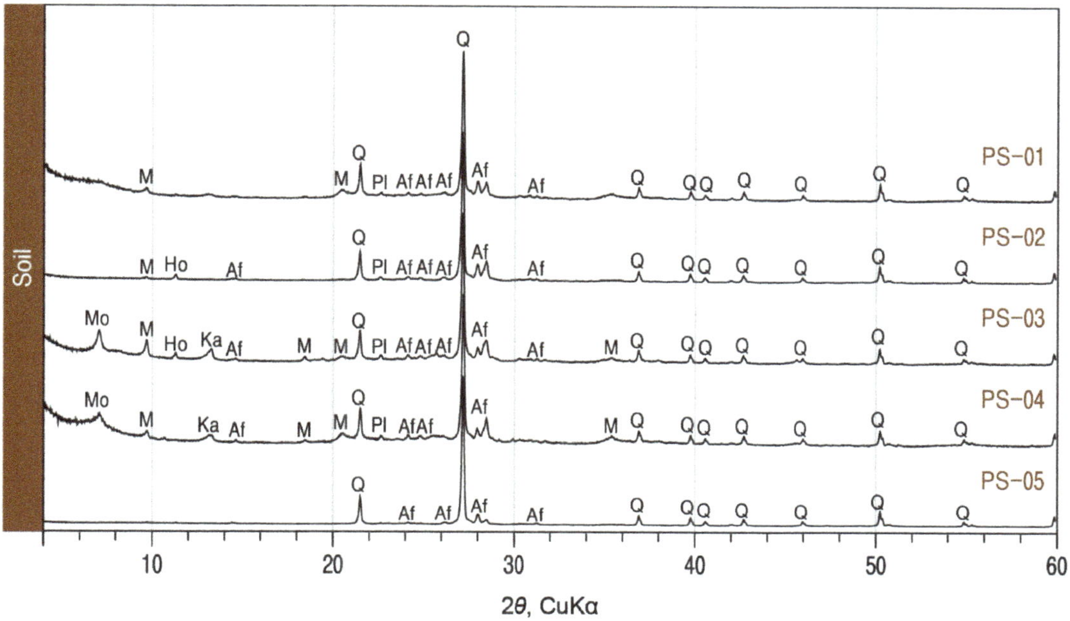

Figure 14. X-ray diffraction analysis of soil samples around the Phaya Thon Zu temple. Mo; montmorillonite, M; mica, Ka; kaolinite, Ho; hornblende, Af; alkali feldspar, Q; quartz, Pl; plagioclase. Sample numbers are the same as those of Figure 13.

Table 5. Summary on X-ray diffraction analysis of calcareous and soil materials from the study area. Sample numbers are the same as those of Table 1.

Sample No.		Mo	M	Ho	Ka	Af	Q	He	Pl	Her	Ca
Soil	PS-1	-	*	-	-	**	***	-	*	-	-
	PS-2	-	*	*	-	**	***	-	*	-	-
	PS-3	*	*	*	*	**	***	-	*	-	-
	PS-4	*	*	-	*	**	***	-	*	-	-
Original Stucco	PST-1	-	*	*	-	**	***	-	*	-	**
	PST-4	-	-	-	-	**	***	-	-	-	**
	PST-6	-	*	-	-	**	***	-	-	-	**
	PST-10	-	-	-	-	*	***	-	-	-	**
Original Mortar	PMO-2	-	*	*	-	**	***	-	*	-	-
	PMO-3	-	*	*	-	**	***	-	*	-	-
New Mortar	PMN-1	-	*	-	-	**	***	-	*	-	**
	PMN-2	-	*	-	-	**	***	-	*	-	**
Original Plaster	PPO-2	-	-	-	-	**	***	-	*	-	**
	PPO-4	-	-	-	-	**	***	-	-	-	***
New Plaster	PPN-1	-	*	*	-	**	***	-	*	-	**
	PPN-2	-	-	-	-	**	***	-	*	-	**

Mo; montmorillonite, M; mica, Ho; hornblende, Ka; kaolinite, Af; alkali feldspar, Q; quartz, He; hematite, Pl; plagioclase, Her; hercynite, Ca; calcite. -: not detected, *: minor, **: moderate, ***: major.

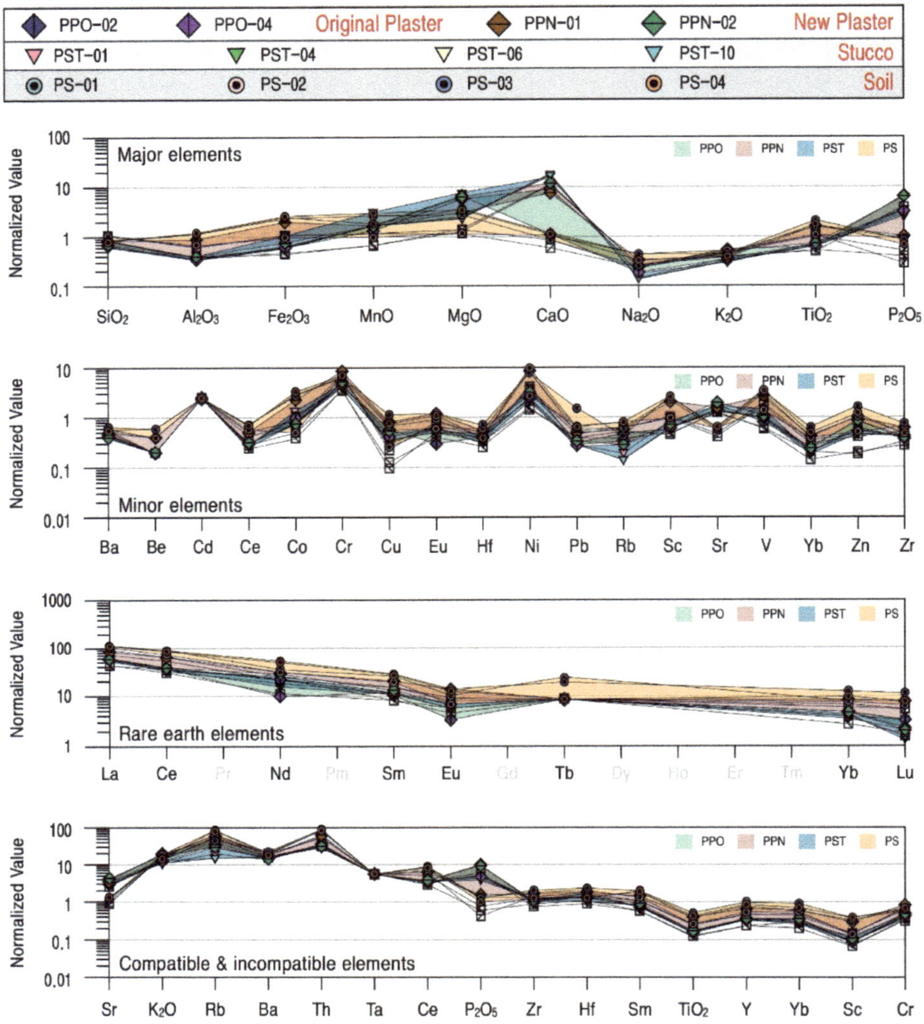

Figure 15. Normalized pattern diagrams showing the enrichment and deficiency factors of major, minor, rare earth, compatible and incompatible elements of calcareous materials around the Phaya Thon Zu temple. Sample numbers are the same as those of Table 1.

As a result, although the behaviors of rare earth elements were the same for all the groups, the main components, compatible and incompatible elements, displayed different levels of enrichment of CaO and P_2O_5 depending on whether lime was included or not. The new mortars, plasters, and stuccos containing lime were interpreted to have a higher CaO content due to the process of their production by mixing lime into the clay. At the same time, the P_2O_5 was believed to be a component generated as the result of organic matter being utilized as an additive in the process.

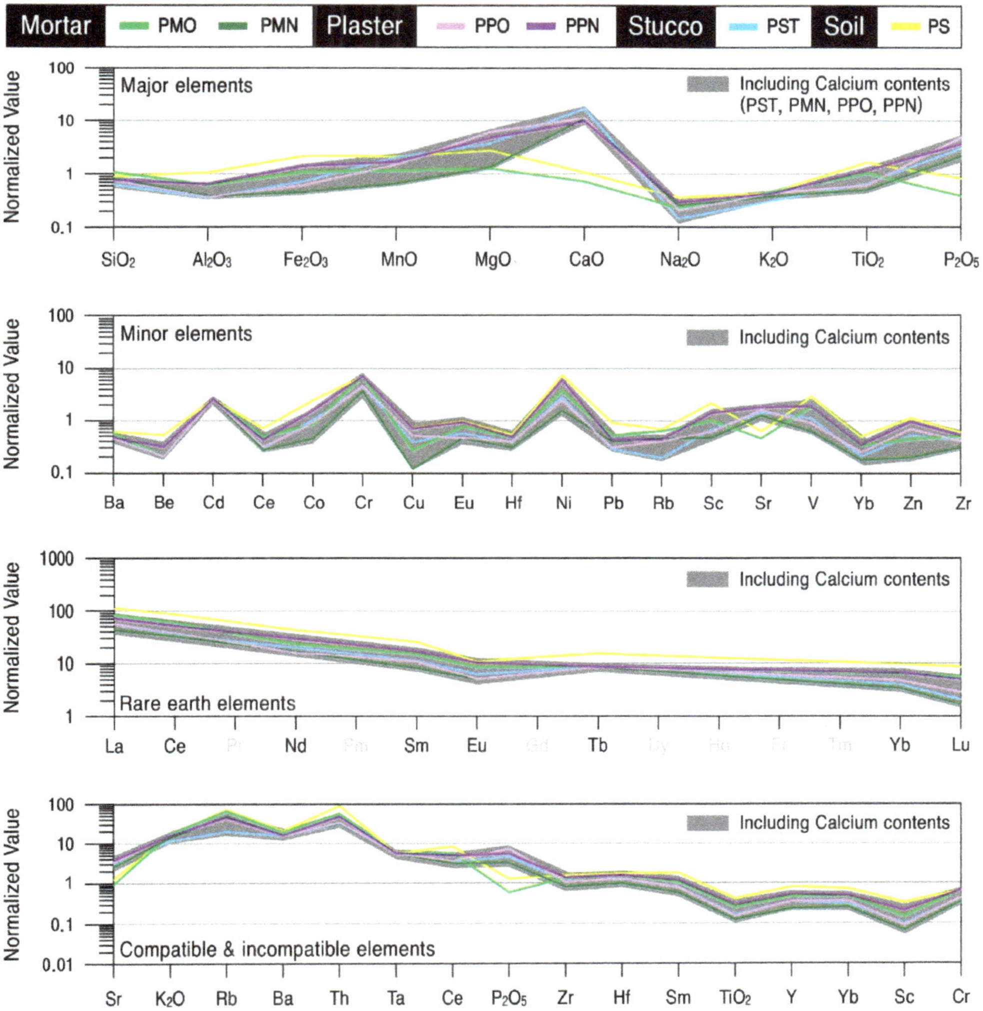

Figure 16. Normalized pattern diagrams showing the enrichment and deficiency factors of the major, trace, rare earth, compatible and incompatible elements by group of calcareous materials and soil around the Phaya Thon Zu temple. Sample numbers are the same as those of Table 1.

The clay minerals detected in the soil mainly dissipated during the production process, and it appears that the lime additives were likely responsible for the detection of a high level of calcite in the new mortar, plaster, and stuccos (Figure 17). Accordingly, it can be concluded that the original materials utilized in the production of the calcareous materials were all soil from the temple sites that were homogeneous in their genesis, and it can be interpreted that the calcareous materials except for the original mortar were composed by adding lime to these clay materials.

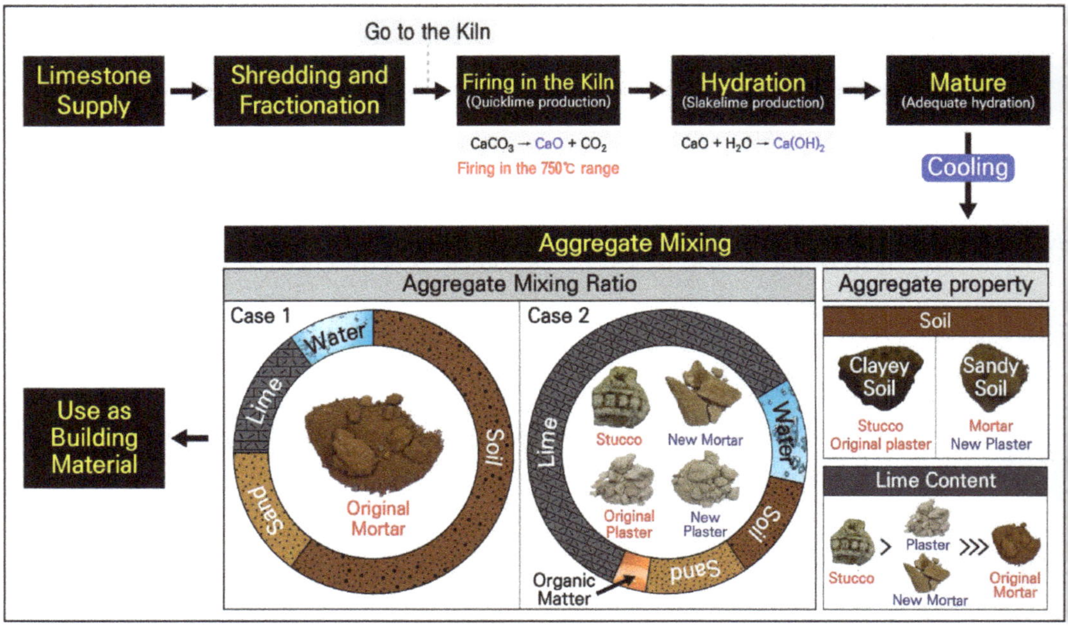

Figure 17. Schematic diagrams of the production technology system of calcareous materials using the construction of the Phaya Thon Zu temple.

4.3. Interpretation of Making Techniques

The calcareous materials exhibited the characteristics of lime mixtures except the original mortars, and thermal analysis confirmed a rapid weight loss in all of the materials at around 700 to 750 °C with an endothermic reaction due to the decarbonization of calcite. The calcite and quartz were mainly identified in the new mortars, plasters, and stuccos, and no detritus was observed in the milky white to white substrate based on their yield state and micro texture.

$CaCO_3$ is the primary component of calcite that composes limestone, and calcite is the most stable form of calcium carbonate. Generally, when limestone is fired at high temperatures, it undergoes decarbonization and converts into quicklime (CaO), which is a steady condition. During this process, calcium carbonate, which is neutral and stable, escapes by decomposing into basic and unstable calcium oxide and acidic carbon dioxide.

Therefore, to obtain the lime needed for the new mortars, plasters, and stuccos, the firing of limestone must have taken place while refining the raw materials. In general, although calcite, a helpful mineral in limestone, begins to undergo phase decomposition at 650 °C, firing at 850 to 1000 °C is required to obtain higher-quality lime.

When limestone is heated at a temperature above 1000 °C, calcite produces quicklime (CaO) due to a calcination reaction involving the dissociation of carbon dioxide. Quicklime reacts with water to form slaked lime ($Ca(OH)_2$), which in turn forms calcite ($CaCO_3$) through a carbonation reaction in which carbon dioxide binds with slaked lime [15–18,21,29].

The utilization of lime as a construction material is based on the carbonization of slaked lime during this limestone cycle, and the calcite formed through this carbonization reaction acts as a mineral adhesive. Therefore, since it is advantageous to have a carbonization reaction to have high strength and durability, high-temperature firing is beneficial to obtain high-quality lime.

As a result of thermal analysis of the calcareous materials collected from the Phaya Thon Zu temple, a maximum endotherm was displayed in the temperature range of 710 to 730 °C due to the decarbonization of calcite, and no carbonation reaction with recrystallization was observed at temperatures higher than 900 °C. As such, it was found that pyrolysis ended at around 750 °C, and it was presumed that the calcareous material used in the temple was relatively low-quality lime that had undergone incomplete decarbonization in obtaining quicklime.

Moreover, the thermogravimetric analysis of each sample displayed a high weight loss of 20.95 wt.% for stuccos and 14.58 wt.% for plaster, and the average LOI was also increased to 20.37 wt.% for stuccos and 14.38 wt.% for plasters. This average LOI is very high compared to the average weight loss of 8.94 wt.% for the studied soil. Thus, although it can be interpreted that organic matter may have been used in the kneading of the lime and soil-based material mixture to produce the plasters and stuccos used in the temple, additional experiments are needed, along with additional review to confirm this.

By integrating the results of the material analysis of limestone and clay materials, which are the main raw materials utilized in the production of the studied calcareous materials, the production technology system was examined (Figure 17). First, the distribution would have been adjusted to make more straightforward utilization of the limestone quarried at the site of origin as construction material. The limestone was then transported to a nearby kiln, fired at a temperature below the range of 700 to 750 °C before undergoing a slaking process through a hydration reaction.

Although the hydration reaction is completed on the surface of the slaked lime particles, there may still be unreacted quicklime remaining inside the particles. Therefore, an additional aging process was possibly implemented to ensure sufficient hydration of all the particles. Aggregates are then mixed to function as a construction material, and the calcareous materials from the Phaya Thon Zu temple showed two different types of aggregate mixing ratios (Figure 17).

While the original mortars displayed characteristics close to normal soils, the other calcareous materials were produced by mixing lime-based materials with clay and organic materials. As a result of the previous comparison of the mineral composition of soils and each construction material, there were some differences in contents and particle sizes. However, the roundness is similar overall (Figure 13).

When the mixing ratio of the clay materials was examined through the results mentioned above, the minerals had relatively large particle sizes of more than 1 mm. In particular, it is likely that the sandy soil in the general region of the Htillominlo temple site was utilized for the production of the new plaster. However, since there is a slight difference in their contents, it is interpreted that some sorting process must have occurred.

Furthermore, since clayey soils appeared to have been used to produce the original plasters and stuccos, it was presumed that the mortar was produced by appropriately mixing clay and sandy soils. A more precise interpretation of the provenance of each material and the mixing ratio of clayey materials will require quantitative examination through experiments, along with the estimated provenance of lime materials.

5. Conclusions

1. The calcareous materials used in the construction of the Phaya Thon Zu temple in Bagan, Myanmar, are mortars, plasters, and stuccos, which were utilized as joint filler, plaster, and decorative elements inside and outside the temple. In the study, specimens of each calcareous material that fell off were collected by classifying their uses and production time, and their material characteristics and making techniques were discussed through mineralogical and geochemical analysis.
2. As a result of the integration of the material scientific characteristics of the studied calcareous specimens, the stuccos and plasters had lime-mixture characteristics due to their high content of CaO along with calcite. At the same time, the mortar was confirmed to have similar characteristics only in the new mortar used for repair. When their geochemical behavioral characteristics are examined, the clayey materials used in the construction were sourced from soils distributed in the temple site.

3. The calcareous auxiliary materials of the Phaya Thon Zu temple are a mixture of soil-based clay and lime paste, to which organic materials have been added. As a result of thermal analysis, in the firing of limestone performed during the process of obtaining the lime used to produce the calcareous materials, pyrolysis was found to have ended at around 750 °C. Therefore, relatively low-quality lime was obtained due to an incomplete decarbonization reaction while securing quicklime.
4. The aggregates mixed in producing the calcareous auxiliary materials were largely of two types. Although the original mortars generally had a composition close to the soil, other calcareous materials were produced with a higher lime content mixed with clay and organic matter. In particular, stuccos had the highest range of lime and organic matter among the calcareous materials and was combined with the most homogeneous selected aggregate. This reflects increased durability against weathering and ease of carving by considering its unique feature of being continuously exposed to the external environment.
5. If research on the clear provenance interpretation of the raw materials for calcareous auxiliary materials and investigation on the mixing ratio of clayey materials were to be carried out in the future based on the results of this study, it would be possible to establish conservational measures to maintain the outstanding universal value, authenticity, and integrity of the Phaya Thon Zu temple as a World Heritage Site.

Author Contributions: Investigation, H.R.Y., G.H.L., D.M.K. and C.H.L.; methodology, H.R.Y., G.H.L. and C.H.L.; analysis, H.R.Y., G.H.L., D.M.K. and C.H.L.; software graphics, H.R.Y. and G.H.L.; data validation, C.H.L. All authors have read and agreed to the published version of the manuscript.

Funding: This research was supported by the Kongju National University Development Project by the Ministry of Education in 2023, Republic of Korea.

Institutional Review Board Statement: Not applicable.

Informed Consent Statement: Not applicable.

Data Availability Statement: The original contributions presented in the study are included in the article, further inquiries can be directed to the corresponding author.

Acknowledgments: This research was supported by the Kongju National University Development Project by the Ministry of Education in 2023, Republic of Korea. Furthermore, we would like to acknowledge that this research is part of a project conducted with the support of the Korea Cultural Agency in 2019. We express our profound gratitude for their assistance.

Conflicts of Interest: Authors (Hye Ri Yang, Gyu Hye Lee, Dong Min Kim and Chan Hee Lee) declare that the research was conducted in the absence of any commercial or financial relationships that could be construed as a potential conflict of interest.

References

1. Barry, M.B. *Bagan: Temples and Monuments of Ancient Burma*; Paths International: Bedford, UK, 2003; pp. 1–76.
2. UNESCO. *Decisions Adopted during the 43rd session of the World Heritage Committee, Baku*; UNESCO: Baku, Azerbaijan, 2019; 2
3. Jeon, B.H.; Kim, Y.J. *Korea-Myanmar cultural heritage ODA project and archaeological significance of Phaya Thone Zu temple excava in Bagan (Myanmar)*; Sogang Institute for East Asian Studies (SIEAS): Seoul, Republic of Korea, 2020; Volume 39, pp. 147– [CrossRef]
4. Cha, J.S. Meet in Beautiful the Myanmar. *Hist. Space* **2013**, *90–109*, 220–359. (In Korean)
5. Amadori, M.L.; Fermo, P.; Raspugli, V.; Comite, V.; Mini, F.M.; Maekawa, Y.; Lin, K. . Integrated scientific investigations constitutive materials from temple n.1205a Bagan valley (Myanmar). In Proceedings of the IMEKO International Conference Metrology for Archaeology and Cultural Heritage, Lecce, Italy, 23–25 October 2017; pp. 387–392.
6. Amadori, M.L.; Fermo, P.; Raspugli, V.; Comite, V.; Mini, F.M.; Maekawa, Y.; Russa, M.L. Integrated scientific investigations constitutive materials from Me-taw-ya temple, Pagan valley Burma (Myanmar). *Measurement* **2019**, *131*, 737–750. [CrossRef
7. Amadori, M.L.; Mengacci, V.; Vagnini, M.; Casoli, A.; Holakooei, P.; Eftekhari, N.; Lin, K.; Maekawa, Y.; Germinario, G. Org matter and pigments in the wall paintings of Me-Taw-Ya temple in Bagan valley, Myanmar. *Appl. Sci.* **2021**, *11*, 11441. [Cross
8. Hudson, B. Restoration and reconstruction of monuments at Bagan (Pagan), Myanmar (Burma), 1955–2008. *World Archaeol.* **2** *40*, 553–571. [CrossRef]

Pichard, P. Pagan, Burma's ancient city of 2000 pagodas damaged by a violent earthquake. *UNESCO Cour.* **1976**, *29*, 14–19. Available online: https://unesdoc.unesco.org/ark:/48223/pf0000049529 (accessed on 24 November 2023).
Weise, K. Safeguarding Bagan: Endeavours, challenges and strategies. *J. Cult. Herit.* **2016**, *1*, 68–84. [CrossRef]
Park, J.S.; Chun, D.Y.; Galloway, C.; Lunsford, R.L. The mural of Phaya Thon Zu temple. *Korea Herit. Agency* **2018**, *5*, 142. (In Korean)
Schwartzbaum, P.M.; Zari, D.; Tint, U.B.; Lazzarini, L. The conservation of the mural paintings and external carving of the temples of Pagan. *Tokyu Congr. Prepr. Conserv. Far East. Art* **1988**, *33*, 103–107. [CrossRef]
Kang, S.H.; Lee, C.H. Interpretation of material characteristics and making techniques for lime soil mixture on tomb barrier of Pyeongtaek Gungri site in Joseon Dynasty. *Econ. Eoviron. Geol.* **2018**, *51*, 49–65. [CrossRef]
Kang, S.H.; Lee, S.O.; Hong, S.G.; Kwon, Y.H. Historical and scientific investigations into the use of hydraulic lime in Korea and preventive conservation of historic masonry structures. *Sustainability* **2019**, *11*, 5169. [CrossRef]
Lee, C.H.; Jo, J.H.; Kim, J.Y. Interpretation of limestone provenance, materials and making characteristics for lime-soil mixture on tomb barrier of the Yesan Mokri site in Joseon Dynasty. *J. Conserv. Sci.* **2016**, *32*, 471–490. [CrossRef]
Lee, C.H.; Eo, E.I.; Shin, S.C. Analysis on composition and strength of lime-soil mixture tomb barrier at Guri Galmae site in Joseon Dynasty. *J. Conserv. Sci.* **2017**, *50*, 40–61. [CrossRef]
Lee, C.H.; Kang, S.H. Evaluation of physical properties and strength interpretation for lime-soil mixture on barrier tomb of Pyeongtaek Gungri site in Joseon Dynasty. *J. Conserv. Sci.* **2018**, *34*, 97–106. [CrossRef]
Lee, G.Y.; Park, J.H.; Lee, C.H.; Lee, S.M.; Lee, K. Mechanical and structural investigation of traditional masonry systems with diverse types of bricks and hydrated lime mortars. *Int. J. Archi. Hert.* **2023**, *18*, 871–893. [CrossRef]
Nam, B.J.; Noh, S.K.; Kim, E.K.; Ahn, S.A.; Kang, S.Y. Initial behavior and shrinkage properties of lime mortars forc restoration of cultural heritage according to the mixing ratio. *J. Conserv. Sci.* **2020**, *36*, 456–474. [CrossRef]
Park, B.Y.; Yang, H.R.; Lee, C.H. Evaluation of physical properties and material characterization for structural frame at the stained glass windows to Gongju Jeil church of the registered cultural heritage in Korea. *Econ. Environ. Geol.* **2023**, *56*, 103–114. [CrossRef]
Choi, I.K.; Yang, H.R.; Lee, C.H. Consideration of procurement system and material homogeneity for lime and clay using the tombs within the King Muryeong and the Royal Tombs in Gongju, Korea. *Econ. Environ. Geol.* **2022**, *55*, 447–463. [CrossRef]
Nockolds, S.R. Average chemical compositions of some igneous rocks. *GSA Bulletin.* **1954**, *65*, 1007–1032. [CrossRef]
Govindaraju, K. Compilation of working values and samples description for 272 geostandards. *Geostand. Newsletter.* **1989**, *13*, 3–16. [CrossRef]
Taylor, S.R.; McLennan, S.M. *The Continental Crust: Its Composition and Evolution*; Blackwell Scientific Publications: Oxford, MI, USA, 1985. [CrossRef]
Pearce, J.A. Role of sub-continental lithosphere in magma genesis at active continental margines. In *Continental Basalts and Mantle Xenolith*; Hawkesworth, C.J., Norry, M.J., Eds.; Shiva: Bristol, UK, 1983; Available online: https://orca.cardiff.ac.uk/id/eprint/8626 (accessed on 24 November 2023).
Hlaing, C.S.S. Scientific analysis and utilization of plant and fruit samples for conservation of ancient decorative arts. In *Proceedings of the Asia Cooperation Program for Conservation Science ACPCS, Daejeon, Republic of Korea, 1 April–31 August 2018*; National Research Institute of Cultural Heritage: Daejeon, Republic of Korea, 2018; p. 2727242023. Available online: https://nrich.go.kr/ichc/programUsrList.do?menuIdx=2727 (accessed on 24 November 2023).
Lou, L.; Zhou, P.; Zhu, H.; Zhang, B.; Hu, Y. A thorough detection of the mortar materials for Buddhist buildings in Bagan, Myanmar. *EPJ Plus.* **2023**, *138*, 151. [CrossRef]
Luján-Lunsford, R. *Guidance Note in Approaches for Conservation of Mural Paintings and Architectural Decorative Works at Pagán*; Internal UNESCO Report; UNESCO: Rome, Italy, 2017.
Noh, J.H.; Lee, H.C. Applied mineralogical characterization for the quick-lime manufactured from fine grained domestic limestones. *J. Miner. Soc. Korea* **2009**, *22*, 261–277.

claimer/Publisher's Note: The statements, opinions and data contained in all publications are solely those of the individual hor(s) and contributor(s) and not of MDPI and/or the editor(s). MDPI and/or the editor(s) disclaim responsibility for any injury to ple or property resulting from any ideas, methods, instructions or products referred to in the content.

materials

Article

Radiocarbon Dating of Mortar Fragments from the Fresco of a Romanian Monastery: A Field Study

Marioara Abrudeanu [1,2,*], Corina Anca Simion [3,*], Adriana Elena Valcea [1], Maria Valentina Ilie [3], Elena Alexandra Ispas [3], Maria Loredana Marin [3], Dragos Alexandru Mirea [3], Dan Cristian Olteanu [3], Cristian Manailescu [3], Alexandru Razvan Petre [3], Denis Aurelian Negrea [4], Sorin Georgian Moga [4], Izabela Mari Dorin Grecu [6], Gheorghe Garbea [6], Flavio Nicolae Finta [1] and Mircea Ionut Petrescu [7]

1 Doctoral School of Materials Science and Engineering, National University of Science and Technology POLITEHNICA Bucharest, 313 Splaiul Independenţei, Sector 6, 060042 Bucharest, Romania; adriana.valcea31@gmail.com (A.E.V.); fintaflavio@gmail.com (F.N.F.)
2 Technical Sciences Academy of Romania, 118 Calea Victoriei, Sector 1, 010093 Bucharest, Romania
3 Horia Hulubei National Institute for R&D in Physics and Nuclear Engineering (IFIN-HH), 30 Reactorului St., P.O. Box–MG-06, 077125 Magurele, Romania; maria.ilie@nipne.ro (M.V.I.); elena_ispas_07@yahoo.com (E.A.I.); marinmarialoredana@yahoo.com (M.L.M.); dragos.mirea@nipne.ro (D.A.M.); cristian.olteanu89@yahoo.com (D.C.O.); cristian.manailescu@nipne.ro (C.M.); alexpetre@nipne.ro (A.R.P.)
4 Regional Center of Research & Development for Materials, Processes and Innovative Products Dedicated to the Automotive Industry (CRCD-AUTO), National University of Science and Technology POLITEHNICA Bucharest, Universiy Center Pitesti, 1 Targu din Vale St., 110040 Pitesti, Romania; aurelian.negrea@upb.ro (D.A.N.); sorin_georgian.moga@upb.ro (S.G.M.)
5 Department of Mineralogy, Faculty of Geology and Geophysics, University of Bucharest, 1 N. Bălcescu Ave., 011401 Bucharest, Romania; izabela@contentlogic.ro
6 Faculty of Theology, Letters, History and Arts, National University of Science and Technology POLITEHNICA Bucharest, University Center Pitesti, 1 Targu din Vale St., 110040 Pitesti, Romania; dorin_grecuro@yahoo.com (D.G.); garbea_59@yahoo.com (G.G.)
7 Department of Engineering and Management of Metallic Materials Casting, Faculty of Materials Science and Engineering, National University of Science and Technology POLITEHNICA Bucharest, Splaiul Independenţei nr. 313, Sector 6, 060042 Bucharest, Romania; ionut.petrescu@upb.ro
* Correspondence: abrudeanu@gmail.com (M.A.); simion_corina_anca@yahoo.com or anke@nipne.ro (C.A.S.)

Abstract: The stone Ensemble from Corbii de Piatră Romania arouses a continuous scientific interest, with the final goal being to obtain an exhaustive and multidisciplinary package of results that will become the support of an extensive restoration project. The cave painting stands out as the most important and most affected by the advanced degradation among the historical monuments in Romania. This article provides for the first time a radiocarbon dating of the first forms of painting by establishing the age of the mortar/plaster used as a pictorial support. Being a very complex context from the point of view of the type of datable material and the disappearance over time through degradation of other elements that would ensure a simpler and more reliable radiocarbon dating (such as the straws used to form the material), it was necessary to use a multidisciplinary approach for the selection of samples and for supporting the radiocarbon results. The set of analyses consisted of visualization techniques through microscopy and compositional analysis, providing information on the similarities/differences between the samples, the degradation mechanisms/impurities and the quality of the calcium carbonate dated by the Accelerator Mass Spectrometry (AMS) technique. The results supported each other, ensured the selection of reliable radiocarbon data and established the most probable moment of the early interventions, namely the two phases corresponding to the 14th century.

Keywords: cave painting; radiocarbon dating; electron microscopy; compositional analysis; accelerator mass spectrometry

1. Historic

Considered to be one of the oldest places of worship, used since the second century AD and sheltered from persecutions and barbarian raids, the current Corbii de Piatră/Stone Crows cave monastery in Romania is built on an old Dacian site [1].

The first mention of the Corbii de Piatră dates back to 15 April 1456, when Vladislav II, the Voivode of Wallachia, issued an act in Old Slavic language in Targoviste, confirming Mogos' ownership of the villages: "All the Corbii, from Aghiş downward, and all the Corbii-de-Piatră, and the pasture from Miceşti, and the mill waters, and half of Mălureni, because they are ancient and rightful pastures of Mogoş (. . .)" [2]. Corbii de Piatră became the first nunnery on the territory of Romania, documented on 23 June 1512, when the nun Magdalina (Muşa in lay life), the hereditary heiress of the estate in Corbi, re-established the monastery, which acquired the status of a princely monastery, donating it to Voivode Neagoe Basarab [3].

The architecture of the Corbii de Piatră cave church (Figure 1A) is typologically derived from hall-type churches, with two altars dedicated to a double rite, characteristic of the Byzantine world from the 10th century (Figure 1B,C) [1–5].

It is considered that the fresco painting, in a purely Byzantine style, was made at the beginning of the 14th century, during the reign of Voivode Basarab I the Founder (1310–1352), having the wall excavated in the sandstone (naturally and/or man-made) as support. Over time, water infiltrations through the sandstone wall, temperature variations and biodegradation processes determined the advanced deterioration of the frescoes.

The previous studies [6–13] carried out for the fresco from Corbii de Piatră provide restorers with data on their execution materials. The recent characterization of two fragments of the church fresco, from areas with different behavior under similar conditions [6,7], posed the problem of their dating and their inclusion in the chronology of the fresco.

(A)

Figure 1. *Cont.*

(B)

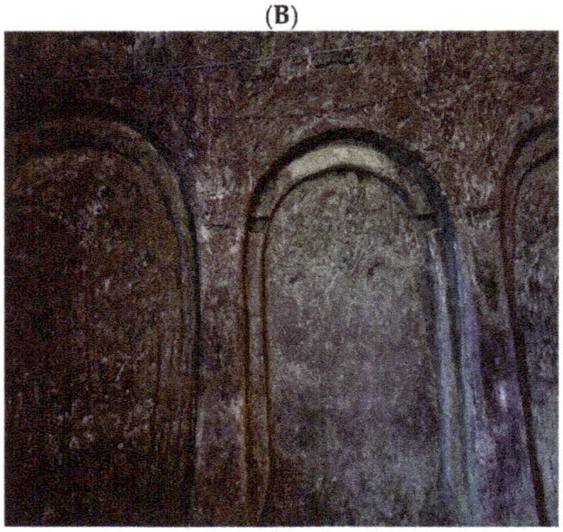

(C)

Figure 1. Corbii de Piatră rupestral church: (**A**) exterior view, tiled wall (Photo: Simion, C. A.); (**B**,**C**) interior view for ship-type structure (Photo: Valcea, A. E.).

2. Geological, Mineralogical and Hydrogeological Considerations

From a geological point of view, the site containing Corbii de Piatră Monastery is located in the Getic Depression (Figure 2), a narrow sedimentary basin in the Southern Carpathian foreland with sediments from the Cretaceous to Miocene ages [14]. The monastery was excavated in massive sandstones of the Oligocene age, which belong to the Corbi Formation [9,15]. These sandstones are characterized by a polymictic composition that includes quartz, alkali feldspar and plagioclase, muscovite granoclasts and metamorphic rock fragments such as gneiss, schists and quartzites [16].

Due to the mineralogy of the grains and the composition of the binder, the Corbi sandstone petrotype is a particular one. The average grain size and the proportion of 40–60% unclogged pores from the total volume of pores determine the permeability, the accumulation of water in the bedrock and the gravitational circulation of water [9]. The research carried out by the team led by Prof. Marin Şeclăman established a percentage of the water stored in the inner walls of the church of 11%, which would correspond to a maximum porosity of 25%, a value much higher than the real porosity. Water from the Pârâul Cascadei/Cascade Stream brings most of the moisture. The Pârâul Cascadei originates from the Brădulet formation and drains the rocks with sulfates from the Corbi sandstone roof, infiltrating the walls of the church, with priority in the northern one. The water has a flow with seasonal fluctuations and variable compositions, containing predominantly Ca^{2+} and Mg^{2+} ions, chlorides and SO_4^{2-} ions, which participate in the degradation process, with the formation of gypsum, $CaSO_4 \cdot 2H_2O$.

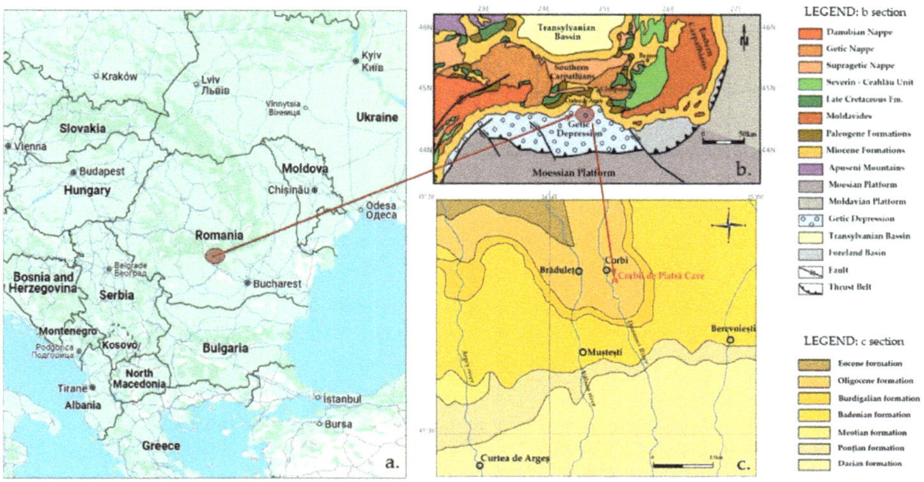

Internal

Figure 2. Map with the researched area, modified after Ghiran et al., 2023 [14]; (**a**) Occurrence of Southern Carpathian (Google Earth source); (**b**) Simplified tectonic map of the Getic Depression, modified from Sandulescu, 1984 [15]; (**c**) The studied area with location of the Corbii de Piatră monastery on Oligocene Corbi Formation, modified after Murgeanu et al., 1967 [16].

The chemical composition established for the wall tiles inside the church also has a high content of uranium (U) and thorium (Th), values that determine a higher natural radioactivity inside the church compared to the radioactivity of the natural background outside it [17]. Measuring the indoor and outdoor dose rate levels and comparing them will express the extent to which the interior is isolated from the exterior [18]. It is possible that radon (with radioactive isotopes produced both by the descendants of thorium and especially by uranium) accumulates inside in higher concentrations than outside. This radioelement is also soluble in infiltration waters, respectively, in the humidity of the air inside the church [19]. In the first constructive forms of the place of worship, when it was insulated from the outside (preserving the walls better) much better than it is today, the radioactivity level of the air inside was most likely much higher. Understanding the risks associated with radon exposure (beyond the states of dizziness, exaltation and other forms that could induce religious trances in the past) and mitigating radon levels through the measures taken will significantly reduce the risk of lung cancer, for example, for those who regularly attend church nowadays [20,21]. Currently, there are kits on the market for testing radon in closed spaces that are simple and relatively cheap. However, the technical solutions for stabilizing the hill and reducing infiltrations should also take into account the long-term minimization of indoor dose rates, with the simplest methods being adequate ventilation but also maintaining a temperature regime specific to the day/night cycles and the succession of seasons. Once the cave complex is stabilized both externally and internally, it is possible to move on to a project to restore the inner walls, including the painting [10].

3. Preliminary Dating Characteristics and Radiocarbon Dating

Materials and Characterization Techniques

The dating of the frescoes can be done from the point of view of identifying the materials and execution techniques used and through the prism of art history but also through physico-chemical analyses to establish their composition (pictorial support, pigments and materials necessary for the creation of frescoes, specific to historical periods).

Intrinsic information can be provided by radiocarbon dating. The result will complement both the conclusions of specialized studies and the investigations that usually accompany them. Thus, compared to the older studies [9–13] or the recent experiences in the field of the RoAMS laboratory from IFIN-HH [22,23], this method provides an answer regarding the age of the investigated materials by means of absolute chronology, in tandem with a series of methods that usually are not associated with radiocarbon dating. This has its limitations, in addition to its advantages [24]. Therefore, it is not important to obtain a radiocarbon date but to interpret it through the prism of all the information held (historical, archaeological, architectural, from the records of the interventions, from the history of art and from the preceding physico-chemical investigations). Based on this preliminary information, the relevant sampling areas can be established from where to extract representative fragments that may contain carbon compounds. Since radiocarbon dating is an invasive technique, the preliminary studies must provide sufficient information to take into account the minimum intervention on the walls/frescoes while obtaining maximum information. A representative sample can be taken, through an appropriate technique, for each wall, which will contain differentiated fragments depending on the studied area.

Based on the characterization through previous studies of the four walls of the church, it was determined that the northern and southern walls remain important in the radiocarbon dating of the first forms of painting on the inner walls [6,7].

Based on the existing information, it was possible to establish the position horizontally in front of the altar and vertically in front of the current floor level from where these samples could be taken, as well as the minimum justified number of required samples. Thus, respecting the deontology of the conservation and restoration of historical monuments [25,26], one sample of mortar/plaster was extracted for each sampling point on each of the two walls, and they were then sub-divided into several fragments, depending on the preliminary results regarding the characterization of the state of preservation of the respective sample, obtained on the remains left after preparing the samples for dating.

The fragments thus obtained for dating, two for the southern wall and three for the northern wall, corresponding to the two chosen areas, were taken from the height of 1.70 m above the floor. The sample on the northern wall was taken from the second square from the altar (Figure 3A), and the one on the southern wall from the corresponding opposite square (Figure 3B).

The remains were prepared for other physico-chemical investigations necessary for the preliminary characterization. The details of these latter fragments are captured in Figure 4A,B.

According to traditional receipts, all mortars may contain carbon residues from the straw inserts used in the preparation [27]. S-type fragments, taken from the southern wall, are very poorly adherent. The N-type fragments, taken from the northern wall, are very adherent. In addition to the S-type fragments, the N-type fragments could contain sheep tallow or mineral oil (Table 1).

Figure 3. Corbii de Piatră church: (A) the second square from the altar on the northern wall from where sample N was taken, divided into fragments N1, N2, N3 for radiocarbon dating (Photo: Simion, C. A); (B) the second square from the altar on the southern wall from where sample S was taken, divided into fragments S1, S2 for radiocarbon dating (Photo: Simion, C. A).

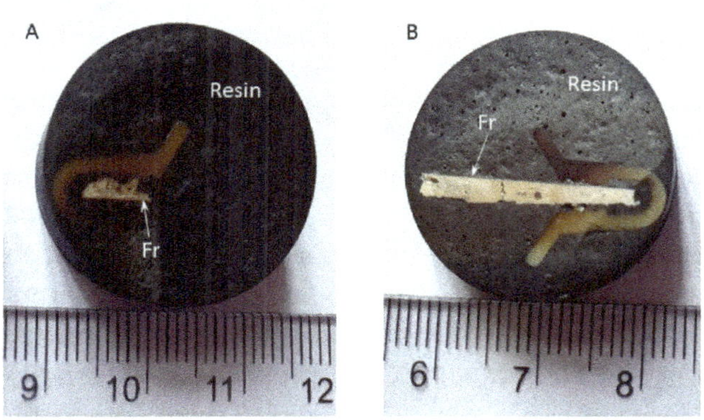

Figure 4. Corbii de Piatră church; details of the leftovers from the samples prepared for radiocarbon dating, used in the preliminary analyses: (A) remains from the sample of the northern wall; (B) remains from the sample of the southern wall.

Table 1. Fragments from the samples used for radiocarbon dating and their masses.

Sample Name	Weight [mg]
S1	265.7
S2	243.5
N1	39.1
N2	196.5
N3	351.0

The sub-samples taken from N-type and S-type remains (Figure 4A,B), one for each side, were investigated before the radiocarbon dating through a series of techniques to reveal the initial pictorial composition/at the intervention in direct context with the plaster that will be dated but also information on the materials with which this plaster came into contact over time, whether it is about the lithic support, whether it is about the fresco itself or about its interaction with the environment inside the church. Thus, a direct connection can be made between the age of the pictorial support and the painting in the respective area, but we can also anticipate contaminants containing exogenous carbon that can affect the radiocarbon date of the calcium carbonate, formed by strengthening the pictorial support after being applied to the walls. The characterization of the fresco fragments was carried out by optical microscopy [28,29], cathodoluminescence [30–32], scanning electron microscopy with energy-dispersive fluorescence spectroscopy mode for elemental chemical analysis [28,29,33], X-ray diffraction [34,35] and X-ray fluorescence [36,37].

The optical microscopy (OM) characterizations were performed in polarized light with a Zeiss Observer A1m optical microscope, which allows magnification up to 1000×, visualization in bright field, dark field, polarized light and the acquisition of images with a Canon camera Zeiss AG, Oberkochen, Germany.

The microscopic analysis with cathodoluminescence (CL) was performed with Nikon E400, (Nikon company, Tokio, Japan) an optical microscope equipped with a cathodoluminescence device with a cold cathode (CL 8200 MK 3A). The images were acquired with a C.OOLPIX 950 digital photomicrograph device (Keyence International, Mechelen, Belgium). The parameters required to perform the technique were the average vacuum value of 0.5 Torr, current voltage on the bundle of 15–17 kV and current intensity on the electron gun of 350–400 mA, according to the standard in use [30,31]. The interpretation of cathodoluminescence results is qualitative.

The characterization of the fresco element by scanning electron microscopy with energy-dispersive fluorescence spectroscopy (SEM-EDS) was carried out using the HITACHI SU5000 electron microscope equipped with a backscattered electron detector and the energy-dispersive fluorescence spectroscopy module for elemental analysis (Hitachi Groupe, Chiyoda, Tokyo, Japan).

The X-ray diffraction analysis (XRD) is used to complete the identification of the crystalline phases of the mortar components and the painting layer. X-ray diffractions (XRD) were performed using an XPERT PRO MPD 3060 facility from Panalytical (Almelo, The Netherlands), with a Cu X-ray tube ($K\alpha$ = 0.154051 nm), a 2 Theta of 20°–70°, a step size of 0.13°, a time/step of 51 s and a scan speed of 0.065651°/s. The phase analysis, qualitative, was carried out using the PDXL2 program (Rigaku Corporation, Chiyoda City, Tokyo, Japan) and the PDF4+ 2022 database (International Center for Diffraction Data).

X-ray fluorescence (XRF): the XRF analysis was performed using a portable Bruker Tracer S1 Titan spectrometer (Billerica, MA, USA). A Rhodium (Rh) anode was used to generate an incident X-ray beam to probe the samples. For safety reasons, the energy used is limited to 50 keV. The incident beam is passed through a collimator, resulting in a circular 8 mm in diameter spot on the sample. The resolution for the medium energy domain (Sc to Zr) is approximately 120 eV, according to the manufacturer. Each sample was analyzed in significant zones. The generated spectra were later analyzed using the Bruker Instrument Tools software 1.7.0.128, supplied by the manufacturer.

Additionally, regardless of the chemical form in which it was present in the samples, the following equipment was used for the identification of total carbon (TC), inorganic carbon (IC) and total organic carbon (TOC): PrimacsMCS in combination with FormacsSERIES and the elemental analyzer EA (VarioMicroCube, Elementar, Hanau, Germany)™.

PrimacsMCS allows us to analyze the total carbon (TC) and inorganic carbon (IC) separately. The total carbon is determined by catalytic oxidation of the sample at 1100 °C, while inorganic carbon is determined by digestion of the sample with phosphoric acid H_3PO_4 in the IC reactor. Carbon dioxide CO_2 is detected by the nondispersive infrared NDIR sensor of the FormacsSERIES analyzer. The TOC concentration of the sample is obtained by subtracting TC − IC = TOC.

The radiocarbon dates were obtained at RoAMS laboratory in IFIN-HH using the 1 MV HVEE Tandetron™ system (High Voltage Engineering Europa B.V., Amersfoort, The Netherlands). Before AMS analysis, the samples were converted into graphite by combustion using the AGE 3 graphitization installation [38], which works in conjunction with the elemental analyzer (VarioMicroCube, Elementar, Hanau, Germany)™. The experimental data obtained for all the samples were normalized against NIST SRM 4990C—Oxalic Acid II (NIST SRM 4990C International Standard Reference Material for Contemporary Carbon-14, 1983) to the modern radiocarbon level, while to estimate the blank level, a fossil coal of Romanian origin was used. By measuring the 13C/12C ratio, the results were also corrected by δ13C parameters determined by AMS data processing software version 4.06 (according to A-4-35-501-7621 Rev. A Operator Manual 1.0 MV Tandetron for AMS B7621 IFIN-HH Magurele, Romania, page 73), representing the cumulative isotopic fractionation of all physico-chemical processes on the analysis chain. The ages were calculated using the BATS software tool (Bats version 4.06 (04.06.2015)) [39], according to Stuiver and Polach (1977) [40]. The dating strategy, after characterization of the pictorial layer, should include the determination of the nature of the material and of contaminants, the limitations of methods on the type of material and the analysis of current literature data, whether they are about the type of material to be radiocarbon dated or whether they are about the most appropriate working methodology.

4. Experimental Results

The microscopic analysis in section of the samples highlighted the differences in the microstructure of the two types of mortars: the friable, slightly adherent mortar has a flat surface with traces of the roughness of the abrasive paper (Figure 5A), which reveals the low hardness, while the adherent mortar presents a raised surface that highlights the particles in the mortar that confer a higher hardness (Figure 5B).

The analysis by cathodoluminescence microscopy of the sample with friable mortar highlighted the degradation process over the entire thickness of the layer and the presence of numerous cracks (Figure 6A).

Adherent mortar presents an unaltered outer layer, which includes the pictorial layer with an unaltered internal substrate. Following the massive dissolutions, vacuolar-type secondary pores were formed in canaliform places (Figure 6B). The pores are large, irregular and interconnected, allowing the circulation of interstitial fluids (infiltration waters). This structure suggests the operation of a settling mortar used since ancient times for walls with infiltrations [41].

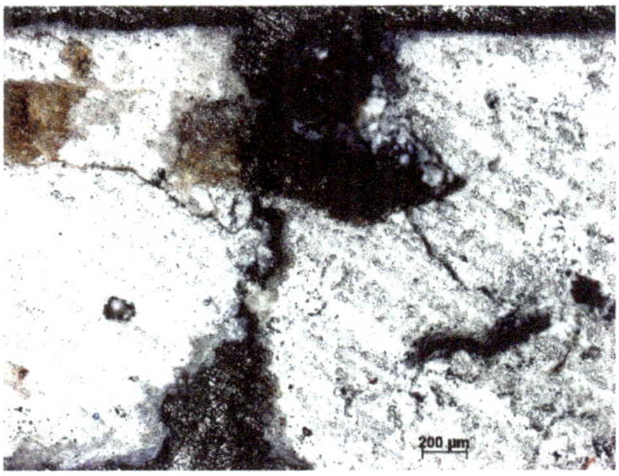

(A)

Figure 5. *Cont.*

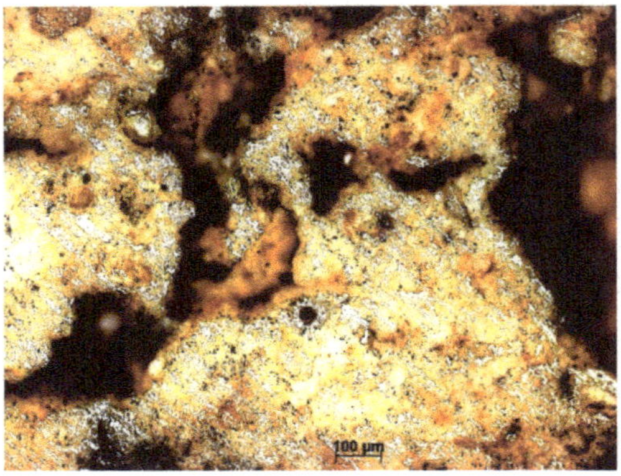

(**B**)

Figure 5. The microstructure of the mortars at the optical microscopic analysis: (**A**) the friable mortar with low adhesion; (**B**) the adhesive mortar with high hardness.

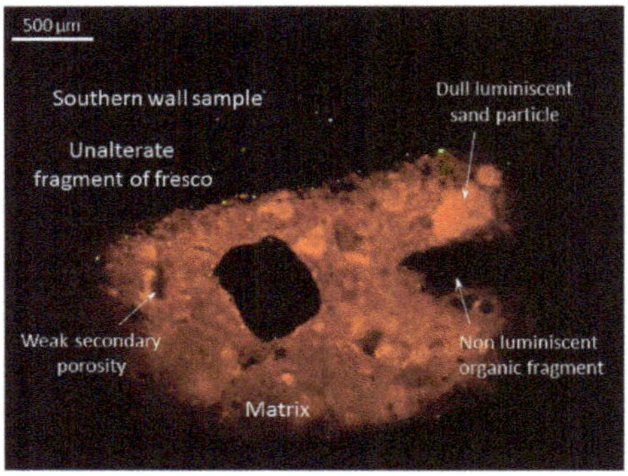

(**A**)

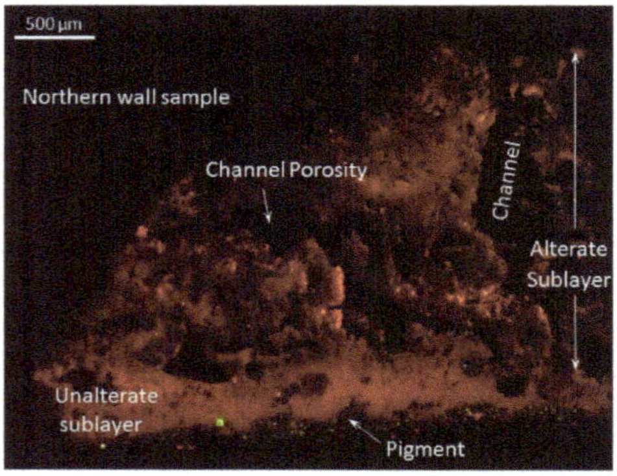

(**B**)

Figure 6. Fragments of fresco under cathodoluminescence microscopy: (**A**) fresco fragment with low adhesion to the sandstone wall; (**B**) fresco fragment with high adhesion to the wall with two distinct areas: an external unaltered sublayer and an internally altered sublayer with high porosity.

Microscopic analysis by cathodoluminescence highlights two fresco fragments with distinct characteristics. The first fragment of the fresco, taken from the southern wall, is made of a carbonate cohesive matrix with weak orange luminescence, which includes sand particles with various grain sizes from 62 to 1000 μm (Figure 6A). The grains embedded in the matrix are mainly represented by carbonate rock fragments and subordinately by muscovite, feldspar and non-carbonate rock fragments. While carbonates have strong or dull orange luminescence, the feldspars show bright green luminescence and the biotite dull blue-violet luminescence. Nonluminescent fragments of carbonized plants and non-carbonate lithoclasts are also distinguished. An incipient secondary porosity is observed represented by small and irregular pores in low concentration.

The second fragment of the fresco was sampled from the northern wall. This has a more complicated structure, being made up of an unaltered outer sublayer and an altered inner one (Figure 6B). The dominant material of fresco is carbonate, with orange colors and variable luminescence, from bright to dull. The unaltered outer sublayer is compact and has pigment infiltrations on the surface that are signaled by bright vivid colors. The altered sublayer contains an active system of interconnected microchannels (channel porosity), formed in time by selective dissolution processes of carbonate bioclast fragments. The dissolution is a continuous process due to the meteoric waters that percolate the monastery wall [32].

Characterization by scanning electron microscopy in the section of the mortar sample [42] with low adhesion revealed a degraded structure, with a granular appearance, for which the stratification in relation to the surface of the fresco cannot be established (Figure 7A). In the areas strongly affected by the infiltration water inside the mortar, the EDS mapping analysis highlighted the presence of sulfur from meteoric water (Figure 7B).

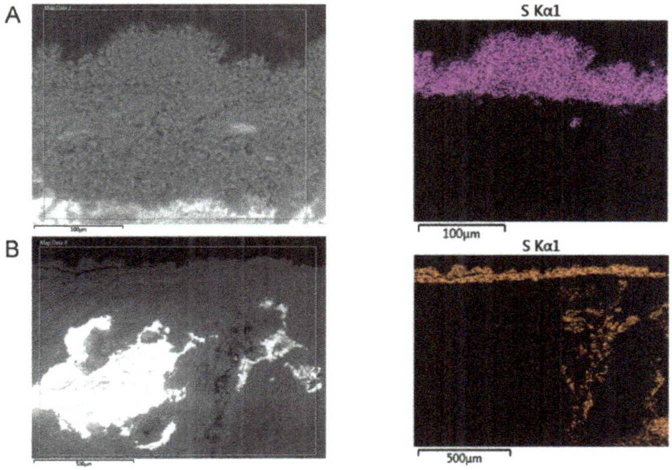

Figure 7. SEM-EDS of the sample with small adhesion in cross-section and elemental mapping: (**A**) surface layer; (**B**) an area where the mortar shows advanced degradation inside.

In the case of the adhesive mortar sample, the scanning electron microscopy analysis in section shows that the fresco has kept the initial layering, and the component layers of the fresco have been highlighted as follows: the superficial carbonation layer, the pictorial layer, the intonaco layer and the support layer (Figure 8).

The qualitative X-ray diffraction analysis of the two fresco fragments [42] highlighted the existing phases in the two mortars and in the pictorial layers. The advanced degradation of the fresco fragment with low adhesion is revealed by the presence of gypsum in the friable superficial layer (Figure 9A) as a result of contact with infiltration waters and diffusion processes towards the surface of the fresco.

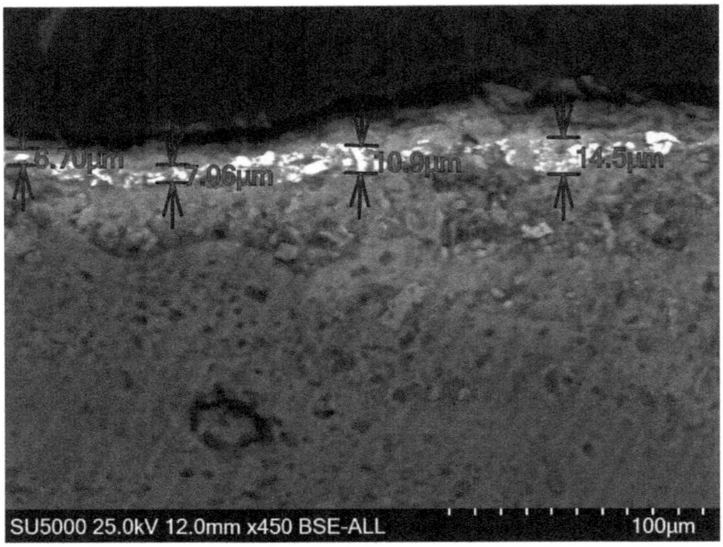

(**A**)

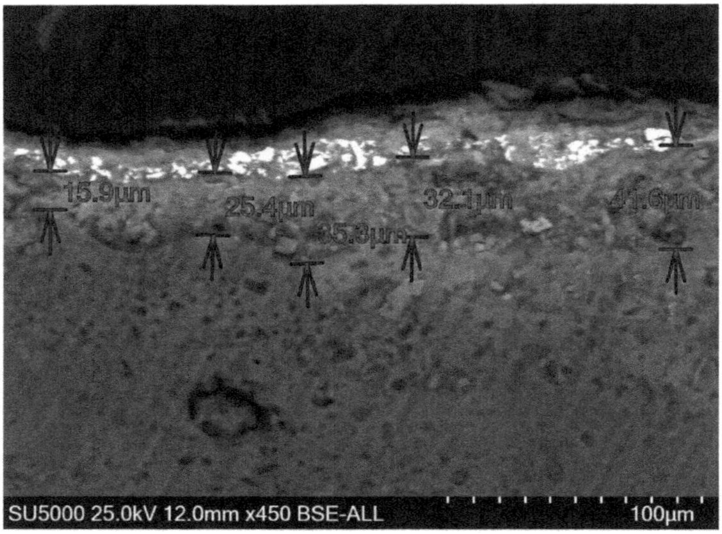

(**B**)

Figure 8. The microstructural analysis in cross-section by scanning electron microscopy of the sample with high adhesion revealed the fresco stratigraphy: (**A**) pictorial layer ×450 WD-10 mm 25 kVcu30si (BSE); (**B**) intonaco layer ×450 WD-10 mm 25 kVcu30si (BSE) [6].

The high durability and adhesion of the fresco element taken from the northern wall (Figure 9B), the most exposed to the action of infiltration waters, in relation to other fresco areas in the Corbii de Piatră church, is explained by the different structure and properties, especially by the way of degradation with the formation of infiltration water drainage channels. The differences can be explained by the realization of this part of the fresco with a different mortar, in a different period of time than the one in which the full painting of the fresco in the church was done, most likely represented by the samples on the southern wall.

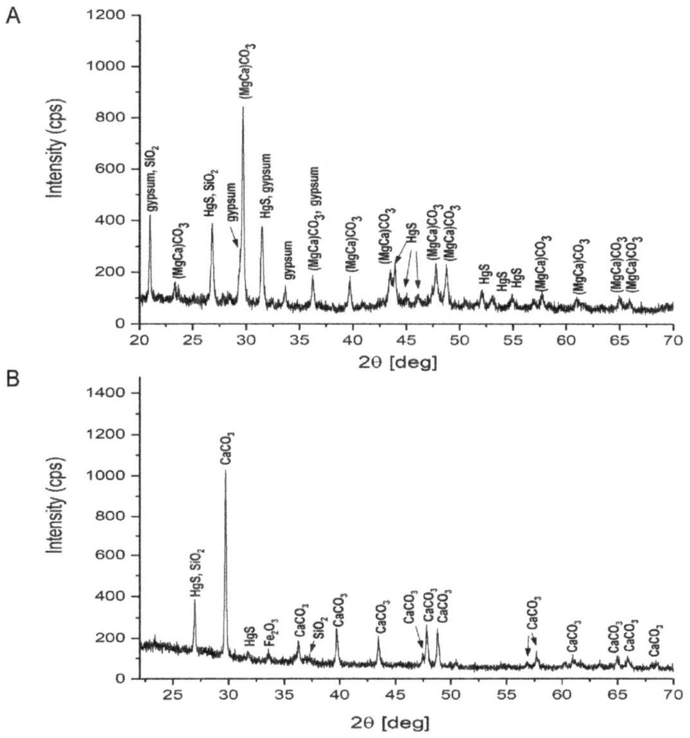

Figure 9. Qualitative phase analysis by X-ray diffraction of pictorial layer of the fresco element: (**A**) with low adhesion to the sandstone wall [7], (**B**) with high adhesion to the sandstone wall [6].

X-ray fluorescence: Compared to previous studies that carried out in situ mapping through dozens of measurement points of the entire surface of the walls and ceiling [10], the current measurements were only performed on the samples that have been taken (Tables 2–4). According to the investigations already published, the previously identifiable elements were Fe, K, Al, Mg, Si, Ca, Hg, S, Pb and Mn.

Table 2. XRF results, without errors, in ppm; all identified elements; brings together types of materials, construction periods/restorations and recipes + sources of raw materials.

Sample Name	Mg	Al	Si	P	S	K	Ca	Ti	V	Cr	Mn
S2	5474	0	165,353	0	2308	0	183,631	135	0	69	197
S1	7543	1055	133,947	2060	55,834	0	162,457	352	33	36	383
N3	7138	7578	277,789	616	1082	1082	128,797	1596	0	92	257
N2	3863	6968	203,358	0	2654	1406	38,027	907	33	15	182
N1	6255	0	137,900	494	2725	0	77,282	122	0	0	136
S2	Fe	Ni	Cu	Zn	As	Sr	Y	Mo	Pb	Th	U
S1	9720	111	33	104	84	197	14	0	19	0	0
N3	24,554	128	0	561	3946	322	74	11	1421	19	176
N2	17,850	43	33	38	0	101	0	0	0	0	0
N1	12,689	127	40	56	0	113	0	0	0	0	0
S2	2498	182	77	75	0	127	0	0	0	0	0

Table 3. XRF results in ppm for elements from Mg through Mn, errors included.

Sample Name	Mg	Err	Al	Err	Si	Err	P	Err	S	Err	K	Err	Ca	Err	Ti	Err	V	Err	Cr	Err	Mn
S2	5474	3077	0		165,353	1170	0		2308	131	0		183,631	325	135	22	0		69	10	197
S1	7543	3299	1055	618	133,947	1085	2060	191	55,834	346	0		162,457	318	352	26	33	14	36	11	383
N3	7138	2955	7578	674	277,789	1472	616	179	1082	110	1082	92	128,797	279	1596	36	0		92	11	257
N2	3863	1945	6968	567	203,358	1163	0		2654	90	1406	59	38,027	137	907	24	33	12	15	8	182
N1	6255	2106	0		137,900	968	494	120	2725	96	0		77,282	192	122	16	0		0		136

Table 4. XRF results in ppm from Fe through U, errors included.

Sample Name	Fe	Err	Ni	Err	Cu	Err	Zn	Err	As	Err	Sr	Err	Y	Err	Mo	Err	Pb	Err	Th	Err	U
S2	9720	61	111	22	33	22	104	14	84	15	197	10	14	8			19	6			
S1	24,554	109	128	23			561	26	3946	64	322	12	74	10	11	5	1421	21	19	2	176
N3	17,850	76	43	16	33	15	38	9			101	7									
N2	12,689	59	127	22	40	19	56	11			113	8									
N1	2498	30	182	28	77	26	75	14			127	11									

This time, 22 elements were identified not only on the pigment traces but also in the mortar/plaster.

The available information [10,42,43] from XRF prior to the current analysis can be summarized as follows and will implicitly help to select the radiocarbon dating strategy:

- Identified pigments: lead minium, Pb_3O_4, applied with the initial addition of an organic binder or during previous interventions (red pigment); white lime; cinnabar (HgS); lead carbonate, $PbCO_3$ as a result of degradation or impurity from the lead mine; red and yellow based on iron oxides and coal black;
- Bluish hue: possible mixing of lime white with coal black;
- Possibly azure or malachite in some paintings; possibly an introduction from the time of painting the pediment, like lead white;
- Scheele green pigment $CuHAsO_3$ and Schweinfurt green pigment $Cu(CH_3COO)_2$-$Cu(AsO_2)_2$, used from the 18th–19th century until the beginning of the 20th century;
- Sulfates and nitrates as salts formed by degradation;
- Alteration products: calcium oxalate; lead carbonate from the alteration of lead minium;
- Analyzes by pigment classes are extensively commented;
- Natural and artificial pigments were used;
- Natural pigments used for ocher and red colors: hematite, limonite, cinnabar;
- Artificial pigments used in white, black, gray and partially red colors;
- Some ingredients are native: limestone, coal, hematite, limonite;
- Cinnabar is "imported".

The XRF results obtained at this stage highlighted, by applying the Principal Component Analysis (PCA) function of the Minitab 17.0 Program, to what extent the five fragments from the two samples are correlated with each other (testing also the degree of homogeneity/the quality of the partition selection of each sample in several fragments) from the same wall and between different walls. If all the elements determined by XRF are used (Tables 2–4), the Score Plot shows a similarity between fragments N2 and N3, respectively, between N1 and S2, with fragment S1 being considered an outlier. The general degree of correlation between all fragments is 92%; it can be an indication of the state of the material from formation to analysis. If only the majority elements are used, and fragments N1, N2 and N3 are separated from S1 and S2, with the general degree of correlation being 86%, it can be an indication of the same type of material but obtained by different recipes. Through the minority elements, N2 and N3 become distinct from S1 and S2, but there is a correlation between S2 and N1; the general degree of correlation is 69%; it shows sources of possibly different raw materials but also the degree of damage by impurities introduced over time from infiltrations from contact with the pictorial material and the surrounding environment. Finally, through trace elements, the following distribution is obtained: N1, N2, N3 and S1 are separated from S2, with the general degree of correlation being 77%.

The interpretation of these results, as well as the information useful for radiocarbon dating, will be presented in the paragraph regarding this method, as in the case of the other analyses.

TC/IC/TOC analysis: The results are shown in Table 5.

Table 5. Total carbon (TC), inorganic carbon (IC) and total organic carbon (TOC) values.

Sample Name	TC [%]	IC [%]	TOC [%]
S1	8.88 ± 0.44	8.56 ± 0.43	0.32 ± 0.01
S2	8.99 ± 0.45	8.45 ± 0.42	0.54 ± 0.03
N2	9.86 ± 0.49	9.86 ± 0.49	Near 0
N3	10.09 ± 0.50	9.55 ± 0.48	0.54 ± 0.03

Being in small quantity, fragment N1 could not be analyzed. The results are expressed as mass percentages.

The association is observed again on each side of the church; those on the northern side contain more inorganic carbon, most likely carbonates, compared to those on the southern side; a correlation may appear with the degree of friability. The very low percentages of total organic carbon do not exclude deposits due to biological attacks, but they show, a fact also demonstrated in [10], that the straws were not preserved; in their place, small degradation products appeared that will not be able to be separated from such quantities of small fragments/samples so that they can be dated separately. If this isolation/purification had been possible, the radiocarbon age could have fixed the moment of application of the material on the wall much better in time, with the straw being a short-life datable material (annual cycle).

Elemental Analysis (EA): This was carried out practically during the graphitization of the samples for subsequent measurement by AMS. The results are presented in Table 6.

Table 6. Elemental analysis results.

No	Weight [mg]	Sample Name	Lab Code (RoAMS)	N Area	C Area	N [%]	C [%]	C/N Ratio
1	34.31	**S1**	G4896_2534.150	8657	92113	0.67	10.13	15.01
2	16.92	**S2**	G4897_2535.150	8238	45221	1.30	10.08	7.76
3	19.09	**N2**	G4898_2537.150	825	52457	0.12	10.36	88.95
4	19.28	**N3**	G4899_2538.150	894	57986	0.13	11.35	89.81
5	17.21	**N1**	G4900_2536.150	1242	44391	0.27	9.73	49.09

The differences can be noticed between the fragments belonging to sample S on the southern side and those on the northern side, N. The total carbon values are somewhat higher compared to the alternative TC/IC/TOC method (Table 5); this is most likely due to the fact that in both methods, the values are in mass percentages and the base value against which the reporting is done in each case makes it count. The average value of the total carbon during graphitization is approx. 10%; then, the actual carbon contained in the samples subjected to graphitization will enter the range below 35 mg, recommended for the initial mass. It will lead to reduced carbon with a mass within the range [1.7–3.4 mg], acceptable for the chance to obtain approx. 1.52 mg of carbon deposited on the iron catalyst, which is the basis for the AMS target. Nitrogen values, a sign of possible organic compounds, are within the limits of the background of the equipment, so they are most likely not present in the dated samples, being another indication of the absence of degraded organic materials, coming from straw (humic, in general, with significant content in nitrogen).

Radiocarbon dating: The work strategy and the primary interpretation of the results, without placing them in a historical and archaeological context, were based on the information provided by the previous methods. They can be summarized as follows:

Imaging: In regard to differences in structure between the sample on the northern side compared to the one on the southern side, the one on the northern side represents a more compact, adherent material, with short cracks and traces of a possible fatty organic additive (this confirms the hypothesis of its use but will not affect the radiocarbon date, being practically of the same age as when the mortar/plaster on the walls hardened); the one on the southern side is friable and shows cracks propagated over long distances (it may be an indication of the introduction of exogenous carbon of a different age than that in the mortar recipe, in support of this hypothesis comes the identification of sulfur/gypsum formed by infiltrations and deposited in the mortar structure). In the friable fresco, we cannot speak of stratification, and therefore, the better-preserved portions cannot be accurately separated (it was still possible to separate two fragments, S1 and S2, based on the observations during the preliminary analyses), while in the adherent fresco, it is possible to identify all layers in their evolution since formation, which can be separated given the degree of preservation (appreciated implicitly by the lack of sulfur/gypsum deposits) as N1, N2 and N3.

Cathodoluminescence: This highlights the degradative processes in the friable mortar and, therefore, a possible appearance of secondary carbonates of different radiocarbon ages; the adherent one presents an unaltered external layer and an altered internal one with interstitial channels through which the water from the infiltrations circulates freely but preserves the initial signature of carbon-14 in this decanting mortar.

EDS: This highlights the presence of sulfur on the southern side; it becomes a poison for the conditioned iron catalyst in the AGE 3 unit in the case of fragments S1 and S2; that is why it is recommended to use thermal combustion at EA for all five sample fragments, with the subsequent separation of the higher sulfur oxides in the hatch dedicated to their retention.

XRD: This indicates the formation of sulfur-containing gypsum in the fragments on the southern side and shows the differences in the recipe and that the mortar on the northern wall in the investigated area appeared after the initial painting of the church, represented by the selected portion on the southern wall.

XRF: The analyzed samples are basically lime mortar plasters, with traces of pigment. There is a difference between the sample on the southern wall and the one on the northern wall. The fragments on the southern wall probably contain lead minium or lead carbonate formed by its carbonation over time, along with traces of green pigments and/or lead arsenate, a biocide. An argument in support of this last hypothesis comes from the very large amount of arsenic, identified in S1, with a higher concentration in these two elements. Compared to the green pigments identified in previous studies [10–12,42,43], used from the 18th–19th century and until the beginning of the 20th century, lead arsenate, introduced later and banned after the 70s in agriculture, was intensively used until the middle of the 20th century and was therefore easy to procure without restrictions on applications, remaining impregnated in the portions of the walls where it was used, which helped to slow down the biocidal attack and preserve the original paintings.

TC/IC/TOC analysis: This confirms the XRF information regarding the lime mortar support, and therefore the possibility of radiocarbon dating. The obtained values show a difference between the north and south sides, which could be revealed by different radiocarbon data. It confirms the practical absence of preserved straw; given the actual situation and the very small quantities of the samples, it will not be possible to separate and purify the inorganic and organic fractions to be dated separately. According to the information from [10], the samples contain calcite and quartz, with mineral impurities originating from the raw materials, from mineral pigments or as a result of aging processes—these will not disturb the radiocarbon dating. The very small amounts of organic carbon identified in the sample on the northern wall come either from traces of straw or from contaminants (possibly sheep tallow or mineral oils for the pictorial support, or as a result of biological degradation); the remains of straw and sheep's wool would have carbon-14 levels close to those of hardened lime and therefore will not disturb the dating; mineral oil, if present, will lead to much too old data, and biological contamination could lead to much too young radiocarbon data.

The experience of the RoAMS laboratory at IFIN-HH established that, due to the nature of the samples and the very small quantities, it is practically impossible to separate and purify the inorganic and organic fractions to be dated separately. Moreover, the extraction and purification of calcite, even if it is the majority in the samples, could become a risk factor in the sense that it is very possible that in the end, the purified quantities will be too small to meet the conditions for graphitization, especially for N1. Knowing the characteristics of the organic part, there are two risks: one or more radiocarbon ages may be too old (mineral oil, geological carbon from pigment or contamination by infiltration with geological carbonates) or too young (biological attack or exchange with atmospheric carbon dioxide). According to previous studies [9–11], it is unlikely that a slow exchange with carbonates from infiltrations or with atmospheric carbon dioxide will occur in the sample from the northern wall, and it is more likely in the sample from the southern wall. Thus, taking into account the risks assumed for objective reasons, the samples were graphitized in raw form in order to measure AMS. The thermal combustion, based on the differential thermal analysis [10], which provided in the diagram a single peak characteristic of calcium carbonate and insignificant peaks for other carbonates or organic compounds, was performed at the elemental analyzer coupled with the graphitization unit. Thus, another set of values was obtained in parallel for the total carbon in the samples, commented on above.

The remaining fragments, after TC/IC/TOC analysis, were finely ground and taken in the quantities indicated in Table 6. The quantity is directly proportional to the carbon/carbonate content established by the other methods. These were subjected to thermal combustion in EA, with the separation, purification and drying of carbon dioxide prior to its introduction into the graphitization unit. The graphitization process, the formation of carbon-containing targets formed by hydrogen reduction on the iron catalyst, the AMS measurements, the calculation of the Conventional Radiocarbon Age (CRA) and the transition from this value to the calibrated age have been extensively exposed in the previous works of RoAMS [44–46]. After all these stages, the calibrated ages are obtained, represented by the probability intervals that the historical age (the age of the material, the mortar in this case) is included in it or, with certain statistical weights, in sub-intervals.

The obtained histograms are represented in Figure 10. The interpretation of the results is done in the next chapter.

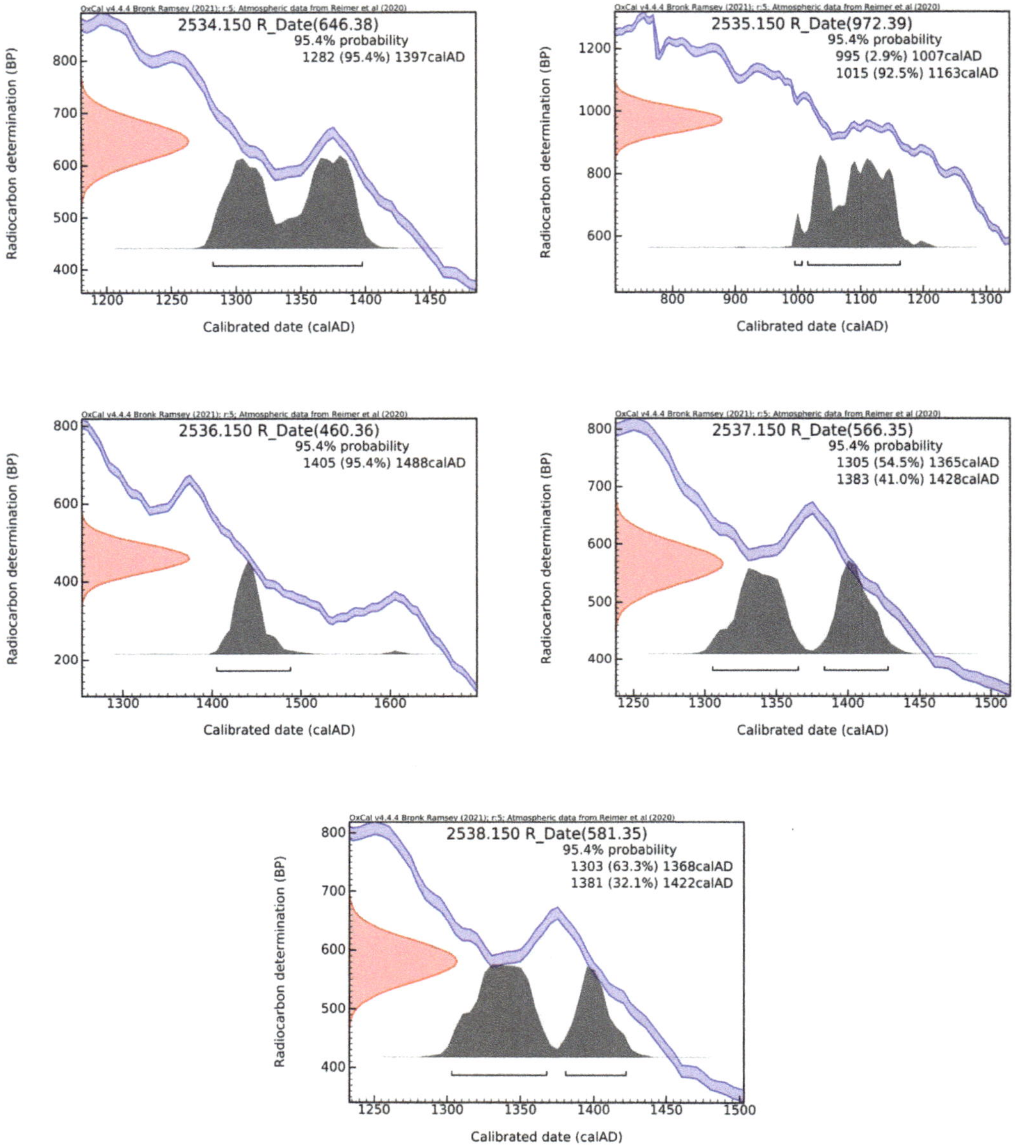

Figure 10. The histograms obtained for the fragments of mortar/lime plasters from the Corbii de Piatră church, separated from the two samples taken, one each for the southern and northern wall, respectively; at the top of each graph, the unique code of the graphitization in RoAMS is indicated (2534.150 for S1, 2535.150 for S2, 2536.150 for N1, 2537.150 for N2, and 2538.150 for N3), as well as the CRA value with the associated measurement uncertainty, both in years BP (according to OxCal version 4.4.4 and [47,48]).

5. Determination of the Chronology

Determining the chronology involves establishing the contact chronology, through which the relative and absolute dating converge towards a consensus.

Preliminary historical-architectural and archaeological information [1–5,9–11]:

15 April, 1456—the oldest documentary attestation through a charter issued by the royal chancellery of the Voivode of Wallachia, Vladislav II, confirming the master Mogoş's right to rule over the "Corbii de Piatră";

Proposed, assimilated or documented founders: Basarab I the Founder (ca. 1310–1352), Neagoe Basarab, Radu de la Afumaţi, Lady Ruxandra and Lady Despina (ca. 1487–1554);

Associations with other religious settlements: cave churches from Cappadocia (7th–12th centuries); the church of St. Procopius from Fasano (11th century); hermitages from Bulgaria (13th–14th centuries); small churches from Macedonia (13th–14th centuries); the dating of the altar (10th–14th centuries);

The cutting of the pilaster from the altar and the formation of an altar table, 1m high, the complete restoration of the west wall, currently made of brick, the porch excavated on the south side of the nave: 19th-century interventions;

It was originally intended for small monastic communities; from the 19th century—the laity church (when the transformations occur at the altar);

Elements of dating through the painting of the church: lacunar, summary, placed in the second half of the 14th century by I. D. Ştefănescu [49]; in 1973, he returns in another work [50], saying that "it could be two or three decades older than the paintings of the church of Sf. Nicolae-Domnesc at Curtea de Argeş originally built at the beginning of the 13th century by Seneslau; developed by Basarb I (1310–1352) and then by Neagoe Basarab (1512–1521); similar to the Corbii de Piatră;

Immediately or shortly after the excavation/anthropization of the church space in the bedrock, it was decorated with painting; a small part was preserved in particularly bad conditions: fragments in the apses of the altar (east), on the east tympanum, on the southern half of the vault of the nave, on the southern wall of the nave; in the northern part, on the vault and walls, the painting is practically destroyed;

Painted areas on the south wall of the nave: on the upper part of the walls and on the vault;

Painted areas on the north wall of the nave: a single painting fragment visible in the niche next to the altar;

The Deisis in the upper part of the apse of the altar, similarities with the cave chapels of Asia Minor, Southern Italy (10th–13th centuries);

The style of the original painter: c. 12th–13th centuries;

There are also elements that belong to a more advanced phase: the second half of the 13th century—the first decades of the 14th century;

With some effort, one could distinguish two or three "hands" in the original pictorial style;

Not many comments can be made on the chromatics: everything is dull, smoky and altered by mold nowadays;

Broadly speaking, from painting and iconography: end of the 13th century or the first decades of the 14th century.

Analysis of historical documents: the existence of an older foundation of the 16th century; for the previous centuries, there is no written information; after 1503–1506, documents appear due to the Magdalina Nun; she worshiped the monastery of Neagoe Basarab in 1512 in order to be completely "a church for nuns"; from a monetary donation made in 1517 for the commemoration, the legend of the refectory table, dated at IFIN-HH in 2015 and which attests that the wood was cut no earlier than the second half of the 16th century, most likely somewhere after 1725–1813; after 1518, he became a satellite church of the Argeș Monastery; in 1809, it became a wedding church for Transylvanian emigrants from over the mountains (the table in the refectory actually originates from this period, thus refuting the legend that it was stipulated to be the table where Neagoe Basarab took counsel); 1814—the excavation of the pronaos with the modification of the south side and the reduction of the trapeze; 1819—the addition of the wall temple; 1887—the restoration of the west wall, after its collapse; 1890—the erection of the wooden belfry;

The conclusions of the 2008 archaeological survey: no archaeological materials were found in surveys older than the 16th century.

Inscriptions were visible in 2023, on the southern wall, near sampling point S (Figure 11); with writing specific to the second half of the 19th century, the year 1881 is mentioned, without any reference in the context of the other historical data, with the year being placed before the restoration of the western wall. However, there is also a writing in Slavonic or possibly with elements in Slavonic alternating with elements in Greek, before the transition to the Latin alphabet (ca. 1860), in a register more erased than that of the inscription 1881. In particular, the inscription of the year 1881 does not seem affected by subsequent works, which can advance the hypothesis that after this year, no notable interventions probably took place in the portion of the wall from where the sample from the southern wall was taken.

Figure 11. Inscription on the southern wall of the Corbii de Piatră church (Photo: Simion, C. A.).

From the previously published information on the state of conservation of the mural painting, on the support of the mural painting, on the damage to the bedrock caused by meteoric infiltration waters in the inner walls of the church and on the damage to the pictorial support [10], the following conclusions can be drawn:

1. Byzantine painting, with lime-based mortar plaster, applied directly to the bedrock wall:

The painting is almost entirely covered with a biological green layer (the north wall entirely, the south wall up to a level of approx. 2 m of pavement);

Pigments more sensitive to humidity lose their bonds with the binder and form powdery layers when touched;

Decohesion: the migration and recrystallization of salts in many places from the inside to the outside, with the formation of efflorescences and crypto-efflorescences;

The advanced degradation is the result of infiltration of meteoric waters, the natural microclimate, human interventions and the loss over time of the properties of pigments and plaster materials through natural aging;

Meteoric water infiltrations on the northern wall are due to the washing of this wall by the waterfall stream, whose flow rate fluctuates depending on the level of precipitation;

The characterization of two fresco samples with different adhesions, from the northern wall and the southern wall, highlighted the different behavior of the two mortars, with higher adhesion for the northern wall.

2. Different types of mortars, identified as belonging to different centuries/periods of intervention:

Mortars A. from the 14th century, with the preservation of the original painting–partially the walls and the vault of the nave and the altar;

Mortars B. from the 19th century, mortars located on the pediment;

Mortars C. from the 19th–20th centuries, mortars considered for repair, on certain portions of the nave, the lower register of the northern wall, near the arch, from the junction of the door with the southern wall;

Mortars A: lime and straw, porous, biological degradation, repeated recrystallization or induced by freeze-thaw; calcite, traces of quartz, rather impure from the limestone source; sometimes gypsum and feldspar (altar and southern wall) as associated or as mortar degradation products; 75% microcrystalline calcite (from the moment of hardening in the masonry), 3% quartz, 4–5% various minerals; from the differential thermal analysis–the samples contain only calcite due to the binder and do not contain limestone added to the recipe; gypsum–due to the limestone and not to the painting; the straw no longer exists; the original calcite led to phenomena of autogenous healing of the lime, filling in time the gaps in the cracks and pores;

The bedrock in the inner walls is inert and has not interacted with the mortar in the fresco/plaster nor with the pigments in the painting layer; it most likely degraded over time, forming a friable layer at the rock-painting support interface that led to its peeling, a sign of the age of the applied layer; water from the sandstone pores: exogenous soluble salts, dominated by sulfates; appearance of plaster;

The study of the infiltration directions through the bedrock/sandstone shows a predisposition that explains why the northern walls are more exposed than the southern ones;

The appeared efflorescences do not contain calcium carbonates;

The organic component of crusts: lichens;

On the northern wall, the biological colonization is intense, being also identified in the mortar; almost exclusive presence of gypsum on the southern wall;

The specific Liesegang carbonation of lime matured, years before use, with the formation of rings rich in calcite succeeded by bands poorer in calcite, precipitated over time, which finally gives different density to carbonated lime.

An interesting hypothesis that deserves to be studied in the future is the accentuation of the state of friability/degradation due to the combined exposure to radon in the presence of oxygen from the air or from the infiltration waters. It is known that in isolated cavities, such as barrow graves, sudden exposure to air/humidity in the presence of high concentrations of radon accumulated over time induces the friability to the pulverulent state of some perfectly preserved archaeological materials until the time of excavations [51,52].

Interpretation of radiocarbon data:

There are differences between the radiocarbon data of the fragments on the southern side compared to the northern one. Those on the southern side offer older radiocarbon data than the northern one. The results are supported by all the previous analyses presented in the summary before but also by those offered by this study. The dates are included in the 11th–15th centuries AD, in accordance with the presented historical-architectural data.

The result for S2 is slightly older (which would support some hypotheses from the preliminary historical-architectural and archaeological information paragraph), and for N1, it is slightly younger (which would support other frameworks from the same paragraph). Although they could represent application sequences of the pictorial layer, the fragments (N1, N2 and N3 and S1 and S2) are too close to each other to have different ages from different historical contexts. Most likely, some contain contaminants with a different level of carbon-14 that affects the radiocarbon result (as explained by some observations independent of radiocarbon dating). Although they do not vary in the same direction, N1 and S2 have in common this modification of the radiocarbon date from the hardening of the mortar in the masonry and therefore form a distinct group in the PCA analysis of the XRF data.

The results for N2 and N3 correlate perfectly; in both cases, the majority of sub-intervals were placed in the 14th century. This result is supported by the PCA analysis of the XRF data, as in the case of S1. For S1, the estimate would be for the end of the 13th century–the beginning of the 14th century.

The 14th century presents a disadvantage in radiocarbon dating, being considered an "imprecise" century [53]; it can be seen from the histograms in Figure 10 how two split signals of relatively equal intensity are obtained. In the particular historical context of the Corbii de Piatră Ensemble, it will be very difficult to decide, for these samples or for others in the future, if the painting segments belong to the first half of the 14th century, therefore associated with Voievod Basarab I (1310–1352), or if there are interventions from a second half of the 14th century. Anyway, the results exclude the interventions of the 16th century (Neagoe Basarab) and highlight the initial forms of the painting.

If we use the R_Date Combination function in OxCal (https://c14.arch.ox.ac.uk/oxcal.html, accessed on 26 January 2024) for N2, N3 and S1, then we will obtain a radiocarbon date with better precision, ±21 years BP, which passes the χ-Test. It gives a clearer situation for interpretation: the paintings on the two walls represent an early pictorial form currently identifiable with the following statistical weight: for $\sigma = 2$, there is a 74.0% probability that the time of execution of the painting is placed in the interval calAD 1305–1365, and for $\sigma = 1$ there is a dominant probability of 56.2% that the interval is restricted to calAD 1321–1358. Most probably, these painting segments belong to the first half of the 14th century and, therefore, are associated with Voievod Basarab I (1310–1352). Anyway, the results exclude the interventions of the 16th century (Neagoe Basarab) and highlight the initial forms of the painting.

Finally, I. D. Ştefănescu correctly defined in 1973 the correspondence with Sf. Nicolae-Domnesc at Curtea de Argeş church, which was originally built at the beginning of the 13th century by Seneslau and developed by Basarab I (1310–1352) and then by Neagoe Basarab (1512–1521).

6. Conclusions

Mortar/plaster samples associated with pictorial layers documented as being from the first forms of painting of the inner walls of the church from the Corbii de Piatră Rupestre Monastery in Romania were analyzed.

During sample collection, it was observed that large sections of the fresco had low adhesion, while small portions were highly adherent to the wall. The specific structural components of the intonaco mortar used include the matrix, hard particles and straw insertions.

The low-adherence, friable mortar sampled from the southern wall is severely degraded, exhibiting large pores and cracks that propagate perpendicularly to the fresco surface due to wetting/drying cycles or developing randomly from the tips of the straw insertions. Areas of advanced degradation contain sulfur carried by rainwater from the slopes, which it washes before infiltrating the church walls. The mineralogical characterization of the mortars is framed within the geological context of the site. From a geological point of view, the site containing Corbii de Piatră Monastery is located in the Getic Depression, a narrow sedimentary basin in the Southern Carpathian foreland with sediments from Cretaceous to Miocene age [14]. The monastery was excavated in massive sandstones of the Oligocene age, which belong to the Corbi Formation [8]. These sandstones are characterized by a polymictic composition that includes quartz, alkali feldspar and plagioclase, muscovite granoclasts and also metamorphic rock fragments such as gneiss, schists and quartzites [9,14–16].

Similar studies have been conducted on fresco paintings in Cappadocia from the 9th–12th centuries [54–58] and on frescoes in Orthodox churches in Bulgaria [42].

The adherent mortar sampled from the northern wall, through which water infiltrates, is compact, with few short cracks and separations of a fatty additive, possibly sheep tallow, which was used in that period to enhance the mortar's properties. In this mortar, rainwater infiltration created a network of large, interconnected pores forming channels that allowed water transfer through the mortar to the middle layer without degrading the masonry–fresco interface or the mortar–pictorial layer interface. This behavior functioned as a settling mortar, which helped the fresco maintain its adhesion to the wall.

To support the radiocarbon dating strategy and the results obtained by this method, a series of physico-chemical investigations were preliminarily carried out on small portions left after the preparation of the two samples for radiocarbon dating (microscopy, elemental analysis, identification of natural compounds or formed as a result of the degradations) with the aim of separating several individual fragments and their separate dating (the consistency of the radiocarbon data originating from the endogenous carbon is monitored).

Techniques were used that cover a wide and complementary palette of analyses, completing the existing studies, and the conclusions are convergent: the fragments represent lime mortars, different in appearance, constitution and behavior.

Regarding the radiocarbon strategy, the results of the preliminary analyses provided the following information: the percentage of calcite in the sample, the content of impurities, the presence/absence of organic materials bearing exogenous carbon (with a carbon-14 level different from that at the time the mortar hardened in the plaster) and the extent to which the last two categories could affect the radiocarbon date.

Thus, there were enough arguments to support direct AMS dating, without pretreatment. The results for the northern wall are different from those for the southern wall. The latter provide older ages but are not old enough to be a clear separation from the results for the northern wall. In both cases, there are even older and newer ages that could be supported by some historical and architectural data, but the PCA analysis of the XRF results shows that the outliers are most likely a sign of carbon-14 contamination levels different from the one at the time of hardening of the mortar on the wall.

From the point of view of historical information, the data up to now support the idea of two close phases of painting: a first phase revealed on the southern wall, which belonged to a painting during the time of Voivode Basarab I the Founder (1310–1352), and a second phase highlighted on the northern wall that was lacunar, brief and framed in the second half of the 14th century by I. D. Ştefănescu.

The fact that the adherent fresco on the northern wall is slightly more recent than the friable one on the southern wall suggests that these are restored areas, following a degradation process, using a mortar with superior properties.

Being extremely degraded, further invasive analyses will no longer be allowed. Therefore, in order to obtain additional information on the areas of the initial painting, a mapping with an XRF/FT-Raman portable instrument [54] can be used, highlighting, compared to previous studies, the traces of the biocide applied to the mural painting 100 years ago. Moreover, with the help of radon determinations, other phenomena can be highlighted that can affect both the painting and the health of the people who come to religious services. This is due to the sandstone from which the bedrock is formed, with the percentages of uranium and thorium from the natural radioactive series being quite significant.

Author Contributions: Agreements, sampling: D.G. and G.G.; OM, SEM-EDS: A.E.V., D.A.N., F.N.F. and M.A.; XRD: S.G.M.; CL: I.M. and M.I.P.; XRF: D.A.M. and C.A.S.; TC/IC/TOC: M.V.I. and D.C.O.; Sample pre-treatment and graphitization: E.A.I. and M.L.M.; AMS measurements, calculation and calibration: M.V.I., C.M. and A.R.P.; Dating strategy, coordination and interpretation: C.A.S. and C.M.; Elaboration/drafting of article text: C.A.S., M.V.I., A.E.V., M.A., G.G. and D.G. All authors have read and agreed to the published version of the manuscript.

Funding: The radiocarbon dating was carried out within the NUCLEU project, project PN 23210201 and PN 23210102, financed by the Ministry of Research, Innovation and Digitization Romania. The experiments were carried out at the Tandetron™ 1MV accelerator from the Horia Hulubei National Institute for R&D in Physics and Nuclear Engineering (IFIN-HH) and were supported by the Romanian Government Program through the National Program for Infrastructure of National Interest (IOSIN funds).

Institutional Review Board Statement: Not applicable.

Informed Consent Statement: Not applicable.

Data Availability Statement: The original contributions presented in this study are included in the article. Further inquiries can be directed to the corresponding authors.

Acknowledgments: Our thanks go to IPS Calinic Argeşeanul, Archbishop of Argeş and Muscel, for the support and the written agreement re-garding the research at the Corbii de Piatră Monastery Church.

Conflicts of Interest: The authors declare no conflict of interest.

References

1. Ilinescu, I.; Gorunescu, I.; Şerban, I.-A. *Monastery of Corbii de Piatră; History Files*; S.C. Zodia Gemenilor S.R.L.: Arges, Ro nia, 2009.
2. Available online: https://viavalahia.ro/en/corbi-arges-county (accessed on 29 September 2023).
3. Available online: https://ro.wikipedia.org/wiki/M%C7%8En%C7%8Estirea_Corbii_de_Piatr%C7%8E (accessed on 29 Sept ber 2023).
4. Available online: https://revistaagressive.ro/cultura/capadocia-de-la-arges (accessed on 29 September 2023).
5. Dumitrescu, C.L. *Corbii de Piatră Rock Church, the Oldest Group of Paintings—Known Today—From Romania*; Art History Studies Research; Romanian Academy Publishing House: Bucharest, Romania, 1975; Volume 22.
6. Vâlcea, A.E.; Grecu, D.; Maris, I.; Negrea, A.D.; Cimpoesu, N.; Giugea, D.; Istrate, B.; Munteanu, C.; Moga, S.G.; Anghel, D et al. The Characterization of a Fragment of a Medieval Fresco from Corbii de Piatră Cave Church. *Appl. Sci.* **2023**, *13*, 4 [CrossRef]
7. Vâlcea, A.E.; Maris, I.; Negrea, A.D.; Cimpoesu, N.; Garbea, G.; Grecu, D.; Moga, S.G.; Istrate, B.; Finta, F.N.; Rizea, A.D.; e Interdisciplinary Research on Medieval Fresco Subjected to Degradation Processes in the Corbii de Piatra Cave Church. *Mate* **2023**, *16*, 5257. [CrossRef]
8. Barzoi, S.C.; Luca, A.C. Significance of studying the petrography and mineralogy of the geological environment of old rupest churches to prevent their deterioration.A case study from the South Carpathians. *J. Cult. Herit.* **2013**, *14*, 163–168. [CrossRef

Corbii de Piatră. *Interdisciplinary Study*; Section 2 "Geological characterization of the site"; Şeclăman, M., Bărzoi, S.C., Luca, A., Roban, R., Eds.; National University of Arts Printing: Bucharest, Romania, 2010.
Corbii de Piatră. *Interdisciplinary Study*; Section 4 "Technological study of the constituent materials of the mural painting": Mohanu, I., Chatzitheodoridis, E., Barbu, O.-H., Niculescu, G., Georgescu, M., Şeclăman, M., Bărzoi, S.C., Luca, A., Roban, R.; Section 5 "Technological milestones regarding the execution of the mural paintings from Corbii de Piatră": Mohanu, D.; Section 6 "Causes and specific forms of degradation": Mohanu, D., Şeclăman, M., Bărzoi, S.C., Luca, A., Roban, R.; Section 7 "Biodeterioration and biodeteriogens": Gomoiu, I., Chatzitheodoridis, E.; National University of Arts Printing: Bucharest, Romania, 2010.
Mohanu, D.; Cincheza-Buculei, E.; Seclaman, M.; Luca, A.; Barzoi, S.C.; Roban, R.-D.; Onete, M.; Gomoiu, I.; Chatzitheodoridis, E.; Niculescu, G.; et al. Interdisciplinary Study. Available online: https://www.researchgate.net/publication/280919986_Corbii_de_Piatra_Studiu_interdisciplinary_Corbii_de_Piatra_Interdisciplinary_study (accessed on 26 January 2024).
Marincas, O. Contribution of the George Enescu Iasi University of Arts to the Interdisciplinary Research Project of Historical Monuments in Romania Undergoing Restoration. Available online: http://www.revistamonumenteloristorice.ro/fisiere/RMI_2008_1-2-010.pdf (accessed on 13 October 2021).
National University of Arts. *Integrated Research Strategy of the State of Conservation of Some Cave Churches in Order to Restore and Enhance Them—Case Study*, 3rd ed.; Corbii de Piatră: Bucharest, Romania, 2008; pp. 8–11.
Ghiran, M.D.; Popa, M.E.; Mariș, I.; Predeanu, G.; Gheorghe, S.; Bălănescu, N.M. Thermal Maturity and Kerogen Type of Badenian Dispersed Organic Matter from the Getic Depression, Romania. *Minerals* **2023**, *13*, 202. [CrossRef]
Săndulescu, M. *Geotectonics of Romania*; Technical Publishing House: Bucharest, Romania, 1984; p. 335. (In Romanian)
Murgeanu, G.; Mihăilă, N.; Stancu, J.; Dessila-Codarcea, M.; Bombiță, G.; Lupu, M. *Harta geologică a României, 1:200 000, Foaia Pitești*; Comitetul de Stat al Geologiei, Institutul Geologic: Bucharest, Romania, 1967. (In Romanian)
El-Taher, A. Assessment of natural radioactivity levels and radiation hazards for building materials used in Qassim area, Saudi Arabia. *Rom. J. Phys.* **2012**, *57*, 726–735.
Özen, S.A.; Celik, N.; Dursun, E.; Taskin, H. Indoor and outdoor radon measurements at lung cancer patients' homes in the dwellings of Rize Province in Turkey. *Environ. Geochem. Health* **2018**, *40*, 1111–1125. [CrossRef]
Calin, M.R.; Ion, A.C.; Radulescu, I.; Simion, C.A.; Mincu, M.M.; Ion, I. Analysis of the radon concentrations in natural mineral and tap water using lucas cells technique. *J. Environ. Eng. Landsc. Manag.* **2022**, *30*, 370–379. [CrossRef]
Muntean, L.E.; Manea, D.L.; Cosma, C. The radioactivity levels in various plastering mortars made with river aggregates from the Alba County. *Rom. J. Mater.* **2012**, *42*, 30–36.
McMillen, M. Radon: How It Can Affect Your Health. WebMD, 7 August 2024. Available online: https://www.webmd.com/lung-cancer/radon-health-effects (accessed on 19 December 2024).
Simion, C.A.; Stanciu, I.M.; Sava, T.B.; Pacesila, D.G. Radiocarbon dating of a sample of hydraulic mortar from Porolissum customs, Revista Română de Materiale. *Rom. J. Mater.* **2022**, *52*, 349–356.
Marcu Istrate, D.; Stanciu, I.M.; Simion, C.A.; Sava, T.B.; Pacesila, D.G.; Ionita, A.; Dobrota, S.O. Feldioara/Marienburg Fortress, in Transylvania, Romania. Traditional Dating and Radiocarbon Dating of Mortar: An archaeometric perspective. *Archaeometry*, 2025; *manuscript in preparation*.
Simion, C.; Stanciu, I.; Gaza, O.; Sava, T.; Pacesila, D.; Ilie, M.; Manailescu, C.; Robu, A. Radiocarbon dating of mortar based on hydraulic lime. Advantages, disadvantages, limitations. *AIP Conf. Proc.* **2019**, *2076*, 050003. [CrossRef]
Romania Laws—UNESCO World Heritage Convention. 2019. Available online: https://fr.unesco.org (accessed on 15 September 2021).
Romania 3.1—Compendium of Cultural Policies & Trends. Available online: https://www.culturalpolicies.net (accessed on 15 September 2021).
Calandra, S.; Barone, S.; Cantisani, E.; Caggia, M.P.; Liccioli, L.; Vettori, S.; Fedi, M. Radiocarbon dating of straw fragments in the plasters of St. Philip church in archaeological site Hierapolis of Phrygia (Denizli, Turkey). *Radiocarbon* **2023**, *65*, 323–334. [CrossRef]
LAMOA Expertise. Available online: https://www.lamoa.fr (accessed on 20 October 2022).
Characterization and Microanalysis of Ancient and Modern Materials by Physico-Chemical Methods. Available online: https://www.archeophile.com/rwcat_129-laboratoires.htm (accessed on 20 October 2022).
Hagni, R.D.; Karakus, M. Cathodoluminescence Microscopy: A Valuable Technique for Studying Ceramic Materials. *MRS Bull.* **1989**, *14*, 54–59. [CrossRef]
Marshall, D. *Cathodoluminiscence of Geological Materials*, 1st ed.; Springer: London, UK, 1988.
Pokrovsky, O.S.; Golubev, S.V.; Schott, J. Calcite, dolomite and magnesite dissolution kinetics in aqueous solutions at acid to neutral pH, 25 to 150 °C and 1 to 55 atm pCO_2: New constraints on CO_2 sequestration in sedimentary basins. *Chem. Geol.* **2009**, *265*, 20–32. [CrossRef]

33. Coacă, E.; Rusu, O.A.; Marin, A.H.; Velciu, L.; Mihalache, M.; Andrei, V.A.; Vişan, T. Microscopic and Electrochemical Charac zation of Alumina Ceramic Films Developed onto 316l Stainless Steel by Microarc Oxidation in Plasma Electrolysis. *U.P.B. Bull. Ser. B* **2018**, *80*, 17–32.
34. Martinetto, P.; Rousselière, H.; Walter, P. Identifier les pigments et comprendre leurs properties à partir de la diffraction rayons X. *L'Actualité Chimique juil-août-sept-oct* **2014**, *387–389*, 170–174.
35. LAMOA: Étude des Peintures Murales Avant la Restauration. Available online: https://www.lamoa.fr/services_analyse_ peintures_murales.html (accessed on 15 October 2022).
36. Dupouy, J.-M. Les rayons X et l'étude des œuvres d'art. *J. Phys. IV* **1996**, *6*, C4-791–C4-808. [CrossRef]
37. Sardos, N.; Velarde, N.; Del, S.; Chapoulie, R.; Castillo Butters, L.J. Caractérisation physico-chimique des pigments des peint murales mochica: San José de Moro (VIII ème—X ème siècles apr. J.-C.). *J. Soc. Am.* **2017**, *103*, 217–235. [CrossRef]
38. Wacker, L.; Němec, M.; Bourquin, J. A revolutionary graphitization system: Fully automated, compact and simple. *Nucl. Inst Methods Phys. Res. B* **2010**, *268*, 931–934. [CrossRef]
39. Wacker, L.; Christl, M.; Synal, H.-A. BATS: A new tool for AMS data reduction. *Nucl. Instrum. Methods Phys. Res. B* **2010**, 976–979. [CrossRef]
40. Stuiver, M.; Polach, H. Discussion: Reporting of 14C data. *Radiocarbon* **1977**, *19*, 355–363. [CrossRef]
41. *Vitruvius, The Ten Books on Architecture*; Translated by Morgan, M.H., with Illustrations and Original Designs Prepared Under Direction of Warren, H.L., and Robinson, N., Jr.; Harvard University Press London: London, UK; Humphrey Milford Oxf University Press: Cambridge, UK, 1914. Available online: https://en.wikisource.org/wiki/Ten_Books_on_Architecture (acces on 16 March 2021).
42. Stamboliyska, B.; Tapanov, S.; Kovacheva, D.; Atanasova-Vladimirova, S.; Ranguelov, B.; Yancheva, D.; Velcheva, E.; Stoyano Guncheva, M.; Fischer, D.; et al. Characterization of art materials and degradation processes in the exterior wall paintings of main church of Rila Monastery, Bulgaria. *Vib. Spectrosc.* **2023**, *128*, 103580. [CrossRef]
43. Pelosi, C.; Agresti, G.; Andaloro, M.; Baraldi, P.; Pogliani, P.; Santamaria, U. The Rock Hewn Wall Paintings in Cappad (Turkey). Characterization of the Constituent Materials and a Chronological Overview. U.S. Pharmacopeia. Available onl https://www.researchgate.net/publication/257365941 (accessed on 26 January 2025).
44. Ghita, D.G.; Moşu, D.V.; Dogaru, M.; Gugiu, M.M.; Popescu, I.V.; Dobrescu, S.; Călinescu, C.I.; Căta-Danil, G.; Enăchescu, Mărginean, N.; et al. Improvements of the research infrastructure at the tandem laboratory in IFIN-HH. *AIP Conf. Proc.* **2** *1525*, 208. [CrossRef]
45. Stan-Sion, C.; Enachescu, M.; Petre, A.R.; Simion, C.A.; Calinescu, C.I.; Ghita, D.G. A new and compact system at the A laboratory in Bucharest. *Nucl. Instrum. Methods Phys. Res. Sect. B—Beam Interact. Mater. At.* **2015**, *361*, 105–109. [CrossRef]
46. Sava, T.B.; Simion, C.A.; Gâza, O.; Stanciu, I.M.; Pacesilă, D.G.; Sava, G.O.; Wacker, L.; Ștefan, B.; Moșu, V.D.; Ghita, D.G.; e Status Report on the Sample Preparation Laboratory for Radiocarbon Dating at the New Bucharest Roams Center. *Radioca* **2019**, *61*, 649–658. [CrossRef]
47. Ramsey, C.B. Radiocarbon calibration and analysis of stratigraphy: The OxCal program. *Radiocarbon* **1995**, *37*, 425–430. [Cross
48. Reimer, P.J.; Austin, W.E.N.; Bard, E.; Bayliss, A.; Blackwel, P.G.; Bronk Ramsey, C.; Butzin, M.; Cheng, H.; Edwards, I Friedrich, M.; et al. The IntCal20 Northern Hemisphere Radiocarbon Age Calibration Curve (0–55 cal kBP). *Radiocarbon* **2020** 725–757. [CrossRef]
49. Ştefănescu, I.D. *La peinture réligieuse en Valachie et Transylvanie (Addenda)*; P. Geuthner: Paris, France, 1932; pp. 421–424.
50. Ştefănescu, I.D. *Iconography of Byzantine Art and Romanian Feudal Painting*; Editura Meridiane: Bucharest, Romania, 1973; p.
51. Thierry, N. Destruction of the sites and mural paintings of Cappadocia: Natural and human causes. In Proceedings of European Symposium: Science, Technology and European Cultural Heritage, Bologna, Italy, 13–16 June 1989; pp. 505–510.
52. Balcázar, M.; Gómez, S.; Peña, P.; Zavala Arredondo, J.; Gazzola, J.; Villamares, A. Presence of a radioactive gas in archaeolog excavations, determination and mitigation. *Appl. Radiat. Isot.* **2014**, *83*, 272–275. [CrossRef] [PubMed]
53. Quirós Castillo, J.A. The radiocarbon dates of historical period sites. Problems and experiences of analysis in medieval cont Munibe. *Anthropol. Archaeol.* **2009**, *60*, 313–324.
54. Sbroscia, M.; Pelosi, C.; Agresti, G.; Paolucci, A.; Pogliani, P.; Ruggiero, L.; Sodo, A. Spectroscopic investigation of Cappad proto-Byzantine paintings. *J. Raman Spectrosc.* **2021**, *52*, 95–108. [CrossRef]
55. Philippova, O.S.; Dmitriev, A.Y.; Tsarevskaya, T.J.; Makarova, A.M.; Grebenshchikova, A.B. Medieval mural painting: A l through the centuries. *J. Cult. Herit.* **2023**, *62*, 460–469. [CrossRef]
56. Shevchenko, D.; Sadovskaya, Y.; Shkondin, V.; Zhurbin, Y.; Ostrovsky, A. Rock-Cut Church Göreme 31 (Cappadocia, Turk New Data from the Study in Situ (2014–2015). *Scrinium* **2021**, *17*, 274–290. [CrossRef]

Available online: https://uchisarcappadocia.org/red-church/ (accessed on 3 February 2025).
Krajcar Bronić, I.; Topić, N.; Radić, I.; Peković, Ž.; Sironić, A. Radiocarbon dating of St. Stephen's in Pustijerna church in Dubrovnik, Croatia. In Proceedings of the Balkan Archaeometry Symposium, Bucharest, Romania, 29–30 September 2012. [CrossRef]

claimer/Publisher's Note: The statements, opinions and data contained in all publications are solely those of the individual nor(s) and contributor(s) and not of MDPI and/or the editor(s). MDPI and/or the editor(s) disclaim responsibility for any injury to ple or property resulting from any ideas, methods, instructions or products referred to in the content.

materials

MDPI

Article

Analysis of the Material and Coating of the Nameplate of Vila D. Bosco in Macau

Liang Zheng, Jianyi Zheng, Xiyue He and Yile Chen *

Heritage Conservation Laboratory, Faculty of Humanities and Arts, Macau University of Science and Technology, Avenida Wai Long, Taipa, Macau 999078, China; zhengliang@must.edu.mo (L.Z.); jyzheng@must.edu.mo (J.Z.); hexiyue@must.edu.mo (X.H.)
* Correspondence: chenyile@must.edu.mo

Abstract: This study focuses on the nameplate of Vila D. Bosco, a modern building in Macau from the time of Portuguese rule, and looks at the types of metal materials and surface coatings used, as well as how they corrode due to the tropical marine climate affecting the building's metal parts. The study uses different techniques, such as X-ray fluorescence spectroscopy (XRF), scanning electron microscopy/energy dispersive spectroscopy (SEM-EDS), X-ray diffraction (XRD), attenuated total internal reflectance Fourier transform infrared spectroscopy (ATR-FTIR), and cross-sectional microscopic analysis, to carefully look at the metal, corrosion products, and coating of the nameplate. The results show that (1) the nameplate matrix is a resulfurized steel with a high sulfur content (Fe up to 97.3% and S up to 1.98%), and the sulfur element is evenly distributed inside, which is one of the internal factors that induce corrosion. (2) Rust is composed of polycrystalline iron oxides such as goethite (α-FeOOH), hematite (α-Fe_2O_3), and magnetite (Fe_3O_4) and has typical characteristics of atmospheric oxidation. (3) The white and yellow-green coatings on the nameplate are oil-modified alkyd resin paints, and the color pigments are TiO_2, $PbCrO_4$, etc. The surface layer of the letters is protected by a polyvinyl alcohol layer. The paint application process leads to differences in the thickness of the paint in different regions, which directly affects the anti-rust performance. The study reveals the deterioration mechanism of resulfurized steel components in a subtropical polluted environment and puts forward repair suggestions that consider both material compatibility and reversibility, providing a reference for the protection practice of modern and contemporary architectural metal heritage in Macau and even in similar geographical environments.

Keywords: metal corrosion; nameplate coating; building materials; modern and contemporary architectural heritage; Portuguese architecture; Macau

1. Introduction

As an important coastal city in China, Macau has a unique historical and cultural background and architectural heritage [1]. Modern and contemporary buildings, especially those during the Portuguese rule, widely use metal components such as iron windows, iron doors, railings, and metal signs [2]. However, due to Macau's special climate, these metal components are prone to corrosion and degradation under long-term exposure to humidity, heat, high salinity, and air pollution [3]. Therefore, studying the corrosion characteristics and influencing factors of metal components in Macau's modern and contemporary buildings is not only crucial for the protection of architectural heritage but also has reference value for the durability of metal materials in subtropical coastal environments.

This study selected Vila D. Bosco as the research object. This building is one of the important representatives of the Portuguese rule in Macau. Numerous environmental factors severely corrode its metal components. By analyzing its location, we can deeply explore the potential impact of the external environment on the corrosion of building metal components.

Research on modern and contemporary metal products has had positive impacts on the development of modern industry in many aspects, such as improving material performance [3,4], innovating manufacturing processes [5,6], optimizing product design [7,8], and sustainable industrial development [9]. For example, through the research on modern and contemporary metal products, researchers have continuously explored the relationship between the microstructure and performance of metal materials and developed a series of high-performance metal materials [10]. Some scholars have found that adding specific alloying elements can significantly improve the strength and corrosion resistance of metals, which has led to the widespread use of high-performance alloys, such as stainless steel and aluminum alloys, in aerospace, automobile manufacturing, and other fields [11]. High-strength aluminum alloys used in aerospace, with their lightweight and high-strength characteristics, help aircraft reduce weight and improve fuel efficiency and flight performance. The industry widely uses modern metal products.

In the field of cultural heritage, recent research on modern and contemporary metal products covers many aspects, such as composition analysis, manufacturing process, corrosion, and protection [12–17]. However, each cultural heritage has its own uniqueness, and the commonalities of many studies may not be similar. Researchers I. Crina Anca Sandu and others studied old "gilded" artworks from the 9th to 19th centuries in European culture using imaging and spectral analysis methods. They looked into how these artworks are preserved, how they change and age over time, and how restoration treatments like cleaning, strengthening, changing, and glazing affect the condition of the original materials [18]. Other scholars have studied the industrial buildings of steel mills that operated in Slovakia from 1800 to 1948, analyzing their architectural and urban characteristics, evaluating the value of metal products in industrial architectural heritage, and formulating protection strategies based on their characteristics [19]. Researchers Matthew Carl and others found that the makeup, thickness, feel, and state of the coating and the main material work well together by studying the coating and base metal of the American silver-plated cultural heritage collection at the Dallas Museum of Art (DMA) [20]. Iranian scholars have studied 14 Achaemenid bronze artifacts in the Persepolis World Heritage Site (550–330 BC) using methods, such as inductively coupled plasma emission spectroscopy, scanning electron microscopy–energy dispersive spectroscopy, and metallographic analysis, and have identified their manufacturing processes, such as casting, coldworking, and annealing. Similar methods can be used to study modern and contemporary metal products, such as iron components and metal decorative parts in modern and contemporary architectural heritage, to explore the inheritance and development of their production processes, as well as the technological level at the time [21]. Based on existing research, we can use modern analytical technology to obtain detailed information on the composition and microstructure of metal products [22]. For example, through a non-destructive portable X-ray fluorescence spectroscopy (XRF), scanning electron microscope equipped with energy dispersive spectroscopy (SEM/EDS) and other technologies, the alloy composition of ancient metal products can be analyzed, which is also of reference significance for understanding the composition of modern metal products. It can help identify the source of modern metal products' materials, the proportion of materials used in production, etc., to better protect and study the heritage metal products.

Existing research mostly focuses on the corrosion of industrial facilities and smaller metal objects in the museum collection, and there is little systematic research on metal components of architectural heritage [23]. Vila D. Bosco in Macau is a typical modern building from the Portuguese rule in the 1960s, reflecting the architectural style and cultural background of Portugal in Macau at that time. The villa nameplate not only has decorative and identification functions but also carries important historical and cultural values. However, the current nameplate has serious metal corrosion, the surface coating has fallen off widely, and the metal substrate is seriously corroded. X-ray fluorescence spectroscopy (XRF) analysis did not find any chlorine; therefore, we initially ruled out the typical direct damage from seawater or sea salt. This finding raises a series of research questions:

(1) What is the specific composition of the nameplate metal material and its process characteristics?
(2) In the absence of chlorine, what environmental factors, pollutants, or other chemical mechanisms may have caused such severe corrosion damage?
(3) Do the metal materials and surface coatings used in modern buildings in Macau during the Portuguese rule have specific era characteristics? How do these characteristics interact with local climate, history, and cultural factors to affect the durability of the materials?

This study aims to better understand how metal materials behave and corrode when protecting Macau's historical buildings, and to offer expert advice for future restoration and protection efforts.

2. Materials and Methods

2.1. Research Subjects: Vila D. Bosco

Vila D. Bosco (鮑思高別墅) is located along the Estrada de Choc Van (竹灣馬路) on Coloane Island in the Macau Special Administrative Region, facing the South China Sea. The building is situated in a region characterized by a typical subtropical marine climate, with high relative humidity (averaging over 80% annually), frequent rainfall, and strong sea breezes, especially during the summer typhoon season [24]. These climatic conditions are known to significantly influence the weather and corrosion of metal materials.

The villa is positioned directly adjacent to the South China Sea and is continuously exposed to the marine atmosphere (Figure 1). The coastal environment, with its high wind speeds and air mobility, can facilitate the transport of atmospheric pollutants such as sulfur dioxide (SO_2) and nitrogen oxides (NO_x) [25]. In humid conditions, these pollutants may form acidic substances that contribute to the corrosion of exposed metal surfaces [26]. Although prior studies have indicated that sea salt aerosols might not be the primary driver of corrosion, the presence of salt can still exacerbate surface deterioration in marine environments [27–29]. Surrounding Vila D. Bosco are extensive nature reserves with dense vegetation cover, which may influence local microclimate factors such as humidity levels [30].

Topographic analysis shows that Vila D. Bosco is located on a southeast-facing hillside on Coloane Island (Figure 2). As seen in the contour map (Figure 2(1)) and slope analysis (Figure 2(2)), the building sits on terrain with moderate to high slopes (15° to 40°), suggesting a varied microtopography. The main axis of the building follows the contour lines, aligned approximately north–south. Vegetation surrounding the site (Figure 2(4)) provides partial shielding from direct marine aerosol deposition, although it may also affect local humidity dynamics.

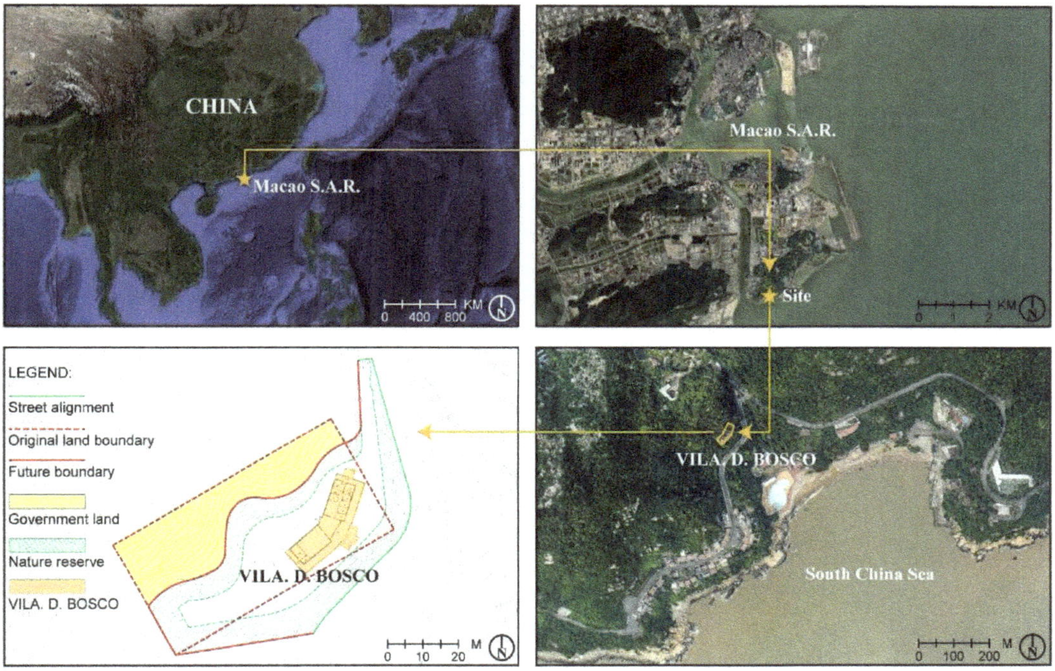

Figure 1. Location and land use conditions of Vila D. Bosco.

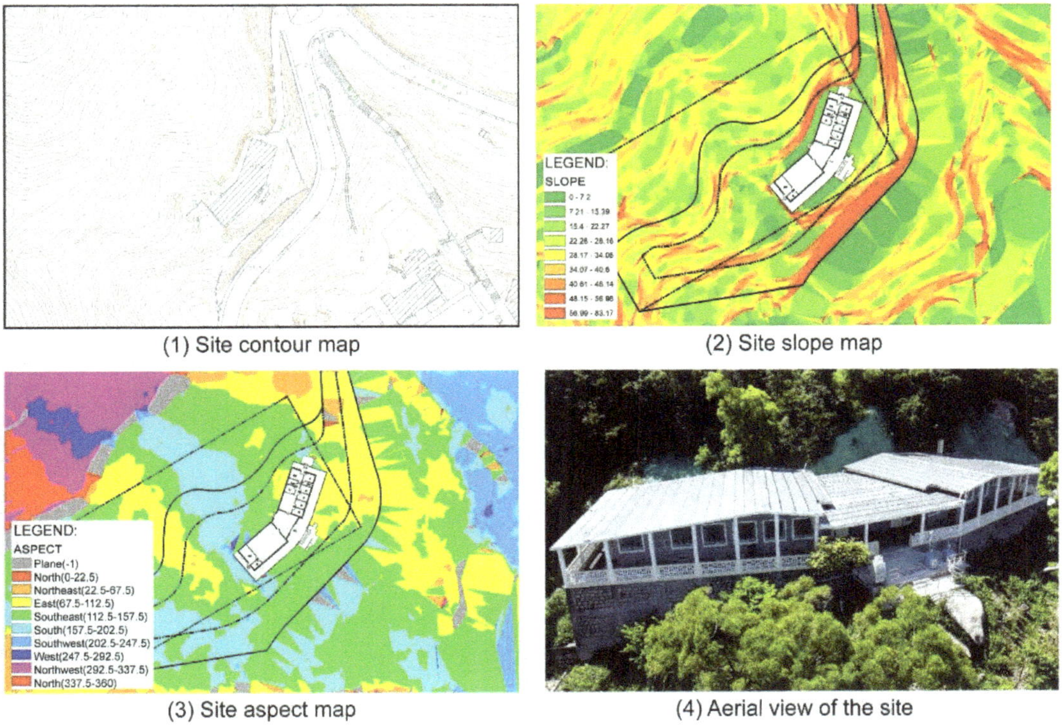

Figure 2. Vila D. Bosco contour lines, aspect, slope, and bird's eye view.

Originally built in the 1960s by Salesian priests of Don Bosco College, Vila D. Bosco served as a holiday camp facility for young people during a time when beach vacations in Taipa and Coloane were popular among affluent Macanese residents [31]. After decades of exposure to the elements, the villa's structure and appearance have deteriorated [32]. Following a restoration project, the site is now managed by the Ministry of Water Resources' Digital Twin Watershed Key Laboratory. The nameplate of Vila D. Bosco is installed under the eaves at the building's main entrance (Figure 3), exposed to a semi-open outdoor environment. Field observation revealed extensive rusting, significant paint peeling, and localized pitting on the nameplate surface, indicating advanced stages of material degradation. Similar signs of deterioration, including peeling paint, concrete spalling, and biological staining, were observed on surrounding structural elements, reflecting prolonged exposure to harsh environmental conditions.

Figure 3. Vila D. Bosco nameplate corrosion status.

The nameplate itself is a rectangular metal plate measuring approximately 173 cm in length and 24 cm in width (Figure 4). Its design features marquee-style three-dimensional letters reading "VILA D. BOSCO", each incorporating sockets for light bulbs. This stylistic approach, popular in early 20th-century signage in Europe and the United States, remains visible in many historical shopfronts within the Macau Peninsula (Figure 5). On the nameplate, the front surface exhibits extensive paint loss, visible rust patches ranging from yellow-brown to dark brown, and isolated greenish stains likely related to the degradation of protective coatings. The back of the nameplate shows more uniform and extensive corrosion, with dark brown rust layers and signs of coating overflow residues. These observations suggest long-term exposure to humid, saline, and pollutant-laden atmospheric conditions.

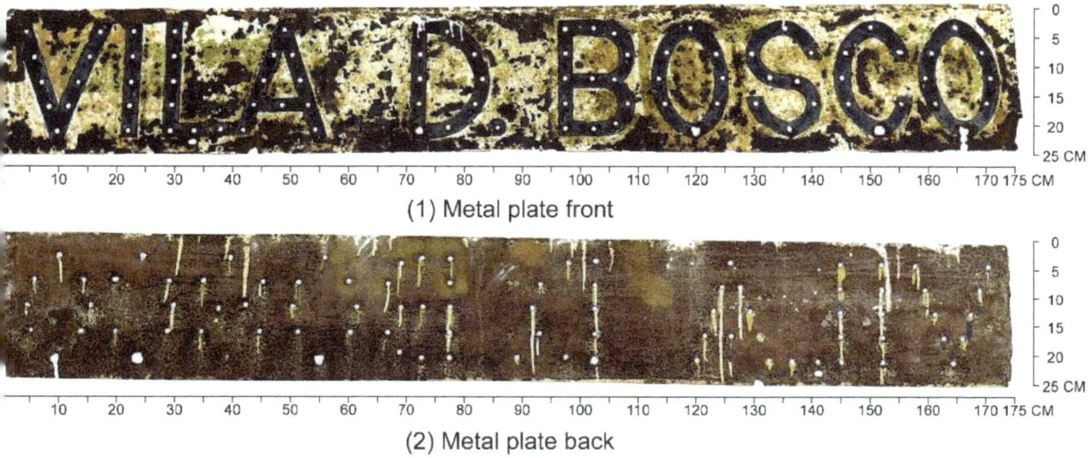

Figure 4. Vila D. Bosco nameplate.

Figure 5. Marquee letters used on restaurant signs on the streets of Macau Peninsula. The Chinese characters in the picture are the names of the signs on the street shops and have no specific meaning.

2.2. Analysis Methods and Processes

The Vila D. Bosco nameplate samples analyzed in this study include on-site non-destructive testing of the entire nameplate and laboratory analysis of selective sampling from the nameplate surface. The specific research methods and steps are as follows:

2.2.1. SEM-EDS Micromorphology and Elemental Analysis

The researchers used a scanning electron microscope (SEM, FEI Quanta 250) combined with energy dispersive X-ray spectroscopy (EDS) to observe the micromorphology and analyze the elemental composition of the small samples. During the experiment, the sample was attached to the stage with conductive glue and covered with carbon to help improve the clarity and quality of the images taken by the electron microscope. The SEM operating voltage was set to 20 kV, the beam current was about 0.5 nA, and the working distance was about 10 mm. The EDS analysis time was 60 s per point, and the data obtained was used to determine the element distribution characteristics of the corrosion area and the possible corrosion mechanism [33–35].

2.2.2. X-Ray Fluorescence Spectroscopy (XRF) Analysis

A portable X-ray fluorescence spectrometer (Bruker Tracer 5i, Karlsruhe, Germany) was used for on-site testing to directly analyze the elemental composition of the nameplate metal substrate and coating in a non-destructive manner [36,37]. The measurement time for each test point was 120 s; the X-ray tube voltage was set to 40 kV, and the current was 35 μA to obtain qualitative and semi-quantitative information about the elements. Small-sized samples (about 5 mm × 5 mm) were selected from areas with different degrees of corrosion on the nameplate for subsequent laboratory testing and analysis. The collected small samples were washed with anhydrous ethanol and dried in a vacuum drying oven at 60 °C for 24 h for use to avoid interference from external contaminants.

2.2.3. XRD Corrosion Product Phase Analysis

X-ray diffraction (XRD) analysis was performed using the Olympus BTX™ III Benchtop XRD Analyzer to identify the crystalline phases present in the corrosion products on the nameplate [38,39]. The instrument uses a Co-Kα ray source (λ = 1.78897 Å), with the tube voltage set at 30 kV and the tube current at 0.33 mA. The scanning was carried out over a 2θ range from 5° to 55°, with a step size of 0.25°. The researchers used Xpowder™ software (http://www.xpowder.com, accessed on 16 March 2025) to compare the obtained XRD spectrum with the Powder Diffraction File (PDF) database, determining the crystalline phases of the corrosion products [40,41]. Since most reference spectra and mineral standards (e.g., those from the RRUFF project) are based on Cu-Kα radiation (λ = 1.5406 Å), the collected Co-Kα diffraction patterns were converted to their Cu-Kα equivalents using Python (version 3.12) to facilitate accurate phase matching. This wavelength transformation ensures consistency when comparing experimental results to the Powder Diffraction File (PDF) and the RRUFF database. Peak positions were cross-checked with standard patterns from the RRUFF project, and corresponding reference spectrum IDs were noted.

2.2.4. Fourier Transform Infrared Spectroscopy (FTIR) Analysis

The organic components of the coating on the surface of the nameplate were analyzed using attenuated total internal reflectance Fourier transform infrared spectroscopy (ATR-FTIR, Perkin Elmer Spectrum III, PerkinElmer Inc, Waltham, MA, USA) [42]. When sampling, a small amount of the sample (about 1 mg) was taken from the coating part and measured non-destructively using ATR. The scanning range was set to 4000–650 cm^{-1}, the resolution was 4 cm^{-1}, and the number of scans was 50. The organic components in the coating, such as resins and pigments, were identified by comparing the FTIR spectrum's characteristic peaks with the standard spectral library.

2.2.5. Cross-Section Microscopic Analysis

Four samples of about 2 mm in length and 1 mm in width (from the surface of the paint to the base of the nameplate) were removed from the white and black paint on the nameplate. These samples were placed in a mold, and Buehler™ epoxy resin was dripped into the mold. The samples were then left to stand and polished. Finally, the sample that has been polished to expose the cross-section is placed under a Motic M230T (Motic Asia, Kowloon, Hong Kong) optical microscope to observe the sample cross-section at a magnification of 50×.

Using the detailed experimental methods mentioned above, it can effectively determine the elemental makeup, microstructure, and crystal structure of rust products, as well as the organic content of the nameplate metal material, allowing for a thorough analysis of the underlying material science reasons for the nameplate corrosion issue.

3. Results

3.1. SEM-EDS Microstructure and Elemental Composition Analysis

The microscopic morphology of different areas of the Vila D. Bosco nameplate was observed using scanning electron microscopy (SEM) combined with energy dispersive spectroscopy (EDS). As shown in Figure 6, multiple typical corrosion features were identified, such as rust layers, micro-cracks, and layered flake structures. To ensure the reliability and representativeness of the analysis, three independent EDS measurements were conducted for each selected region (A–D), and the results were averaged and summarized in Table 1.

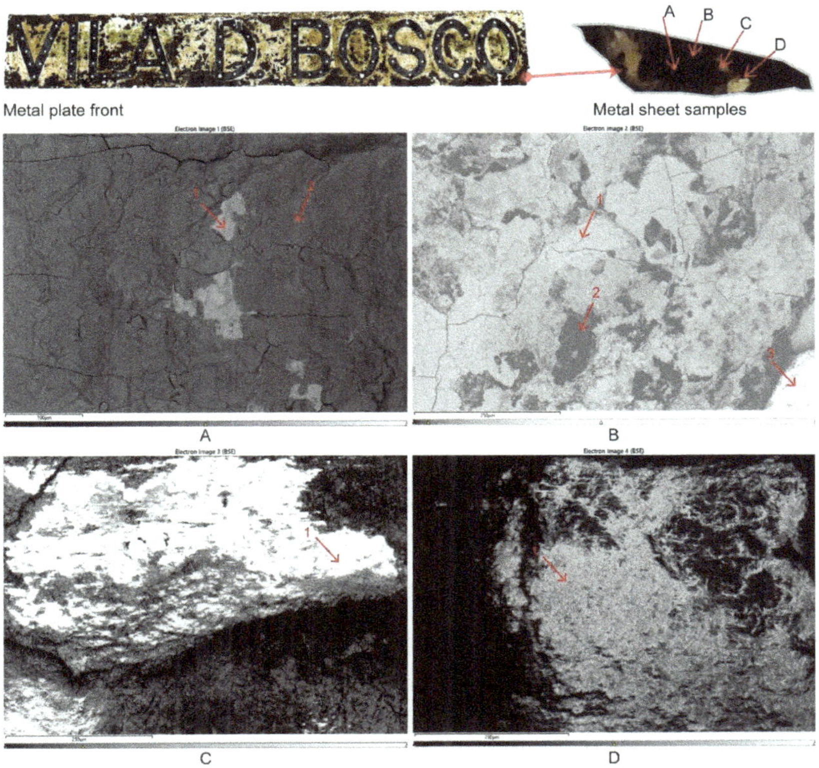

Figure 6. SEM micrographs of different regions (**A**–**D**) of the Vila D. Bosco nameplate and sampling points for EDS analysis.

Table 1. Elemental composition results of the Vila D. Bosco nameplate surface based on EDS point analysis (average of three points per sample).

Element	Sample A [%]			Sample B [%]			Sample C [%]			Sample D [%]		
	1	2	3	1	2	3	1	2	3	1	2	3
O	3.08	19.85	5.66	39.91	50.66	54.53	38.09	32.18	31.79	38.36	38.79	36.4
Mg	0.00	0.00	0.00	0.00	0.82	3.06	0.95	1.59	0.98	0.00	0.00	0.0
Al	0.13	0.00	0.18	1.81	0.30	5.84	4.19	2.27	2.74	2.42	4.11	4.4
Si	0.00	0.00	0.00	10.87	0.36	3.95	0.00	2.15	3.51	2.51	4.88	5.4
S	0.00	0.58	0.00	9.29	0.32	1.18	11.96	5.00	3.34	12.78	11.03	7.2
K	0.00	0.00	0.00	4.13	0.00	0.34	0.72	0.00	1.22	4.54	4.30	2.5
Ca	0.00	0.00	0.00	0.27	0.38	2.43	16.94	5.62	21.56	0.00	0.62	1.6
Ti	0.00	0.00	0.00	3.02	0.16	0.00	4.59	3.39	1.37	3.99	3.13	4.7
Cr	0.00	0.00	0.00	0.00	0.00	0.00	1.26	4.62	0.85	0.00	0.00	0.0
Fe	95.58	78.64	92.86	26.89	46.42	28.67	0.00	14.92	18.77	26.92	27.50	31.4
Co	1.21	0.93	1.30	0.00	0.58	0.00	0.00	0.00	0.68	0.00	0.00	0.0
Ni	0.00	0.00	0.00	0.00	0.00	0.00	1.80	0.78	0.00	0.77	0.00	0.7
Zn	0.00	0.00	0.00	0.00	0.00	0.00	19.50	15.01	6.12	0.00	0.00	0.0
Zr	0.00	0.00	0.00	3.81	0.00	0.00	0.00	12.47	7.07	7.71	5.64	5.2

In sample A (Figure 6A), the surface shows relatively limited corrosion. EDS point 1 indicates a high iron (Fe) content (95.58%) and a low oxygen (O) content (3.08%), suggesting the presence of the exposed metallic substrate. Points 2 and 3 reveal slight increases in oxygen (19.85% and 5.66%, respectively), accompanied by minor amounts of sulfur (0.58% at point 2) and cobalt, reflecting the initial stages of oxide formation and environmental deposition.

In sample B (Figure 6B), the corrosion features are more prominent. EDS results show that oxygen content varies significantly from 39.91% to 54.53%, while iron content drops to between 26.89% and 46.42%. High concentrations of silicon (up to 10.87%), sulfur (up to 9.29%), potassium, and zirconium were also detected. These findings suggest sulfate corrosion and possible environmental dust deposition. The considerable presence of sulfur and other non-metallic elements indicates that this area is particularly vulnerable to chemical attack under humid and polluted atmospheric conditions.

Sample C (Figure 6C) mainly represents areas associated with paint residues and corrosion layers. The EDS points reveal an oxygen content around 31.79–38.09%, with substantial enrichment in zinc (up to 19.50%), calcium (up to 21.56%), and sulfur (up to 11.96%). Trace levels of titanium, aluminum, nickel, and chromium were also detected. This elemental composition implies that this area includes residues from zinc-rich coatings and the accumulation of environmental dust. The combination of sulfur and calcium suggests the possible formation of sulfate salts on the surface.

Sample D (Figure 6D) exhibits a complex, highly deteriorated morphology. EDS results indicate oxygen contents between 36.49% and 38.79%, and iron levels between 26.92% and 31.45%. Sulfur remains significantly high (7.28–12.78%), along with considerable amounts of zirconium, potassium, titanium, aluminum, and silicon. These results confirm the formation of corrosion products mainly composed of iron oxides and sulfate compounds, combined with environmental pollutant deposition. Severe coating peeling and corrosion-product accumulation were observed in this area.

Overall, the SEM-EDS analysis confirms that the Vila D. Bosco nameplate has undergone multiphase corrosion processes, characterized by localized oxidation, environmental particle deposition, and surface deterioration associated with complex atmospheric exposure. Due to equipment limitations, the SEM images shown reflect the maximum achievable resolution during testing.

In this study, an elemental surface scan (Figure 7) was performed on sample D (Figure 6D). The carbon, oxygen, titanium, and iron in the nameplate were evenly distributed, indicating that the nameplate was oxidized to form evenly distributed rust. Carbon (C) was found in higher amounts in certain spots (Figure 7, blue area), likely due to the buildup of organic materials in the coating (like alkyd resin or polyvinyl alcohol) and organic dirt from the outside environment. This finding further confirms the significant effect of aging and shedding of the coating on the element distribution in the corrosion area of the nameplate. The titanium element should come from the residue of titanium dioxide pigment in the coating. It is worth noting that the bright area in sample D is white paint, which should be rich in titanium; however, its EDS analysis results indicate that the titanium content in the bright area is not much different from that in the dark rust area. It is speculated that the weak conductivity of the coating itself causes the weak EDS signal. In addition, the distribution of sulfur elements throughout the surface of the nameplate is closely related to the fact that the nameplate's original material is resulfurized steel. A certain proportion of sulfur is added to resulfurized steel during the smelting process to improve the material's cutting performance. However, in a humid subtropical environment, this type of steel is more likely to rust due to sulfur, leading to a large increase in sulfur in the rust. It should be noted that although the carbon detected in the SEM-EDS surface analysis was primarily attributed to external organic contamination (such as aged coatings and environmental deposits), due to the limited sensitivity of EDS to light elements like carbon, a small amount of intrinsic carbon from the metal matrix cannot be entirely excluded. Based on the comprehensive analysis of iron and sulfur content and historical background, the base material is more consistent with resulfurized steel rather than raw iron (pig iron) [43].

3.2. XRF Elemental Analysis

In order to further clarify the overall chemical composition of the metal matrix of the nameplate and support the conclusion of the SEM-EDS analysis, this study conducted XRF tests on 5 points in the front of the nameplate (Figure 8 and Table 2). The test results indicate that the main component of all samples is iron (Fe), with a mass fraction ranging from 96.34% to 98.22% and an average of 97.30%, indicating that the nameplate is an iron alloy. The Fe content fluctuates slightly between different points; however, it is maintained at a high level overall, indicating that the overall material of the nameplate is relatively uniform, and there is no material replacement due to later maintenance or replacement. A certain proportion of sulfur (S) was detected in all samples, with a content ranging from 0.67% to 1.98%. The content in sample 4 was the highest, significantly higher than the other points. This sulfur content is much higher than the impurity content of sulfur in general low-carbon steel. Combined with the phenomenon of widespread sulfur enrichment in the rusted area in the aforementioned SEM-EDS results, it is speculated that the nameplate uses resulfurized steel. Resulfurized steel was common in industrial products in the mid-20th century and was widely used in nameplates, signboards, and other precision-processed metal components due to its excellent cutting performance [44,45]. However, high sulfur content also reduces the corrosion resistance of steel. In a humid, high-humidity, and pollutant-rich atmosphere, it is more likely to generate sulfate corrosion products, thereby accelerating metal surface oxidation and spalling.

All samples contained trace amounts of silicon (Si), manganese (Mn), copper (Cu), aluminum (Al), and molybdenum (Mo). The contents of Si and Mn were slightly higher, ranging from 0.65–1.13% and 0.34–0.39%, respectively. These elements may be deoxidizers and alloys added during the steel manufacturing process, which can improve the structural stability and strength of the steel to a certain extent. The contents of Cu, Al, and Mo are all at trace levels, which may be impurities in the raw materials, residual pollutants on the surface, or interference from coating components. The nameplate's corrosion morphology shows that the presence of alloying elements has not significantly changed the material's overall corrosion resistance.

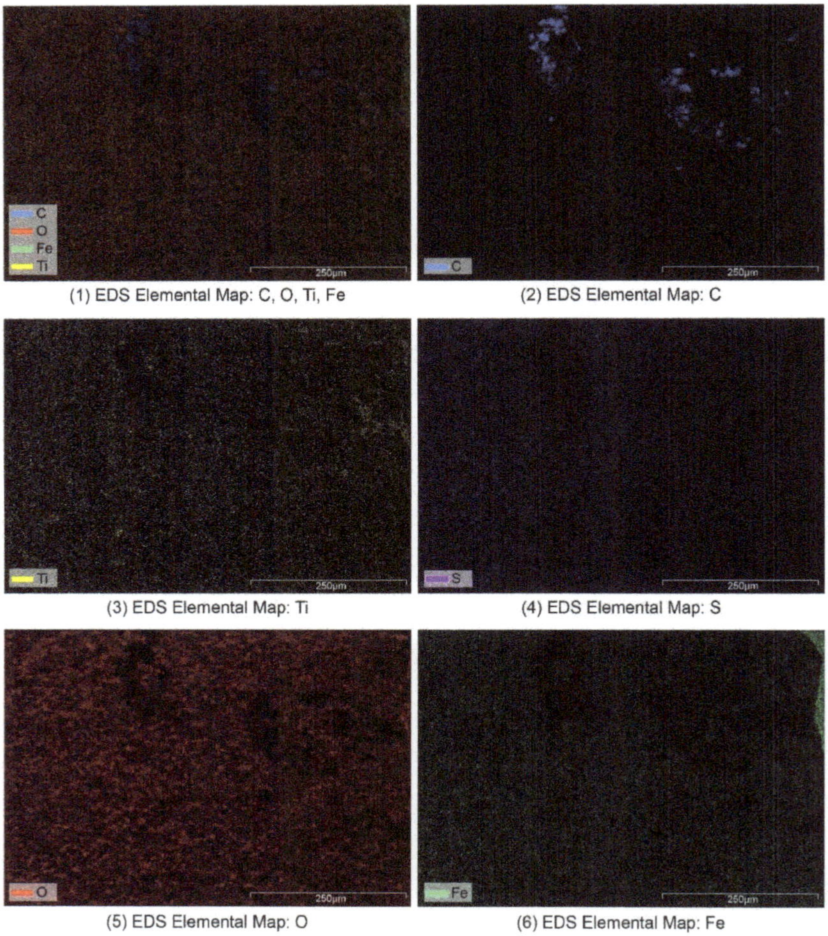

Figure 7. SEM-EDS elemental surface scanning image of the Vila D. Bosco nameplate.

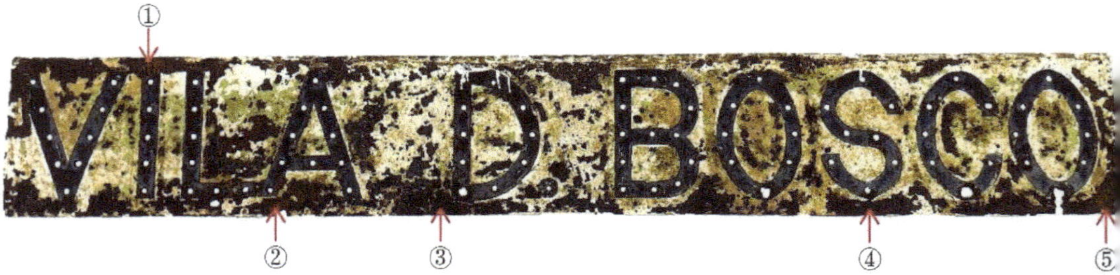

Figure 8. Five detection points of Vila D. Bosco nameplate iron-based XRF analysis.

Table 2. Vila D. Bosco nameplate iron-based XRF analysis.

Element Symbol	Value (%)				
	Point ①	Point ②	Point ③	Point ④	Point ⑤
Fe	98.02	98.22	97.80	96.34	98.13
S	0.85	0.70	0.78	1.98	0.67
Si	0.65	0.67	0.92	1.13	0.76
Mn	0.37	0.34	0.39	0.35	0.37
Cu	0.05	0.03	0.04	0.09	0.04
Al	0.03	0.00	0.04	0.08	0.00
Mo	0.03	0.04	0.03	0.03	0.03

To determine the material composition and pigment source of the nameplate surface coating [46,47], XRF elemental analysis was performed on a total of 7 points of white, yellow-green, and black coatings on the nameplate (Figure 9 and Table 3).

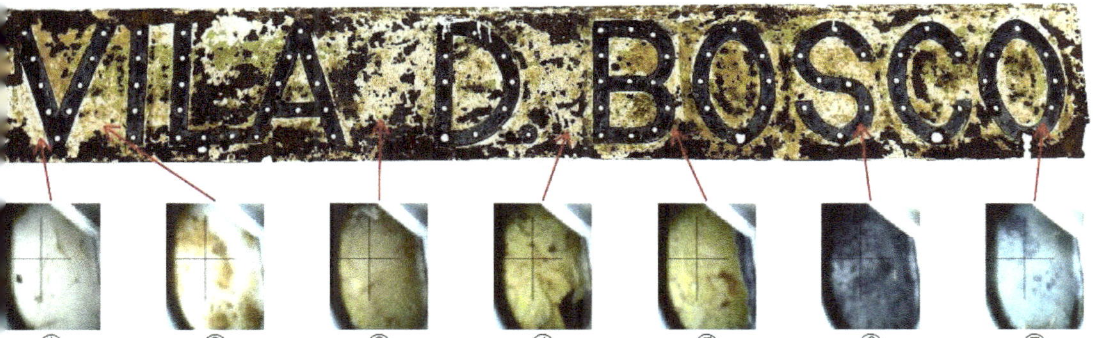

Figure 9. Seven detection points of XRF analysis in the coating on the Vila D. Bosco nameplate.

Table 3. XRF analysis of the coating on the Vila D. Bosco nameplate.

Element Symbol	Value (%)						
	Point ①	Point ②	Point ③	Point ④	Point ⑤	Point ⑥	Point ⑦
Ti	84.70	54.47	10.02	17.92	4.56	54.81	60.56
Fe	8.19	37.90	15.60	46.30	34.39	39.65	32.31
Cr	3.58	3.77	11.92	7.12	12.13	1.17	2.03
Al	1.87	1.45	0.00	1.17	2.07	0.64	1.05
Ta	1.13	0.71	0.00	0.00	1.44	0.32	0.64
Si	0.00	1.34	1.41	1.03	0.06	3.26	3.13
W	0.53	0.30	0.00	0.00	4.59	0.15	0.27
V	0.00	0.05	1.66	1.00	3.14	0.00	0.00
Zn	0.00	0.00	12.11	5.68	0.00	0.00	0.00
Pb	0.00	0.00	47.27	12.84	24.32	0.00	0.00
S	0.00	0.00	0.00	6.55	11.78	0.00	0.00
P	0.00	0.00	0.00	0.40	1.53	0.00	0.00

(1) White coating area (samples 1–2). Samples 1 and 2 mainly consist of titanium (Ti), with amounts of 84.70% and 54.47%, respectively, showing that titanium dioxide (TiO_2) is used as a color pigment in the white coating. Titanium dioxide has excellent hiding power and weather resistance and is widely used in modern buildings and industrial products. Also, a little bit of aluminum (Al), chromium (Cr), and tantalum (Ta) were found here, likely from impurities or additives in the pigment. Since the white coating has a strong hiding power, XRF cannot detect the metal matrix; therefore, the iron (Fe) content in this area is low.

(2) Yellow-green coating area (samples 3–5). The elemental composition of this area is relatively complex and should be a mixture of multiple color pigments. In addition to the medium contents of Ti (4.56–17.92%) and Fe (15.60–46.30%), relatively high contents of Cr (7.12–12.13%), Pb (12.84–47.27%), and Zn (5.68–12.11%) were also detected. This indicates that the yellow-green pigment is most likely chrome yellow ($PbCrO_4$) and zinc yellow ($ZnCrO_4$). The mid-20th century saw widespread use of this type of pigment in industrial and architectural coatings due to its strong durability and corrosion resistance. This paint is more common in the metal parts of Portuguese colonial buildings, which is consistent with the nameplate's production background.

(3) Black coating area (samples 6–7). The Ti content in the black area is still high (54.81% and 60.56%), indicating that the black paint may also be mixed with TiO_2 as a covering filler. The Fe content is moderate (32.31–39.65%), which matches what we expect from common iron black pigments (like iron manganese oxides); samples 5 and 6 also showed some sulfur (6.55–11.78%), which could come from sulfur-containing black organic or inorganic pigments, or it might be left over from environmental pollution.

3.3. XRD Phase Analysis of Corrosion Products

To further investigate the mineral composition and crystalline structure of the corrosion layer on the nameplate, an X-ray diffraction (XRD) analysis was conducted on rust samples collected from its surface. The experiment employed a Co-Kα radiation source (λ = 1.78897 Å), and the resulting spectra were converted to Cu-Kα equivalent values (λ = 1.5406 Å) to allow direct comparison with standard reference data.

As shown in Figure 10, the diffraction pattern of the experimental sample (converted to Cu-Kα) is compared with reference spectra from the RRUFF database. Specifically, goethite (RRUFF ID: R120086), hematite (RRUFF ID: R040024), and magnetite (RRUFF ID: R060191) are included in the comparison. Major peak positions are annotated based on this matching process. The experimental spectrum shows several distinct peaks at approximately $2\theta \approx 21.3°$, 33.3°, 36.7°, etc. Notably, the peaks around 33.2° and 35.5° are characteristic of hematite (α-Fe_2O_3) and magnetite (Fe_3O_4), confirming the presence of these phases. A peak at 21.3°, corresponding to goethite (α-FeOOH), further suggests a typical mixed iron oxide rust layer. This combination of phases indicates that the corrosion developed in a humid, oxygen-rich environment, with possible exposure to mildly acidic airborne particles—a scenario consistent with the subtropical climate of southern Macau. The detection of stable phases like goethite implies that the corrosion occurred under alternating wet and dry conditions, allowing for both active oxidation and passive stabilization processes over time.

The XRD results of the white coating (Figure 11) indicate that the sample predominantly consists of synthetic rutile (TiO_2, RRUFF ID: R040049). The experimental pattern—converted from Co-Kα to Cu-Kα radiation—shows strong diffraction peaks at $2\theta \approx 27.5°$, 36.1°, and 41.3°, which align closely with the reference peaks of rutile. These results confirm that the white pigment is crystalline and primarily composed of TiO_2 in its rutile form, commonly used in industrial coatings as a white pigment.

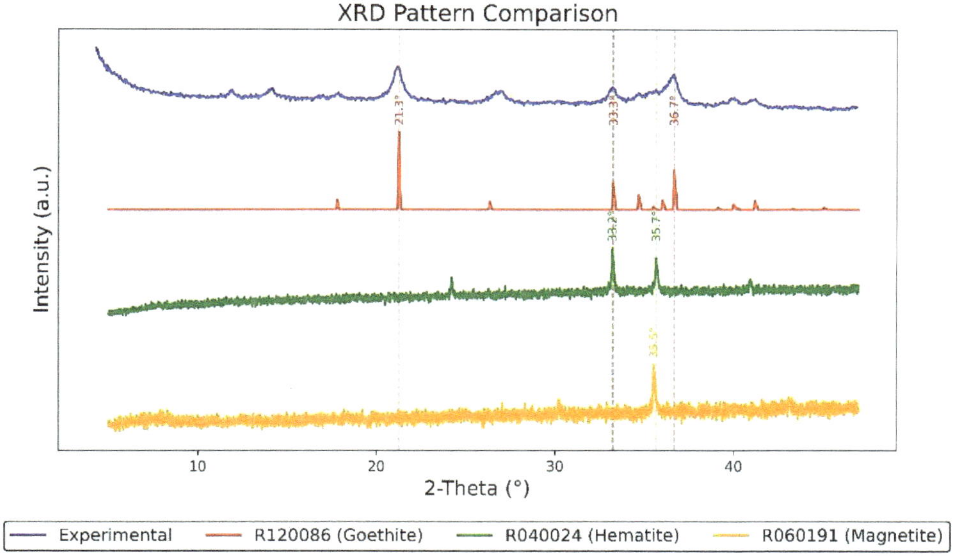

Figure 10. XRD analysis results of rust on the Vila D. Bosco nameplate.

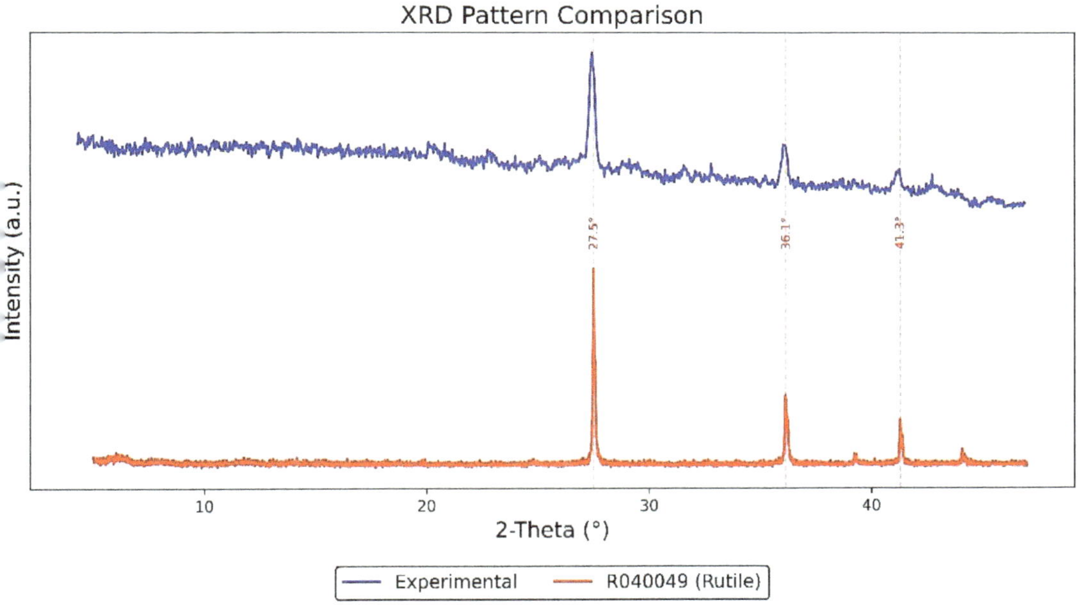

Figure 11. XRD analysis results of the Vila D. Bosco nameplate's white coating.

In contrast, the XRD analysis of the black and yellow-green coatings on the nameplate revealed broad background signals and baseline drift, with no discernible sharp diffraction peaks. This suggests that these layers are either amorphous or possess an extremely low degree of crystallinity. Such results are typical of complex or degraded organic coatings or corrosion products where crystalline phases are poorly developed.

3.4. FTIR Spectroscopy Analysis of Coatings

In order to identify the organic components in the surface coating of the nameplate, ATR-FTIR tests were performed on white, yellow-green, and black coating samples. Both white and yellow-green coatings contain alkyd resin (polyester alkyd), and the comparison with the standard spectrum is shown in Figure 12. The sample shows a strong C–H stretching vibration absorption peak at ~2925 cm^{-1}, corresponding to the vibration of the –CH$_2$– (methyl) group of the aliphatic hydrocarbon chain, indicating that the sample contains a segment of a long fatty acid chain. The strong absorption peak near 1730 cm^{-1} is the C=O carbonyl stretching vibration of the ester, which is a typical feature of alkyd, polyester, or other ester-based resins. Additionally, a faint peak at 1638 cm^{-1}, which may be attributed to C=O stretching and/or C=C stretching vibrations, is consistent with the typical spectral features of alkyd resins. Several absorption peaks were observed in the 1150–1000 cm^{-1} region, which can be attributed to C–O–C and C–O stretching vibrations. These peak shape characteristics are also consistent with the alcohol ester structure of alkyd paint. There is also a C–H out-of-plane bending vibration peak at approximately 720–740 cm^{-1}, indicating the presence of some hydrocarbon side chains in the sample.

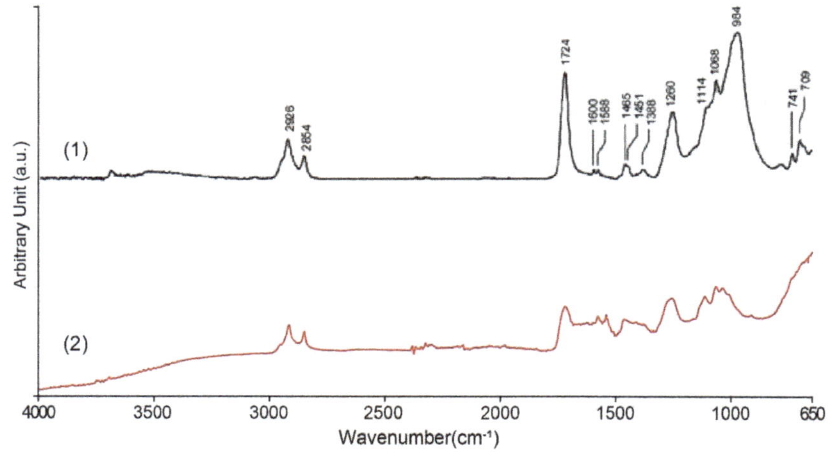

Figure 12. FTIR analysis of white and yellow-green coating on the Vila D. Bosco nameplate: (1) alkyd binder in mixture with inorganic pigments [48]; (2) the results of this study.

The FTIR spectrum results of the black coating sample show that its main component is polyvinyl alcohol (Figure 13). The broad and strong absorption peak at about ~3300 cm^{-1} in the spectrum corresponds to the stretching vibration of the hydroxyl group (–OH); the absorption peak near about 2920 cm^{-1} corresponds to the asymmetric stretching vibration peak of the methylene group (–CH$_2$–) in the aliphatic chain segment; the weak peak at about 1700 cm^{-1} should be attributed to the stretching vibration peak of the carbonyl group (C=O) contained in the small amount of polyvinyl acetate (PVAc) component in the hydrolysis process of the residual polyvinyl alcohol in the sample. In addition, the strong peak at about 1080 cm^{-1} shows the stretching vibration of the alcohol group (C–O) in the polyvinyl alcohol chain. These absorption peaks match very well with the known features of standard polyvinyl alcohol, confirming that the black coating material is indeed polyvinyl alcohol. At the same time, due to incomplete hydrolysis during the preparation process, a small amount of polyvinyl acetate components remains in the coating. The significant differences in elemental composition between the coatings and the metal substrate suggest that the substrate was initially painted, and the decorative letters were installed afterward.

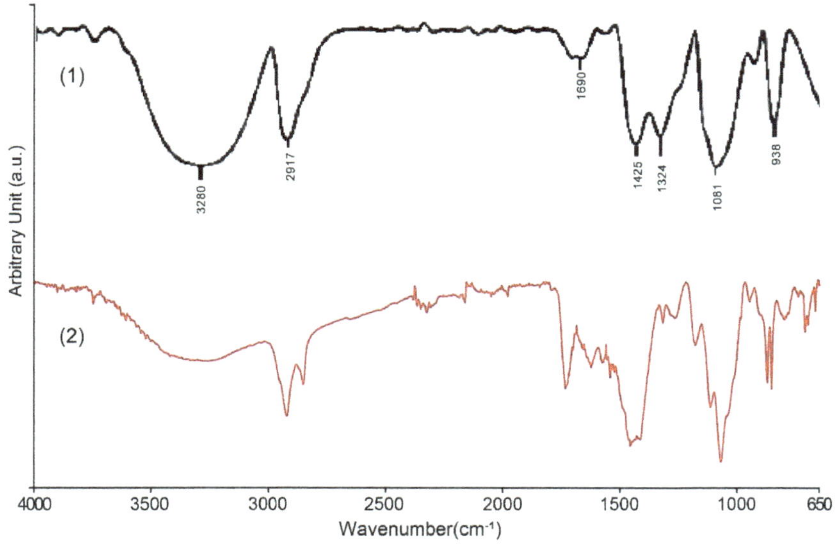

Figure 13. FTIR analysis of black coating on the Vila D. Bosco nameplate: (1) PVA [49]; (2) the results of this study.

3.5. Cross-Section Microscopic Analysis

The microphotographs (Figure 14) show that the cross-sections of the four Vila D. Bosco nameplate coating samples have different numbers of layers; however, they all have a white-green-white overlay structure. The compositional interpretation of the observed layers is based on the correlation of morphological features with the results from XRD and FTIR analyses, rather than the microscopic images alone. Sample 1 is located on a white background, with titanium white resin paint on the top layer, a yellow-green lead yellow resin paint on the second layer, titanium white paint on the third layer, black paint on the fourth layer, a zinc powder layer on the fifth layer, and a rust layer on the sixth layer. Sample 2 features black letters, black paint on the top layer, titanium white resin paint, lead yellow resin paint, titanium white resin paint, black paint, and a rust layer underneath. Sample 3 is like sample 2, but it has a white background and no black surface layer. In the microphotograph of sample 4, there is no titanium white coating above the rust layer, but a small amount of it can be observed on the right side of the picture. We speculate that either the paint application was uneven or the paint was loose during production, leading to the white layer falling off.

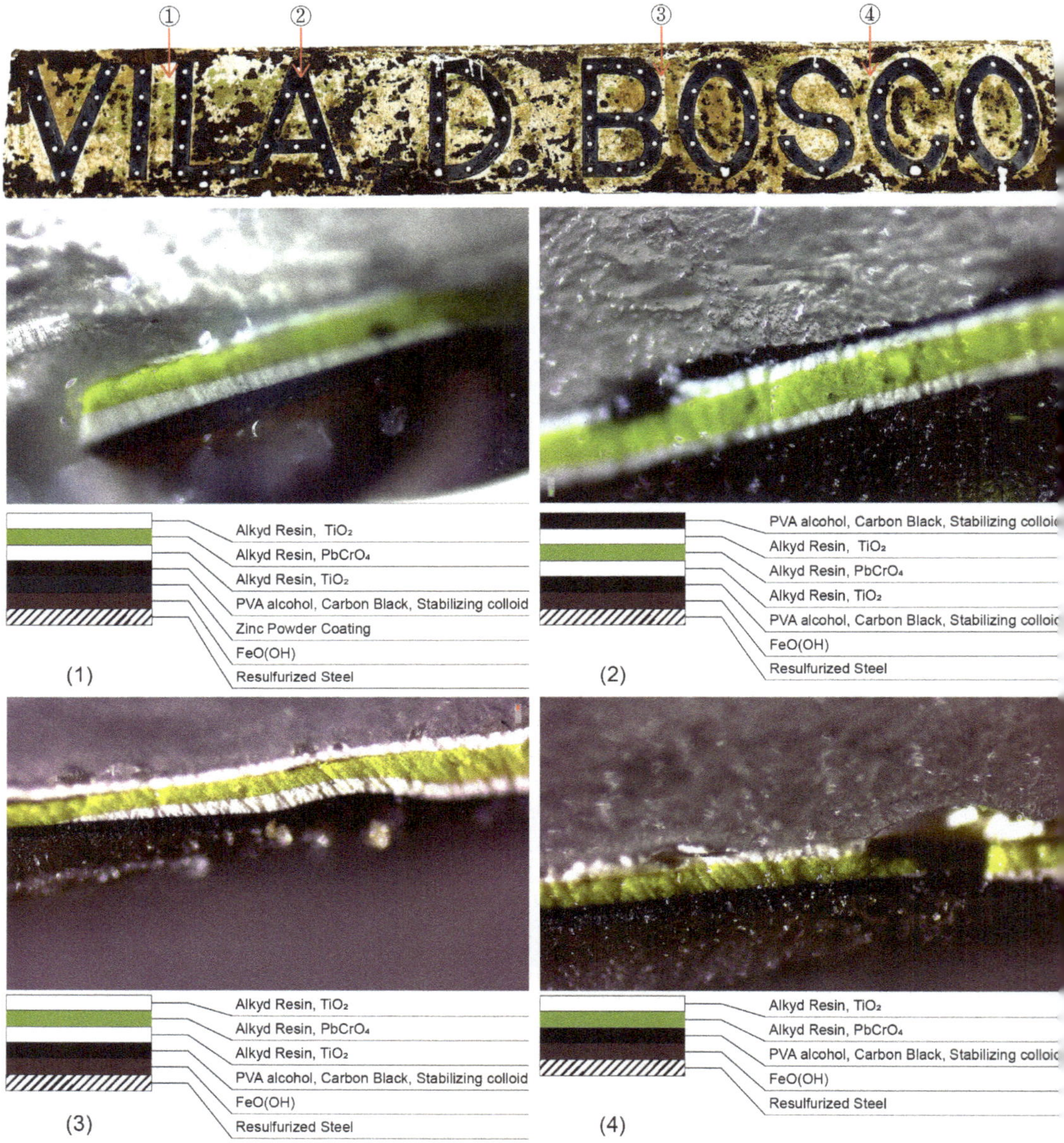

Figure 14. Microscopic observation of the cross-section of the Vila D. Bosco nameplate.

4. Discussion

Based on the results of XRF, SEM-EDS, XRD, ATR-FTIR, and cross-section microscopic analysis, this study systematically revealed the material composition, corrosion mechanism, and surface coating structure characteristics of the Vila D. Bosco nameplate.

XRF testing shows that the nameplate is an iron alloy with an iron content of 96–98% and about 0.7–2.0% sulfur (S). Combined with its historical background and the application characteristics of common metal materials in the mid-20th century, it is speculated that the nameplate is made of resulfurized steel. This type of steel can improve cutting performance due to its sulfur content and is often used for precision-processed components such as nameplates. However, its high sulfur content also makes it easy to generate sulfate corrosion products in hot and humid atmospheres, thereby reducing corrosion resistance.

SEM-EDS and XRD analysis show that the nameplate's rusted area is rich in Fe, O, S, Ti, and Zn. The main types of rust found are goethite (α-FeOOH), hematite (α-Fe_2O_3), and magnetite (Fe_3O_4), which are common results of rusting in the air and are usually seen in Macau's humid, salty, and polluted environment.

FTIR testing confirmed that the white and yellow-green coatings on the nameplate were oil-modified alkyd resin pigments. XRF and XRD tests showed that the coating used synthetic rutile titanium dioxide, chrome yellow, zinc yellow, and other pigments, which together made a strong industrial coating that hides well and lasts outdoors. Unlike the matte white background, the black coating of the letters has a glazed gloss, and the FTIR results are polyvinyl alcohol. Since polyvinyl alcohol is rarely used directly as a pigment binder, it is speculated that polyvinyl alcohol is a transparent protective layer on the surface of the black coating, and the binder of the black coating itself is unknown.

The cross-sectional microscopic analysis of the samples indicated that the nameplate coating method was to apply black paint as a primer on the zinc powder coating, then apply it on the black base layer in the order of white-green-white, and finally draw black letters on the white background layer. Only the bottom layer of sample 1 displayed the presence of the zinc powder layer among the four samples. The reasons for this phenomenon are as follows: (1) the zinc coating layer separated from the surface coating after deterioration; (2) considering that there was no gap between the titanium dioxide layer and the rust layer in samples 2, 3 and 4, the initial process may have been immature and caused part of the nameplate surface to be ungalvanized. The coating thickness in the samples was uneven, the surface was not smooth, and the titanium dioxide layer was relatively thick in some places, which also indicated that a spray gun was not used during production, but was painted by hand. The uneven surface coating made the anti-rust performance of the nameplate surface uneven, resulting in varying degrees of deterioration. In particular, the lack of the zinc powder layer weakened the cathodic protection of the alloy matrix. Since the resulfurized steel is prone to corrosion from sulfur, some areas in a wet and polluted environment made the problem worse and allowed it to spread.

After systematically analyzing the metal material, corrosion products, coating composition, and layered structure of the Vila D. Bosco nameplate, one of the ultimate goals of this study is to provide a scientific, operational, and academically based protection and restoration strategy for this type of modern architectural metal component. The nameplate material is mainly resulfurized steel, the corrosion products are mainly mixed iron oxides, the white coating is TiO_2 alkyd resin paint, and the yellow-green coating is $PbCrO_4$ alkyd resin paint. From the perspective of material chemistry, the deterioration mechanism it faces includes not only typical atmospheric oxidation corrosion, but also local failures caused by interlayer peeling, pigment aging, and uneven original construction. Relying solely on a single material or restoration process cannot achieve long-term, stable protection for these different types of problems. A "multi-material-multi-mechanism" collaborative restoration method must be adopted to consider the structural, functional, and cultural expression integrity.

Therefore, combined with the experimental results and cross-sectional structure analysis in the previous article, this paper summarizes a comprehensive protection plan including rust removal, functional material recoating, organic layer reconstruction, and reversible sealing, and the summary is shown in Table 4. Each technical treatment suggestion clearly points to the specific problem identification basis and selects modern materials with good compatibility based on the material properties. For example, in response to the problem of partial loss of the zinc powder protective layer, it is recommended to use high-content zinc-rich epoxy primer to restore its cathodic protection ability; for the reconstruction of the multi-layer organic alkyd resin structure, it is recommended to give priority to alkyd resin with a highly similar chemical structure to the raw material and strictly control the PVC value of the pigment to ensure its visual restoration and material stability [50,51]. In addition, for the repair of the blurred area of the nameplate text, a non-invasive micro-brush finishing process should be adopted as much as possible, supplemented by a standard method for controlling color differences to ensure that the historical information of the text is accurately restored. At the same time, all repair measures follow the principle of "reversibility" and try to avoid irreversible reactions and material cross-linking in the selection of bonding materials, coating components, and construction processes to ensure the flexibility of sustainable maintenance in the future. The entire restoration system advocates a strategy of zoning treatment and adapting to local conditions. Different treatment measures should be applied to different structural parts and degrees of deterioration, rather than integrated coverage, to respect the continuity of the historical layered information of the original components.

Table 4. Recommendations for the preservation and restoration of the Vila D. Bosco nameplate.

Repair Projects	Problem Identification Basis	Materials/Methods Recommendations	Technical Description
Surface rust removal	The presence of corrosion products such as FeOOH and Fe_2O_3, and the rust layer is relatively thick	Use a neutral pH chelating agent (such as EDTA-2Na solution) + mechanical micro-grinding	Avoid using acidic cleaning agent to prevent aggravation of corrosio reaction on resulfurized steel; micro-grinding is limited to non-inscription font areas
Zinc powder protective layer re-coating	The zinc powder layer is partially missing, and the cathodic protection fails	Apply high-purity zinc powder epoxy primer ($\geq$95% Zn content)	Two-component zinc-rich primer can be used to enhance the anti-corrosion performance of the substrate and rebuild the structur of the original zinc powder layer
Black base layer reconstruction	The PVA alcohol + carbon black interface layer is a key stable structure, and it is partially peeled off	Self-prepared polyvinyl alcohol/carbon black dispersion material or commercially available stable carbon black paint	Surface tension and adhesion nee to be adjusted to match the origina coating thickness (recommended t be controlled at 20–30 μm)
Multi-layer organic paint repair	The organic coating peels off, the layers are uneven, and the light aging is serious	Use alkyd resin adhesive (optionally alkyd resin) + TiO_2/$PbCrO_4$ simulated coating	Apply in layers according to the original structure, first yellow laye and then white layer, and pay attention to controlling the pigme volume fraction (PVC) to ensure hiding power and weather resistan
Protective transparent coating	After aging, the coating has strong hygroscopicity, the surface is slightly cracked, and it needs to be protected from moisture and UV	Apply UV-stable acrylic sealant or fluorocarbon transparent protective film	It is recommended to use product with UV aging resistance of QUV-5000h (Q-Lab, Westlake, OH, USA or above, and control the thicknes to 5–10 μm to avoid excessive interference with the apparent colo
Nameplate text regeneration	The paint on the edge of the font is seriously peeled off, and some letters are blurred	Use micro-pen repair method + UV-visible color difference comparison and refinement	Fonts are simulated with original black pigment (PVA + carbon blac and redrawn with fine templates when necessary

Table 4. *Cont.*

Repair Projects	Problem Identification Basis	Materials/Methods Recommendations	Technical Description
Reversible repair logo	Ensure the operability of later repairs	Use reversible adhesive for all repair materials, and fully record the repair position	Material records must include brand, ratio, batch number, and use date to meet the principle of reversible protection

5. Conclusions

This study systematically investigated the material characteristics and degradation mechanisms of the Vila D. Bosco nameplate, a representative metal component from modern and contemporary architecture in Macau during the Portuguese rule in the 1960s. Through a combination of XRF, SEM-EDS, XRD, FTIR analyses, and cross-sectional microscopy, the following conclusions were drawn:

(1) Metal Matrix Composition: The nameplate substrate is composed of resulfurized steel with an iron (Fe) content of up to 97% and sulfur (S) content between 0.7 and 2%. The high sulfur content contributes significantly to internal sulfate corrosion under humid and polluted atmospheric conditions.

(2) Corrosion Products: The main corrosion products include goethite (α-FeOOH), hematite (α-Fe_2O_3), and magnetite (Fe_3O_4), forming a multiphase mixed structure indicative of prolonged and progressive corrosion processes.

(3) Coating Materials: The surface coatings were identified as based on alkyd resin, incorporating pigments such as titanium dioxide (TiO_2), lead chromate ($PbCrO_4$), and zinc chromate ($ZnCrO_4$), reflecting typical mid-20th-century material usage.

(4) Layered Degradation: Cross-sectional analysis revealed significant stratification, including complete coating structures with zinc-rich primer layers and areas of coating disorder caused by aging, maintenance interventions, and environmental factors.

(5) Based on these findings, this study proposes conservation strategies emphasizing material compatibility, structural integrity restoration, and reversibility. The outcomes provide scientific guidance for the protection of similar modern metal components, balancing structural functionality with cultural significance.

Author Contributions: Conceptualization, L.Z., J.Z., X.H., and Y.C.; methodology, L.Z., X.H., and Y.C.; software, L.Z., X.H., and Y.C.; validation, L.Z., X.H., and Y.C.; formal analysis, L.Z., X.H., and Y.C.; investigation, L.Z., J.Z., X.H., and Y.C.; resources, L.Z., J.Z., X.H., and Y.C.; data curation, L.Z., X.H., and Y.C.; writing—original draft preparation, L.Z., X.H., and Y.C.; writing—review and editing, L.Z., J.Z., X.H., and Y.C.; visualization, L.Z.; supervision, J.Z.; project administration, J.Z.; funding acquisition, J.Z. All authors have read and agreed to the published version of the manuscript.

Funding: This research was funded by the Macau Science and Technology Development Fund (FDCT)'s Funding for Innovation and Technology Promotion (0080/2023/ITP2).

Institutional Review Board Statement: Not applicable.

Informed Consent Statement: Not applicable.

Data Availability Statement: The original contributions presented in this study are included in the article. Further inquiries can be directed to the corresponding author.

Conflicts of Interest: The authors declare no conflicts of interest.

References

1. Liang, R. Protection of Macau's Historic Architectural Heritage. In Proceedings of the 2019 International Conference on A tecture: Heritage, Traditions and Innovations (AHTI 2019), Moscow, Russia, 25–27 February 2019; Atlantis Press: Paris, Fra 2019; pp. 474–479. [CrossRef]
2. Ng, M.N. *Pilgrimages: Memories of Colonial Macau and Hong Kong*; Hong Kong University Press: Hong Kong, China, 2009.
3. Yu, X. Growth and degradation in the Orient's 'Las Vegas': Issues of environment in Macau. *Int. J. Environ. Stud.* **2008** 667–683. [CrossRef]
4. Zhong, C.; Gasser, A.; Kittel, J.; Wissenbach, K.; Poprawe, R. Improvement of material performance of Inconel 718 forme high deposition-rate laser metal deposition. *Mater. Des.* **2016**, *98*, 128–134. [CrossRef]
5. Horton, P.M.; Allwood, J.M.; Cleaver, C. Implementing material efficiency in practice: A case study to improve the mate utilisation of automotive sheet metal components. *Resour. Conserv. Recycl.* **2019**, *145*, 49–66. [CrossRef]
6. Lakshmanan, V.I.; Roy, R.; Ramachandran, V. *Innovative Process Development in Metallurgical Industry*; Springer: New York, USA, 2016; Volume 10, ISBN 978-3-319-21598-3. [CrossRef]
7. Mc Clelland, M. Evaluating Sectoral Innovation System Functional Performance in the Additive Manufacturing Sector: mented Tungsten Carbides Case Studies. Ph.D. Thesis, Stellenbosch University, Stellenbosch, South Africa, 2023. Avail online: http://hdl.handle.net/10019.1/127013 (accessed on 30 March 2025).
8. Birkhofer, H.; Fügenschuh, A.; Günther, U.; Junglas, D.; Martin, A.; Sauer, T.; Ulbrich, S.; Wäldele, M.; Walter, S. Optimizatio Sheet Metal Products. In *Operations Research Proceedings 2005: Selected Papers of the Annual International Conference of the Ger Operations Research Society (GOR)*; Springer: Berlin/Heidelberg, Germany, 2006; pp. 327–336. [CrossRef]
9. Bonte, M.H.A.; Boogaard, A.H.V.D.; Huétink, J. An optimisation strategy for industrial metal forming processes: Model screening and solving of optimisation problems in metal forming. *Struct. Multidiscip. Optim.* **2008**, *35*, 571–586. [CrossRef]
10. Romanova, O.A.; Sirotin, D.V. Metal Industry Development in the Conditions of Formation of New Technological and Ins tional Trends. *KnE Mater. Sci.* **2019**, *5*, 15–28. [CrossRef]
11. Sheng, Y.; Hua, Y.; Wang, X.; Zhao, X.; Chen, L.; Zhou, H.; Wang, J.; Berndt, C.C.; Li, W. Application of High-Density I tropulsing to Improve the Performance of Metallic Materials: Mechanisms, Microstructure and Properties. *Materials* **2018** 185. [CrossRef]
12. Eggert, G.; Fischer, A. The formation of formates: A review of metal formates on heritage objects. *Herit. Sci.* **2021**, *9*, 26. [Cr Ref]
13. Sandu, I. Modern aspects regarding the conservation of cultural heritage artifacts. *Int. J. Conserv. Sci.* **2022**, *13*, 1187–1208.
14. Macchia, A.; Colasanti, I.A.; Rivaroli, L.; Favero, G.; de Caro, T.; Munoz, L.P.; Campanella, L.; La Russa, M.F. Natural ba products for cleaning copper and copper alloys artefacts. *Nat. Prod. Res.* **2023**, *37*, 1177–1184. [CrossRef]
15. Eggert, G.; Fischer, A. Curious Corrosion Compounds Caused by Contact: A Review of Glass-Induced Metal Corrosion Museum Exhibits (GIMME). *Corros. Mater. Degrad.* **2022**, *3*, 553–565. [CrossRef]
16. Filopoulou, A.; Vlachou, S.; Boyatzis, S.C. Fatty Acids and Their Metal Salts: A Review of Their Infrared Spectra in Light of T Presence in Cultural Heritage. *Molecules* **2021**, *26*, 6005. [CrossRef] [PubMed]
17. Ershad-Langroudi, A. *Protective Material Coatings for Preserving Cultural Heritage Monuments and Artwork*; Bentham Science I lishers: Oak Park, IL, USA, 2022.
18. Zhao, Q.; Sun, Q.; Xin, S.; Chen, Y.; Wu, C.; Wang, H.; Xu, J.; Wan, M.; Zeng, W.; Zhao, Y. High-strength titanium alloys aerospace engineering applications: A review on melting-forging process. *Mater. Sci. Eng. A* **2022**, *845*, 143260. [CrossRef]
19. Sandu, I.C.A.; de Sá, M.H.; Pereira, M.C. Ancient 'gilded'art objects from European cultural heritage: A review on diffe scales of characterization. *Surf. Interface Anal.* **2011**, *43*, 1134–1151. [CrossRef]
20. Belláková, E. Analysis of industrial architectural heritage–iron and steel plants as a development potential. *Procedia Eng.* **2** *161*, 1926–1931. [CrossRef]
21. Carl, M.; Young, M.L. Complementary analytical methods for analysis of Ag-plated cultural heritage objects. *Microchem. J.* **2** *126*, 307–315. [CrossRef]
22. Oudbashi, O.; Hessari, M.; Bahadori, S. An archaeometallurgical study of Achaemenid copper-base artefacts from the Persep World Heritage Site, Iran. *J. Cult. Herit.* **2023**, *60*, 114–121. [CrossRef]
23. Soffritti, C. Metals in Heritage Science. *Heritage* **2024**, *7*, 1822–1825. [CrossRef]
24. Zhou, M.; Wang, R.; Guo, Y. How urban spatial characteristics impact surface urban heat island in subtropical high-density ci based on LCZs: A case study of Macau. *Sustain. Cities Soc.* **2024**, *112*, 105587. [CrossRef]
25. Muhammad, M.A.; Sani, S.; Sulaiman, M.; Shuaibu, A.M.; Kura, A.U.; Rabiu, S. Impact of climate change on environme chemistry: A review. *Curr. Res. Interdiscip. Stud.* **2024**, *3*, 28–49. [CrossRef]
26. Oesch, S.; Faller, M. Environmental effects on materials: The effect of the air pollutants SO_2, NO_2, NO and O_3 on the corros of copper, zinc and aluminium. A short literature survey and results of laboratory exposures. *Corros. Sci.* **1997**, *39*, 1505–1 [CrossRef]

Thomas, E. The effects of the marine environment on the built environment of Llandudno: Comparing the effects from coastline to inland properties. *J. Build. Surv. Apprais. Valuat.* **2016**, *5*, 113–143. Available online: https://hstalks.com/article/1299/ (accessed on 30 March 2025). [CrossRef]
Ijsseling, F.P. General guidelines for corrosion testing of materials for marine applications: Literature review on sea water as test environment. *Br. Corros. J.* **1989**, *24*, 53–78. [CrossRef]
Chandler, K.A. *Marine and Offshore Corrosion: Marine Engineering Series*; Elsevier: Amsterdam, The Netherlands, 2014.
Gracia, A.; Rangel-Buitrago, N.; Oakley, J.A.; Williams, A. Use of ecosystems in coastal erosion management. *Ocean Coast. Manag.* **2018**, *156*, 277–289. [CrossRef]
Vong, L.T.N.; Lam, C.C.C. Holiday-making and the leisure space of the Macao people. In *Domestic Tourism in Asia*; Routledge: London, UK, 2009; pp. 301–314. Available online: https://www.taylorfrancis.com/books/edit/10.4324/9781849770057/domestic-tourism-asia-shalini-singh (accessed on 30 March 2025).
Colónia de Férias Nas Ilhas em 1972 (I)—Colégio de D. Bosco. Available online: https://nenotavaiconta.wordpress.com/2016/08/04/colonia-de-ferias-nas-ilhas-em-1972-i-colegio-de-d-bosco/ (accessed on 16 April 2025).
Zhang, M.; Fan, J.; Liu, J.; Chen, Y.; Lu, Y.; Lei, Y.; Kaya, M.G.A.; Tang, K. A comprehensive evaluation of a historical leather armor from Yanghai Cemetery, Turpan. *Herit. Sci.* **2024**, *12*, 162. [CrossRef]
Saleh, S.M.; El-Badry, A.E.-H.A.; Abdel-Karim, A.M. Evaluation of the corrosion resistance of bronze patina or/and protective coating on the surface of the archaeological coins. *Sci. Rep.* **2025**, *15*, 2361. [CrossRef]
Fruth, V.; Todan, L.; Codrea, C.I.; Poenaru, I.; Petrescu, S.; Aricov, L.; Ciobanu, M.; Jecu, L.; Ion, R.M.; Predoana, L. Multifunctional Composite Coatings Based on Photoactive Metal-Oxide Nanopowders (MgO/TiO_2) in Hydrophobic Polymer Matrix for Stone Heritage Conservation. *Nanomaterials* **2021**, *11*, 2586. [CrossRef]
Silveira, P.; Falcade, T. Applications of energy dispersive X-ray fluorescence technique in metallic cultural heritage studies. *J. Cult. Herit.* **2022**, *57*, 243–255. [CrossRef]
Asvestas, A.; Chatzipanteliadis, D.; Gerodimos, T.; Mastrotheodoros, G.P.; Tzima, A.; Anagnostopoulos, D.F. Real-Time Elemental Analysis Using a Handheld XRF Spectrometer in Scanning Mode in the Field of Cultural Heritage. *Sustainability* **2024**, *16*, 6135. [CrossRef]
Campos, B.; Marco, A.; Freire-Lista, D.M.; Durães, N.; Silvestre-Albero, J.; da Silva, E.F.; Vieira, E.; Pintado, M.; Moreira, P.R. Rare Biogeochemical Phenomenon Associated to Manganese Patinas on Mural Painting and Granite Ashlars. *Coatings* **2021**, *11*, 917. [CrossRef]
Gajić Kvaščev, M.; Milojević, P.; Kapuran, A.; Ljuština, M.; Maksimović, D.; Bulatović, A. Pottery from the Bubanj Archaeological Site: A Study on Provenance and Production Technology. In *Interaction, Transmission, Transformation. Long-Distance Connections in Metal Ages of South-Eastern Europe*; Institute of Archaeology: Belgrade, Serbia, 2025; pp. 187–198.
Elert, K.; Monasterio-Guillot, L.; Cultrone, G. Effect of iron and organic matter on mineralogy and texture of replacement bricks for heritage conservation: The case of the Alhambra Formation soil (Granada, Spain). *J. Eur. Ceram. Soc.* **2024**, *44*, 4294–4306. [CrossRef]
Tomás, R.; Cano, M.; Pulgarín, L.; Brotóns, V.; Benavente, D.; Miranda, T.; Vasconcelos, G. Thermal effect of high temperatures on the physical and mechanical properties of a granite used in UNESCO World Heritage sites in north Portugal. *J. Build. Eng.* **2021**, *43*, 102823. [CrossRef]
Liu, G.L.; Kazarian, S.G. Recent advances and applications to cultural heritage using ATR-FTIR spectroscopy and ATR-FTIR spectroscopic imaging. *Analyst* **2022**, *147*, 1777–1797. [CrossRef] [PubMed]
Lou, D.; Cui, K.; Jia, Y. Study on the machinability of resulfurized composite free-cutting steels. *J. Mater. Eng. Perform.* **1997**, *6*, 215–218. [CrossRef]
Bhattacharya, D. Machinability of steel. *JOM* **1987**, *39*, 32–35. [CrossRef]
Roper, C.R., Jr.; Strattan, J.K.; Hetherington, R., Jr.; Hiller, A.J. Process for Making Resulfurized Machinable Steel. U.S. Patent US4373967A, 15 February 1983. Available online: https://patents.google.com/patent/US4373967A/en (accessed on accessed on 12 March 2025).
Fontana, D.; Alberghina, M.F.; Barraco, R.; Basile, S.; Tranchina, L.; Brai, M.; Gueli, A.; Troja, S.O. Historical pigments characterisation by quantitative X-ray fluorescence. *J. Cult. Herit.* **2014**, *15*, 266–274. [CrossRef]
Costa, T.G.; Ritcher, F.A.; Correia, M.D.D.M.; Escorteganha, M.R.; Santiago, A.G.; Gonçalves, S.; Spudeit, D.A.; Micke, G.A.; Miranda, F.S. Multi-technical analysis as a tool to investigate structural species in the "replica" of First Mass in Brazil painting by Sebastião Vieira Fernandes. *J. Mol. Struct.* **2016**, *1120*, 196–204. [CrossRef]
Pagnin, L.; Calvini, R.; Wiesinger, R.; Weber, J.; Schreiner, M. Photodegradation kinetics of alkyd paints: The influence of varying amounts of inorganic pigments on the stability of the synthetic binder. *Front. Mater.* **2020**, *7*, 600887. [CrossRef]
Kharazmi, A.; Faraji, N.; Hussin, R.M.; Saion, E.; Yunus, W.M.M.; Behzad, K. Structural, optical, opto-thermal and thermal properties of ZnS–PVA nanofluids synthesized through a radiolytic approach. *Beilstein J. Nanotechnol.* **2015**, *6*, 529–536. [CrossRef]

50. Juhl, M.; Hauschild, M.Z.; Dam-Johansen, K. An eco-strategy for development of more sustainable coatings. *Prog. Org. C* **2024**, *197*, 108781. [CrossRef]
51. Buyondo, K.A.; Kasedde, H.; Kirabira, J.B. A comprehensive review on kaolin as pigment for paint and coating: Recent trenc chemical-based paints, their environmental impacts and regulation. *Case Stud. Chem. Environ. Eng.* **2022**, *6*, 100244. [CrossR

Disclaimer/Publisher's Note: The statements, opinions and data contained in all publications are solely those of the individua thor(s) and contributor(s) and not of MDPI and/or the editor(s). MDPI and/or the editor(s) disclaim responsibility for any injur people or property resulting from any ideas, methods, instructions or products referred to in the content.

materials

MDPI

icle

ntifungal Hybrid Graphene–Transition-Metal ichalcogenides Aerogels with an Ionic Liquid Additive as nnovative Absorbers for Preventive Conservation of Cultural eritage

orge Gorgolis [1,2,*], Maria Kotsidi [2], Elena Messina [3], Valentina Mazzurco Miritana [4], Gabriella Di Carlo [3,*], Elsa aria Nhuch [5], Clarissa Martins Leal Schrekker [5], Jeniffer Alves Cuty [5], Henri Stephan Schrekker [5,*], orge Paterakis [1], Charalampos Androulidakis [6], Nikos Koutroumanis [2] and Costas Galiotis [1,2,*]

1 Institute of Chemical Engineering Sciences, Foundation of Research and Technology-Hellas (FORTH/ICE-HT), Stadiou Street, Platani, 26504 Patras, Greece
2 Department of Chemical Engineering, University of Patras, 26504 Patras, Greece
3 Institute for the Study of Nanostructured Materials (ISMN), National Research Council (CNR), SP35d, 9, 00010 Montelibretti, Italy; elena.messina@cnr.it
4 Department of Energy Technologies and Renewable Sources, Italian National Agency for New Technologies, Energy and Sustainable Economic Development (ENEA), Via Anguillarese 301, 00123 Rome, Italy
5 Laboratory of Technological Processes and Catalysis, Institute of Chemistry, Federal University of Rio Grande do Sul, Av. Bento Gonçalves 9500, Porto Alegre 91.501-970, RS, Brazil
6 Skeletal Biology and Engineering Research Center, Department of Development and Regeneration, KU Leuven, O&N1, Herestraat 49, PB 813, 3000 Leuven, Belgium
* Correspondence: ggorgolis@iceht.forth.gr (G.G.); gabriella.dicarlo@cnr.it (G.D.C.); henri.schrekker@ufrgs.br (H.S.S.); c.galiotis@iceht.forth.gr (C.G.)

Abstract: The use and integration of novel materials are increasingly becoming vital tools in the field of preventive conservation of cultural heritage. Chemical factors, such as volatile organic compounds (VOCs), but also environmental factors such as high relative humidity, can lead to degradation, oxidation, yellowing, and fading of the works of art. To prevent these phenomena, highly porous materials have been developed for the absorption of VOCs and for controlling the relative humidity. In this work, graphene and transition-metal dichalcogenides (TMDs) were combined to create three-dimensional aerogels that absorb certain harmful substances. More specifically, the addition of the TMDs molybdenum disulfide and tungsten disulfide in such macrostructures led to the selective absorption of ammonia. Moreover, the addition of the ionic liquid 1-hexadecyl-3-methylimidazolium chloride promoted higher rates of VOCs absorption and anti-fungal activity against the fungus *Aspergillus niger*. These two-dimensional materials outperform benchmark porous absorbers in the absorption of all the examined VOCs, such as ammonia, formic acid, acetic acid, formaldehyde, and acetaldehyde. Consequently, they can be used by museums, galleries, or even storage places for the perpetual protection of works of art.

Keywords: cultural heritage; volatile organic compounds; graphene; transition-metal dichalcogenides; aerogels; ionic liquids

1. Introduction

The air quality to which cultural objects are exposed is highly important for maintaining the appropriate conditions regarding the conservation of an artwork collection. In museums, galleries, and archives, the construction and decoration materials of the building and the technical equipment itself can be important sources of harmful chemical emissions that interact with the artworks exhibited or stored in closed environments [1,2]. The interactions between these harmful chemical substances, which are transferred by air, and the materials of art objects occur in different ways, mainly induced by temperature and relative humidity. Along with the aesthetic effect of pollutants and dust, such reactions cause material degradation and eventually result in disfiguring phenomena on the surfaces of the objects. Extra attention should be paid to containers, such as display cases, storage crates, cabinets, and drawers, where the emission of volatile chemical compounds from constituent materials can affect very badly the containing objects. In all the aforementioned enclosures, the air exchange amount is constrained in the sense that the local emissions will be easily accumulated in the closed environment until reaching concentration levels able to cause alterations to the works of art. Current museum guidelines consider that well-designed display cases are not only designed to provide an object's visibility and physical protection but are also tools for the primary environmental control of the objects [3].

Chemically, the volatile compounds originating from display case materials and other containers used for the needs of museums are mainly organic and categorized as volatile organic compounds (VOCs) [4]. VOCs can be divided according to the boiling points at standard atmospheric pressure of 1 atm of their emissions. Thus, very volatile organic compounds (VVOCs) show boiling points <100 °C, volatile organic compounds (VOCs) are those with boiling points up to 250 °C, and semi-volatile organic compounds (SVOCs) stand for boiling points >250 °C, up to 380–400 °C [5]. These substances can be either present inside the materials as residues from their synthesis or otherwise can be produced after chemical reactions (oxidations and other degradation processes). Some of the most commonly reported VOCs that are detected inside museums are acetaldehyde, formaldehyde, acetic acid, formic acid, and ammonia [6–10]. The control of VOCs inside museums is one of the most effective preventive conservation methods [11], although it is hard to achieve a delay of the more invasive techniques of conservation of artworks [12]. Museums trying to keep their environment stable use various porous materials in their facilities and inside storage crates [8]. Currently, these materials are not so effective and have a limited operational time.

Except for VOCs, mold and fungus growth is one of the least controllable parameters that cause aesthetic and structural alterations to artworks and historical monuments [13]. The fungal contamination of art objects is mostly airborne, with significant seasonal variations. Water availability and high temperature are the most significant factors favoring mold growth [14]. Aside from deteriorative fungi, which affect only the objects, pathogenic fungi can cause health issues to people working with or visiting the contaminated objects [15]. The reason why the art objects are so appealing for fungal thriving is that artworks contain a diversity of organic materials, namely proteinaceous materials (like eggs, bovine milk, animal skin, and bones), polysaccharides, and oils, which in total, comprise the perfect source of nutrients for microorganisms [16]. Within this context, easel paintings of Giorgio Martini from the late nineteenth century showed the attack of different species of microorganisms at different parts of the paintings [16]. For example, *Cladosporium* and *Ulocladium* species, which produce cellulolytic enzymes, were responsible for the damage of canvas, and other species like *Aspergillus* and *Penicillium* attacked the paint binder and caused chromatic alterations and detachment of the support. Fabrics are another category that are mostly affected by microorganisms. Fabrics consist mainly of fibers, and based on the origin of the fibers, they can be divided into natural and chemical [17]. Plant fibers like cotton, linen, hemp, and jute or animal fibers like wool, silk, or leather are considered natural, while chemical fibers consist of modified natural or synthetic molecules like viscose, polyester, acryl, and PVC based [18]. Some microorganisms that have been reported to colonize fabrics of significant historical value are Arthrobacter, Microbispora, Sporocytophaga, Cellulomonas, Bacillus, Cellvibrio, Clostridium, Cytophanga, Pseudomonas, Nocardia, Streptomyces, Aspergillus, Chaetomium, Mnemoniella, Stachybotrys, Verticillium, Penicillium, Mucor, Trichoderma, Myrothecium, Rhizopus, Alternaria, Fusarium, Aureobasidium, and Cladosporium [19]. Mold inside the museums can be prevented by complying with and following the revised environmental guidelines that accept relative humidity between 40–60% for more sustainable storage [20]. Also, conservators and mycologists should collaborate to establish targeted detection and prevention practices for heritage repositories.

Graphene-based aerogels have caught scientific attention for being highly effective in VOC absorption [7–9]. Thanks to their unique structure and high surface area, such materials have been used as superior absorbers of toxic pollutants [21]. Graphene is a hydrophobic material, and its theoretical specific surface area is equal to 2600 m^2/g [22]. Hence, it possesses remarkable adsorption ability for hydrophobic organic compounds and can be exploited as an excellent absorber for air purification. The precursor material for such structures is graphene oxide (GO), which is a sheet of graphene with carboxylic groups at its edges and epoxide/phenolic hydroxyl groups on its basal plane. Reduced graphene oxide (rGO) can be obtained after chemical treatment or thermal annealing, eliminating such functional groups on GO [21]. Compared with GO, rGO can reach a higher absorption capacity for aromatic pollutants thanks to its lower oxygen content, higher hydrophobicity, and higher surface area. Chen et al. [23] examined the absorption of *m*-dinitrobenzene, nitrobenzene, and *p*-nitrotoluene onto GO and rGO, showing that rGO nanosheets have higher absorption capacities for nitroaromatic compounds. Apart from hydrophobicity, π–π interactions also contribute to the strong adsorption of organic molecules onto graphene-related materials [24]. Wang et al. [25] examined the adsorption of phenolic compounds on rGO and concluded that the adsorption depends on π–π interactions between the aromatic molecules and rGO. The degree of rGO reduction and the chemical structure of the phenols can affect the π–π interactions.

Another category of two-dimensional materials in tandem with graphene is the transition-metal dichalcogenides (TMDs), which are known to be able to adsorb environmental pollutants [26–28]. The fascinating characteristics of the TMDs, like the layered structure, tunable bandgap, and unique optical, thermal, and electrical properties, are the main reasons why the potential of these two-dimensional materials is explored for pollution prevention [26]. Graphene can be combined with TMDs and create three-dimensional macrostructures like aerogels with enhanced properties. The as-resultant structures are usually named as *'hybrid'* aerogels [8,29]. Worsley et al. [29] have reported molybdenum disulfide (MoS_2)/graphene aerogels with a high surface area of ~700 m^2/g and an electrical conductivity of 112 S/m. Also, Zhu et al. have explored graphene/MoS_2 aerogels for the absorption of water-soluble organic contaminants [30]. The 3.2 wt.% polydopamine addition resulted in a composite structure with small MoS_2 nano-crystallites homogeneously dispersed over the graphene surface without aggregation.

Ionic liquids (ILs) are a very interesting category of materials and consist of a discrete cation and anion pair. The research on these salts with low melting temperatures was intensified thanks to their tunable properties, high ionic conductivity [31], good thermal stability [32], low flammability [33], negligible vapor pressure [34], and tunable polarity and surface activity [35]. Regarding their biological activity, various ILs have been studied as potential molecules for antitumoral [36] and antimicrobial agents, due to their relatively low toxicity [37,38]. Such biological activity of the ILs strongly depends on the charge, size, alkyl chain, electronegative group, charge distribution on the ions, and small changes in the shape of covalent bonds with protein or micro-molecules interaction, and their thermal properties as well [39–41]. In the case of the imidazolium ILs, 1-hexadecyl-3-methylimidazolium chloride ($C_{16}MImCl$) has been identified as a strong antimicrobial [37,38] and antitumoral agent [36]. The incorporation of this IL in materials represents an effective strategy for the preparation of biologically active materials, including poly(L-lactide)- [42] and high-density polyethylene-based [43] biomaterials, and calcium phosphate-based bionanocomposites [44]. The advantages of using ILs in such processes can be listed as the following. (1) It is considered an easily applicable and cost-efficient way to create tunable, diverse libraries of biologically active compounds with a countless number of combinations of anions and cations. (2) ILs result in control of the ion formation in a solution and the adjustment of the solvation properties in water and biological fluids to provide a reliable solution for solubility and bioavailability. (3) Their hydrophobic/hydrophilic properties, nature of the ionic core, covalent/ionic binding, linker size, and characteristics can be finely tuned via simple organic synthesis procedures. On the other hand, some challenges that are currently faced are the following. (1) Mechanisms of action of many biologically active compounds with ionic nature are not well understood and require further examination. (2) Systematic comparative studies on different types of IL-based drug development systems have not yet been reported, since most of the reported works focus on specific types of IL systems. (3) The relationship between the molecular structure and nano-/microscale arrangement of molecular properties into self-organized structures is at the first stages of understanding [45].

Our research consortium has recently reported on graphene-based aerogels with an IL (C_{16}MImCl) additive, exhibiting anti-fungal and VOC absorption properties [7]. Several IL contents were evaluated, and the graphene aerogel containing 10 wt.% IL was found to have the best anti-fungal activity, preventing the aerogel contamination with *Aspergillus niger*. When exposed to a VOC-saturated micro-environment, this aerogel is highly suitable for the absorption of acetaldehyde, formic acid, acetic acid, and formaldehyde. Since the combined anti-fungal and VOC absorption properties are prerequisites for bringing preventive conservation to a higher level, herein, graphene was combined with MoS_2, tungsten disulfide (WS_2), and C_{16}MImCl to prepare *'hybrid'* absorbent materials for enhanced selective absorptions of pollutants that present, at the same time, anti-fungal activity. It is strongly believed that, apart from the context of the preventive conservation of cultural heritage, these results are of much general importance, including the chemical industry and research laboratories.

2. Experimental Procedure

For the materials, an aqueous solution of GO was prepared by the modified Hummer's method [45,46] and was subsequently diluted in water to obtain a concentration of 1 mg/mL. The commercial WS_2 (Sigma-Aldrich, Steinheim, Germany, 2 µm, 99%) and MoS_2 (Sigma-Aldrich) powders were used as received. C_{16}MImCl was synthesized as described in the literature [47–49].

For the preparation of the *'hybrid'* graphene—TMDs aerogels, the approach presented by Hong et al. [50] for the simultaneous assembly and reduction of GO (the steps for the synthesis of neat rGO aerogel are described in the Supplementary Materials) was adapted. The freeze-drying technique was exploited for solvent sublimation and obtaining the aerogels. Ethanol–water (50/50% *w/w*) solutions of 1 mg/mL of 2D material bulk platelets (MoS_2 and WS_2) were prepared and bath-sonicated for 10 min. The GO solution was mixed with one of the two 2D materials in contents of 90/10%, 70/30%, and 50/50% *w/w* and stirred for 10 min, and then, the same steps were followed as for the preparation of the neat GO (see Supplementary Materials). For the *'hybrid'* aerogels with the IL additive, at the stage of the reducing agents' addition (H_3PO_2 and I_2, see Supplementary Materials), the amount of C_{16}MImCl was added based on the IL:GO weight ratio of 1:10 *w/w*. The as-prepared samples were correspondingly coded as rGO, rGO10IL, rGO10IL/MoS_2, and rGO10IL/WS_2.

For scanning electron microscopy (SEM), SEM images were obtained using a LEO SUPRA 35VP with a maximum resolution of 1.5 nm and 2 nm at high and low vacuum, respectively.

For Raman spectroscopy, Raman spectra were collected using a Renishaw InVia Raman Spectrometer with a 1200 grooves/mm grating for the 785 nm laser excitation and several lenses, such as 20×, 50×, or 100×. The power of the laser beam was kept lower than 1 mW to avoid heating of the specimens.

For the attenuated total reflectance Fourier transform infrared spectroscopy (ATR-FTIR), ATR-FTIR spectra were collected using a Nicolet iS50 spectrometer (Thermo Fisher, Waltham, Massachusetts-United States) equipped with an ATR accessory. The measurements were recorded using a germanium crystal cell using, typically, 32 scans at a resolution of 4 cm^{-1}. No ATR correction has been applied to the data. The range of wave numbers within which the measurements were performed was equal to 450–1650 cm^{-1}.

For X-ray diffraction (XRD), a Bruker D8 Advance X-ray diffractometer was used for performing the XRD measurements. The specimens were examined under ambient conditions.

For X-ray photoelectron spectroscopy (XPS), the surface-analysis measurements were performed in a UHV chamber (P ~5 × 10^{-10} mbar) equipped with a SPECS Phoibos 100-1D-DLD hemispherical electron analyzer and a non-monochromatized dual-anode Mg/Al X-ray source for XPS. The XP Spectra were recorded with MgKa at 1253.6 eV photon energy and an analyzer pass energy of 10 eV giving a full width at half maximum (FWHM) of 0.85 eV for the Ag3$d_{5/2}$ line. The analyzed area was a spot with a 3 mm diameter. The atomic ratios were calculated from the intensity (peak area) of the XPS peaks weighted with the corresponding relative sensitivity factors (RSF) derived from the Scofield cross-section, taking into account the electron transport properties of the matrix (the inelastic mean free path (IMFP). λi and the elastic-scattering correction factor Q depend mainly on the corresponding electron kinetic energy (KE)) and the energy analyzer transmission function. For spectra collection and treatment, including fitting, the software SpecsLab Prodigy with version No 4.12.0r49869 (Specs GmbH, Berlin, Germany) was used. The XPS peaks were deconvoluted with a sum of Gaussian–Lorenzian peaks after a Shirtey-type background subtraction.

For the volatile organic compounds (VOCs) absorption tests, the gas absorption tests were conducted under static conditions in a closed glass desiccator with an excess of pollutant using a saturated vapor stream at room temperature (Figure S2A,B), as described also in detail in the Supplementary Materials file. The absorption capacity [51] of the prepared graphene aerogels was determined by calculating the percentage of weight change:

$$A\% = \frac{last\ weight\ measurement - initial\ weight\ measurement}{initial\ weight\ measurement} \times 100\%$$

Since the graphene-based aerogels reached the saturation point in the volatile gas absorption, which corresponds to the maximum weight change, the weight of the aerogels was stabilized. For the establishment of a baseline for the saturation of the aerogels, after two consequent gravimetric measurements similar to the maximum observed value, the samples were considered saturated and were submitted to the regeneration process. The regeneration was carried out by drying the absorber and removing the absorbed pollutant above its boiling temperature. The regeneration of the hybrid aerogels after their saturation in the VOCs absorbance was performed by heating them (60 °C) with a common electric hair dryer for 24 h (Figure S2C), as published elsewhere [52]. Alternatively, overnight drying at 120 °C in an oven with a vacuum can be also efficiently used. All samples were tested for three absorption–desorption cycles.

For the anti-fungal tests, these tests were conducted in triplicate, placing the aerogels in Petri dishes with a malt agar extract as the culture medium, followed by incubation in the dark at 25 °C. Finally, the fungal growth and, thus, the percentage of Petri dishes colonized by *Aspergillus niger* were verified after 24 h and quantified by image analysis using ImageJ software with version No 1.8.0.

3. Results and Discussion

The two-dimensional materials were kept in bulk form to maintain their ability to trap gaseous pollutants [53], avoiding also the elevated cost of the exfoliation process. Ethanol was added to ensure a good dispersion of the bulk 2D materials, which also significantly stabilized the solutions. It is reported elsewhere that a mixture of 50/50% w/w of ethanol–water results in good dispersion of two-dimensional nanosheets and that ethanol can be removed from hydrogels during the freeze-drying process [53,54]. When only water was used, the platelets tended to separate from the solution, and even though gelation still occurred, there was a non-uniform distribution of the platelets in the formed aerogels. The density of the prepared aerogels was found to lie in the range of 13.9–23.4 mg/cm^3, which is within the desirable range of values, without significant variation between samples. In fact, these values (less than 30 mg/cm^3) are a prerequisite for classifying the 3D porous structure as an aerogel [30].

Since the 2D materials do not contain reactive groups on their surfaces, they were randomly dispersed in the starting GO solution. During the formation of the conjugated network of rGO sheets, the bulk nanosheets of 2D materials got wrapped in the 3D rGO structure, which served as a rigid template for the final 3D porous hydrogel [8]. The result was a 3D rGO foam with two-dimensional nanosheets randomly attached to its pores, as shown in the SEM images of the produced samples. To examine the size and structure of the pores of the prepared aerogels, SEM images for all the IL-free samples with various mixed ratios were also acquired (Figure 1). The platelets of the 2D materials are clearly distinguished and seem to be either attached to the surface of the rGO layers or wrapped around them, creating a robust macro-scale interconnected porous network. Also, the magnitude of the pores tended to increase for the samples with a higher content of the nanosheets, and the prepared aerogels are considered mainly macro-porous structures (because the size of pores exceeded 50 nm). Also, some meso-pores [55] are detectable. It is known that the porosity of carbon aerogels can be controlled by the size and shape of the sheets and the concentration of the precursor GO solution [56]. Since the same GO stock solution was used for the preparation of all the samples, this observation can be explained by considering the concentration of the initial GO solutions for each set of aerogels (90/10, 70/30, and 50/50 rGO/nanosheets). In the preparation of the pure rGO aerogel, the concentration of the used GO was 1 mg/L, while for the rGO/TMD hybrids, the concentration of GO was lower. Even though the starting GO solution for hybrid aerogels had lower GO content, the final volume of these aerogels was the same as that of the neat rGO [8], which means that the porous network inside them should be less dense.

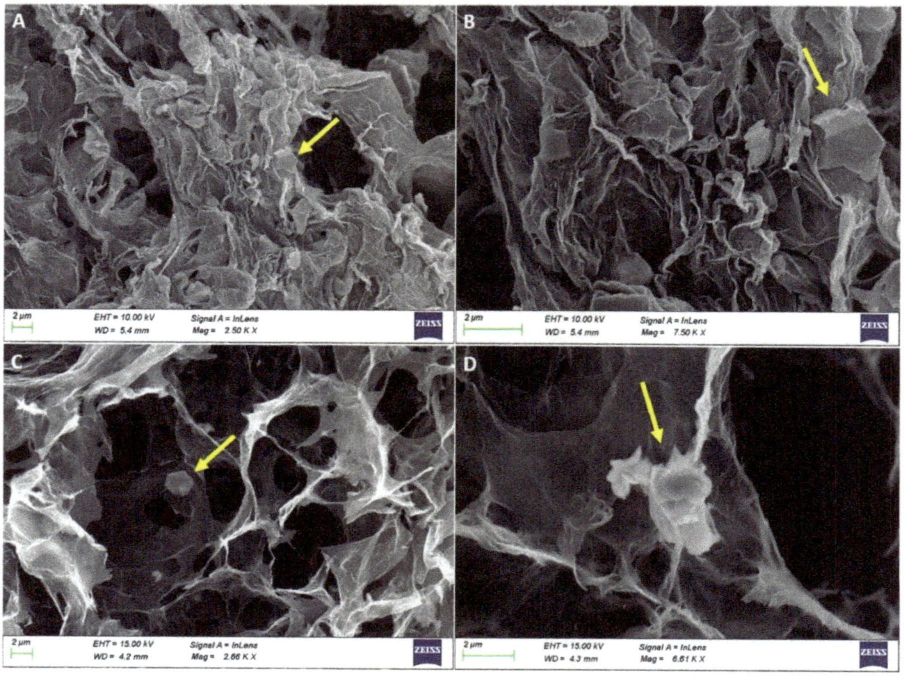

Figure 1. SEM images of rGO/MoS_2 *50/50* (**A**,**B**) and rGO/WS_2 *50/50* (**C**,**D**) aerogels with different scale bars. The yellow arrows point to flakes of MoS_2 or WS_2, correspondingly.

The survey XPS scan (Figure 2) of the rGO/MoS_2 *50/50* aerogel showed, except for the signals of the C, O, and P atoms related to rGO, a spin-orbit doublet at 229.2 eV

($Mo3d_{5/2}$) and 232.4 eV ($Mo3d_{3/2}$), which is attributed to the Mo^{4+} of MoS_2. Furthermore, two S_{2p} peaks at 162.2 eV ($S2p_{3/2}$) and 163.3 eV ($S2p_{1/2}$) were observed, assigned to S^{2-} of MoS_2 [29]. The XPS survey spectrum of the rGO/WS_2 *50/50* aerogel (Figure 3) demonstrated the coexistence of C, O, W, and S atoms. The binding energy of the S_{2p} peak due to S^{2-} appears at 162.2 eV and 163.4 eV, corresponding to $S2p_{3/2}$ and $S2p_{1/2}$, while $W4f_{7/2}$ and $W4f_{5/2}$ are observed at 32.5 eV and 34.7 eV, respectively, indicating the existence of W^{4+} in WS_2 [57]. The sulfur-to-tungsten ratio was calculated to be 2.2, which is slightly S-rich. Additionally, to find the reduction level of the hybrid aerogels with TMDs and compare it to the reduction level in the pure rGO sample, the C:O atomic ratio was calculated for each specimen, subtracting the oxygen concentration due to the presence of P_2O_5 (resulting from the addition of hypophosphorous acid (H_3PO_2), see Supplementary Materials). These levels were equal to 10.4 and 8.3 for the rGO/MoS_2 and rGO/WS_2 samples, respectively. The calculated C:O ratio indicates that the higher the percentage of two-dimensional material contained in a sample, the lower its reduction level. This observation was further assessed by Raman measurements, which are presented below.

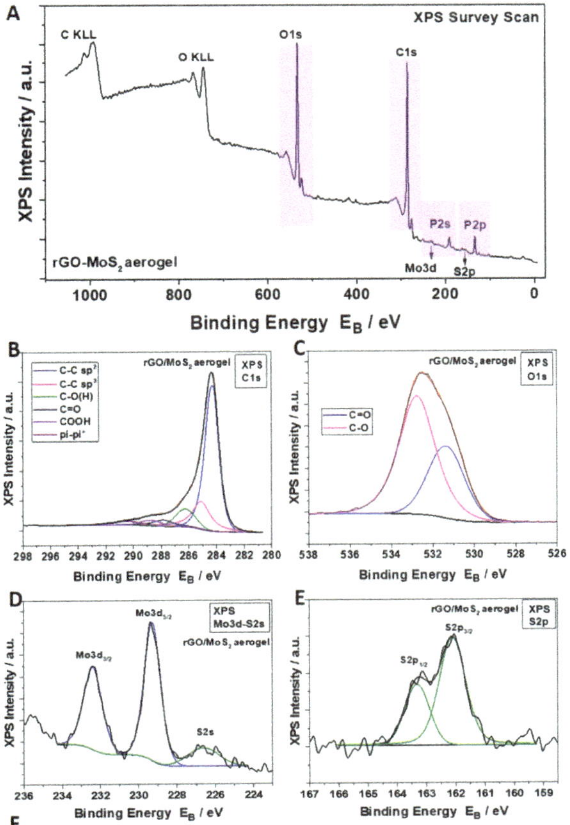

F

sample	% component of C bonds					Relative atomic ratio
rGO/MoS_2	C-C sp^2	C-C sp^3	C-O(H)	C=O	COOH	C:O:TMD:S
	81.1	18.0	9.8	5.1	3.0	74.77:24.58:0.21:0.44

Figure 2. (**A**) XPS survey scan of the rGO/MoS_2 *50/50* hybrid aerogel. Deconvolution of its C_{1s} (**B**) and O_{1s} (**C**) peaks. XPS peaks attributed to its Mo^{4+} of MoS_2 (**D**) and S^{2-} of MoS_2 (**E**). (**F**) Percentages of C_{1s} components derived from the C_{1s} peak deconvolution and relative atomic ratio C:O:TMD:S of this hybrid aerogel.

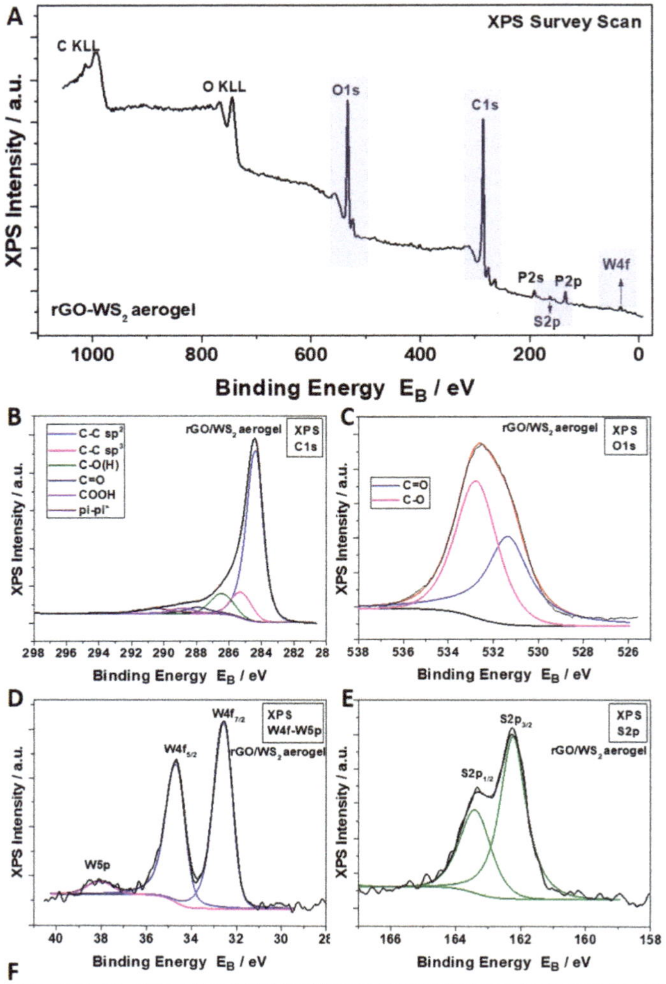

F

sample	% component of C bonds					Relative atomic ratio
rGO/WS_2	C-C sp^2	C-C sp^3	C-O(H)	C=O	COOH	C:O:TMD:S
	70.9	11.3	10.1	4.9	2.7	77.5:21.4:0.34:0.76

Figure 3. (**A**) XPS survey scan of the rGO/WS_2 *50/50* hybrid aerogel. Deconvolution of its C1s (**B**) and O1s (**C**) peaks. XPS peaks attributed to its W^{4+} of WS_2 (**D**) and S^{2-} of WS_2 (**E**). (**F**) Percentages of C1s components derived from the C1s peak deconvolution and relative atomic ratio C:O:TMD:S of this hybrid aerogel.

Figure 4 shows the results of XRD and Raman measurements for the hybrid samples. Both analyses proved the presence of rGO and TMDs in the hybrid aerogels. The XRD plot of a neat rGO aerogel shows a broad peak at $2\theta = 26°$, which corresponds to the (002) plane of graphite [58]. For the rGO/TMD *50/50* hybrid aerogels, the obtained XRD spectra confirm the crystalline nature of MoS_2 and WS_2 [56,58], and a lower reduction level of GO as the peak of rGO shifted to lower 2θ values compared with the 26° of neat rGO [59]. The Raman spectra of a neat rGO aerogel comprise several frequency bands, each one assigned to a specific structural configuration [60]. For the main Raman peak, the so-called G peak is located at ~1590 cm^{-1} and is related to the E_{2g} mode that corresponds to the in-plane stretching of C=C bonds [61]. Another prominent Raman peak that has been associated with the presence of disorder has been termed the D peak, and it is present at about 1356 cm^{-1}. It corresponds to a breathing vibrational mode, and its intensity is associated with the number of lattice defects or discontinuities, such as flake edges [51,61]. The two main Raman peaks of GO, G and D, are very sensitive to small sp^3/sp^2 ratio changes. Also, the intensity ratio of these two peaks $I_{D/G}$ is correlated to the changes in the sp^3/sp^2 ratio and can be used to confirm the reduction of GO. From the calculated ratio of D/G peaks, it is evident that the highest reduction occurred in the neat rGO aerogel, while the lowest ones occurred in hybrid aerogels with the highest 2D materials content. The main Raman peaks found for the MoS_2 aerogel, except the peaks of rGO, are those at 408 cm^{-1} (first-order mode A_{1g}) and 373 cm^{-1} (second-order mode E_{2g}), and for the WS_2 aerogel, there are those at 420 cm^{-1} and 355 cm^{-1} [62,63]. These peaks' positions are in good agreement with similar values from the literature for nanoparticles that are composed of multiple layers [29].

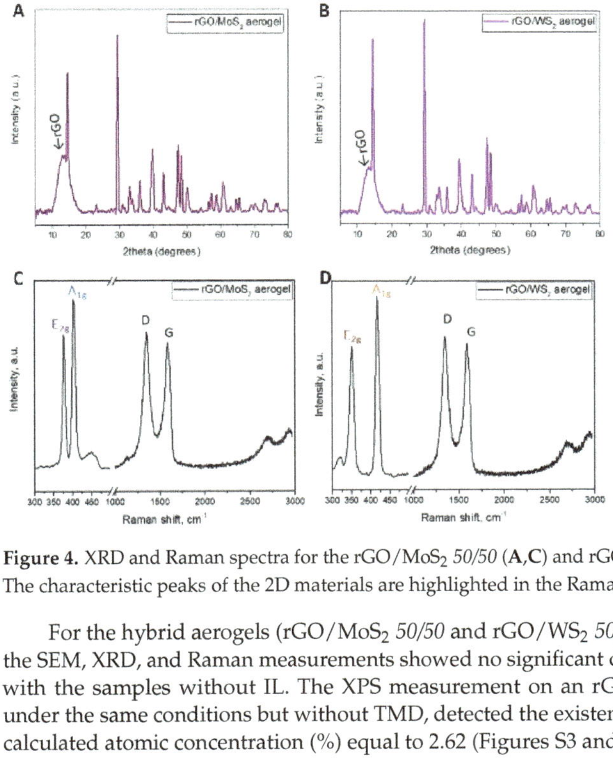

Figure 4. XRD and Raman spectra for the rGO/MoS_2 *50/50* (**A**,**C**) and rGO/WS_2 *50/50* (**B**,**D**) aerogels. The characteristic peaks of the 2D materials are highlighted in the Raman spectra.

For the hybrid aerogels (rGO/MoS_2 *50/50* and rGO/WS_2 *50/50*) with the IL additive, the SEM, XRD, and Raman measurements showed no significant differences in comparison with the samples without IL. The XPS measurement on an rGO/IL aerogel, prepared under the same conditions but without TMD, detected the existence of nitrogen (N) with a calculated atomic concentration (%) equal to 2.62 (Figures S3 and S4, Table S1).

VOCs Absorption

The prepared hybrid aerogels were tested for sorption of formaldehyde (CH_2O), acetic acid (CH_3COOH), formic acid (CH_2O_2), acetaldehyde (CH_3CHO), and ammonia (NH_3), as well as water vapor (humidity), which are the most common pollutants in places where artworks are exhibited.

The regeneration ability was examined, while the significance of this process lies in the necessity of recyclability of the 3D material from an environmental, economic, and practical perspective. Ideally, absorbent material is desired to maintain its absorption ability after many regeneration cycles. From the absorption results summarized in Figures 5 and 6, the incorporation of TMDs in rGO aerogels favored the selective absorption of ammonia, while the absorption capacity increased with increasing TMD content. Thus, the most efficient ammonia absorbers were the hybrid aerogels with the highest amount of MoS_2 and WS_2, namely the *50/50* samples. For the assessment that the higher increase in weight of the samples with TMDs was not due to enhanced humidity absorption, similar sets of samples were tested under 55% RH (inside a desiccator with a saturated aqueous salt solution [64]), and did not show any weight increase. This means that the weight increase of the aerogels was caused only by ammonia sorption. Meanwhile, for the cases of formaldehyde, acetic acid, formic acid, and acetaldehyde, the opposite trend was observed for the aerogels containing MoS_2, with the rGO/MoS_2 *90/10* sample being the most efficient absorber. At the same time, the hybrid aerogels with WS_2 showed a dramatic decrease in absorption capacity for the examined VOCs (apart from ammonia), particularly for the *50/50* sample, and a 40% decrease for both formaldehyde and acetic acid was recorded. An important note is that the humidity absorption at 75% RH was insignificant, in the range of 4–6%, while at 99% RH, the absorption capacity was decreasing with the increasing TMD content (rGO/MoS_2 *90/10*: 32.4%, rGO/MoS_2 *70/30*: 22.2%, rGO/MoS_2 *50/50*: 6.7%, GO/WS_2 *90/10*: 31.8%, rGO/WS_2 *70/30*: 16.6%, and rGO/WS_2 *50/50*: 5.6%, compared with the neat rGO aerogel). This behavior can be ascribed to the hydrophobic nature of TMDs. As a result, water molecules are physiosorbed with a low degree of charge transfer on TMDs [65]. When the performance of the proposed *'hybrid'* aerogels of this study is compared with the corresponding ones of neat rGO aerogels (Figure S5) [7], the absorbance values are found to be two-to-three (2–3) times lower on average for all the examined VOCs, apart from ammonia. For the case of ammonia, the neat rGO aerogel shows an absorption of only 84% (Figure S5), while the corresponding values for rGO/MoS_2 and rGO/WS_2 are equal to 273% and 232%. This difference of one order of magnitude in the absorption performance reveals the selectivity for ammonia of the *'hybrid'* aerogels.

There is extended research on ammonia detection using TMDs, so their selective absorption of ammonia after incorporation in rGO aerogels was maintained [64,65]. In the case of neat rGO aerogels, the absorption of ammonia takes place only on oxygen-containing functional groups, such as epoxy and hydroxyl, through hydrogen bonding [66,67], while when adding MoS_2 and WS_2 into the rGO structure, practically active adsorbing sites with selectivity to ammonia are added [68–70]. Owing to the existence of a lone pair of electrons, ammonia behaves as a charge donor to provide electrons to get physiosorbed on TMDs. Due to the strong electronegativity of the elemental sulfur layer of the TMD layers, NH_3 molecules are easily adsorbed to the edge sites of the few-layered and bulk TMDs and transfer electrons to the TMD layers to form NH_4^+. Thus, NH_4^+ ions adhere to the edges of two adjacent elemental sulfur layers and expand the interlayer space. Subsequently, more NH_3 molecules enter deeper inside, expand the interlayer space, and eventually fill the entire interlayer space [64,65,71–73]. Another observation from the absorption-capacity results (Figures 5 and 6) is that MoS_2 samples exhibited higher ammonia absorption relative to the WS_2 samples. This can be related to the larger radius of W^{4+} in WS_2, resulting in a weaker bonding ability to ammonia than for Mo^{4+} in MoS_2, which is also supported by the calculated absorption energy for ammonia of -216 eV for WS_2 and -250 eV for MoS_2. That practically means that the desorption of ammonia from WS_2 becomes easier compared to that from MoS_2 at room temperature. This phenomenon also has been found to contribute to the better recovery feature of the WS_2-based ammonia sensors, when compared to the MoS_2 ones [53].

Regarding the other VOCs, the addition of TMDs into rGO aerogels resulted in lower absorption capacities, which has been also reported elsewhere [53]. TMDs cannot interact with VOCs through π–π stacking or hydrogen bonding, but only via electrostatic attraction forces. The flakes of TMDs (*p*-type), which were dispersed in the three-dimensional pores of rGO aerogels, interact with some active adsorption sites, oxygen-containing groups (*n*-type), and form *p*–*n* junctions, leading to less active sites of rGO aerogels for absorbing VOCs. Furthermore, the TMDs are hydrophobic, especially tungsten disulfide, and this is not favorable for attracting molecules of formaldehyde, formic acid, acetic acid, and acetaldehyde. Despite this drawback, there is a positive aspect to the absorption of formic acid under high RH. As the hydrophobicity of TMDs disfavors the competitive adsorption of water molecules, the hybrid aerogels favor the absorption of formic acid molecules. The hybrid aerogels with TMDs maintained similar absorption capacities of ammonia through all the absorption–desorption cycles (Figures 5 and 6). This behavior can be attributed to the reversible interaction of ammonia with TMDs. Moreover, Raman spectroscopy revealed that ammonia promoted a further reduction of the hybrid aerogels with TMDs, which did not occur with the other examined VOCs. More specifically for the rGO/WS_2 *50/50* aerogel, the ratio of the intensities of characteristic D and G peaks was increased from 1.01 $\pm$ 0.06 to 1.05 $\pm$ 0.016. A red shift of the G peak was detected from 1591 cm^{-1} to 1587 cm^{-1}, proving the further reduction of rGO/TMDs aerogels. Contrarily, the absorption capacities and saturation times of formaldehyde decreased after regeneration. This can be justified by the fact that TMDs do not exhibit selective adsorption of formaldehyde, so all the formaldehyde molecules were attached to rGO active sites, leading to degradation. Meanwhile, the acquired Raman spectra detected no further reduction of rGO.

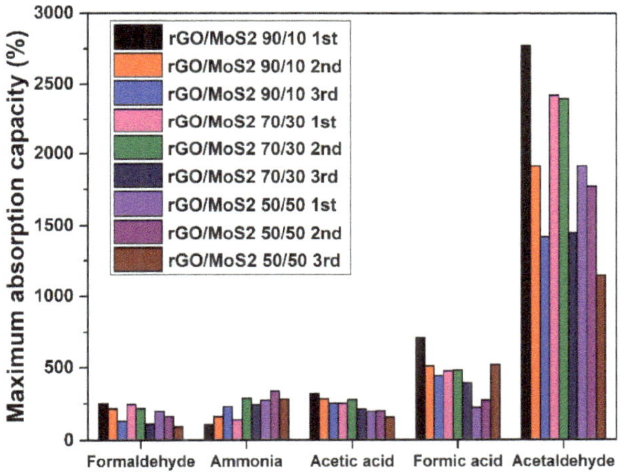

Figure 5. Maximum absorption capacities of rGO/MoS_2 aerogels for three absorption-desorption cycles with VOCs.

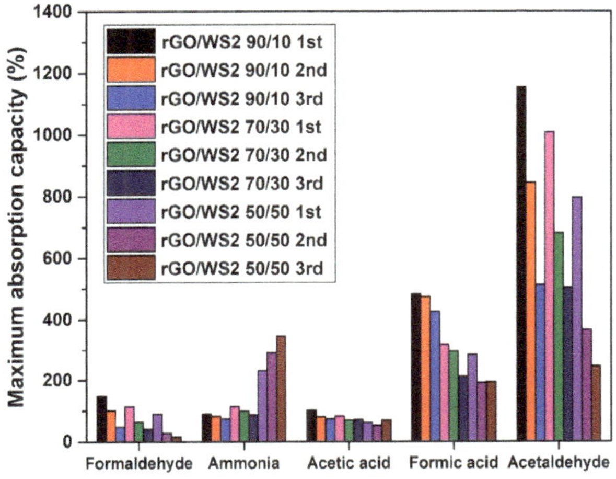

Figure 6. Maximum absorption capacities of rGO/WS_2 aerogels for three absorption–desorption cycles with VOCs.

Although TMDs are prone to oxidation, WS_2 is considered to be more stable than MoS_2. When water and pollutant molecules are physiosorbed on TMDs, they can be transformed into semiconducting metal oxides (MoO_3, WO_3). Yet, they maintain their sensing and selectivity properties [68]. In the case of acetic acid, some fluctuations of the acquired data were observed, and these can be attributed to slight oxidation after exposure to the acid, as shown from Raman measurements. Regarding the saturation time of the samples, in general, the trend remained the same for all the VOCs except ammonia, which showed better performance.

As shown in Figures 7 and 8, the addition of the IL (10% w/w related to GO) resulted in increased absorption capacities for all examined VOCs. ILs have been reported to exhibit remarkable affinities for VOCs [74]. C_{16}MImCl should have induced a significant increase in hydrophobicity, enhancing the surface potential of the GA-based surfaces [75] and turning the interaction with the VOC molecules stronger [76]. Furthermore, such IL favors the formation of strong hydrogen bonds with VOCs [77]. According to molecular interactions between the VOCs and the ions of the ILs, the VOCs do have preferential adsorption near the IL interfacial region compared to the bulk region [78]. The VOC molecules have been reported to prefer to reside near the IL/VOC interface, while, also, the longer the length of alkyl substituents of the imidazolium cation is, the higher the solubility of VOCs in ILs [79]. Acetaldehyde was absorbed the most, and this could be attributed to its high volatility, which depends on its boiling point and saturation vapor pressure [80]. In addition, for acetaldehyde, there is a synergistic action of dipole–dipole and hydrogen bonding interactions with the GA surface [81].

Even though some of the aerogels showed lower performances after the first absorption–desorption cycle, the samples exhibited significantly higher absorption capacities than the other relevant reported materials [8]. Similar experiments have been performed for testing reduced graphene oxide-based porous materials as gas absorbers for various VOCs, including benzene, toluene, carbon dioxide, ammonia, acetic acid, and formaldehyde [8]. In addition to rGO aerogels, various other materials have been studied as pollutant removers, ranging from activated carbons to zeolites and from metal–organic frameworks (MOFs) to hyper-crosslinked polymeric resins (HPRs). Thanks to their large surface areas, these porous materials are capable of trapping volatile pollutants either by physical or chemical absorption [81,82]. Nevertheless, several drawbacks of commercially available porous materials have been discovered when it comes to VOC removal. For example, activated carbon cannot capture VOCs with low molecular weight like formaldehyde, and other materials may not be effective in capturing polar VOCs like ammonia [83]. Hence, the selectivity, as well as the reusability of these materials, needs improvement. Also, the manufacturing costs of the materials play a crucial role. In fact, to make the materials attractive for practical applications in museums and art galleries, where the real needs are, for thousands of display cases or storage boxes, the production costs should be as low as possible. Activated carbons, natural porous materials, and natural zeolites represent reasonable compromises between production costs and environmental impact. Indeed, activated carbons have tremendous potential thanks to their low manufacturing costs and thanks to the biomass nature of the raw materials, which could be coconut shells, walnut shells, woods, and many more. Also, some natural porous materials, such as diatomite, stellerite, and vitric tuff, have low environmental impact and could be used as VOC removers after easy activation steps. Despite this, both natural porous materials and activated carbons have low VOC absorption capacities, and even more importantly, their powdered nature makes their practical usage extremely limited (e. g. they need filters, boxes, etc.).

Some of the most common commercial absorbent materials were used for the benchmarking study, like activated carbon, silica gel, and polyurethane. These materials were dried and placed inside the desiccators with the same VOCs, and the calculated absorption capacities and kinetics are presented in Figure S6. This confirmed that commercial absorbent materials exhibit significantly lower absorption performance than graphene-based aerogels without any particular selectivity. Comparing the best commercial absorbent with the best aerogel sample for each examined VOC, the rGO/MoS_2 and rGO/WS_2 *50/50* samples showed about 530% more weight increase for ammonia absorption than polyurethane. Also, it is worth mentioning that the superior performance of the presented materials was achieved by using much less weight (a few milligrams) compared with the commercial absorbents (a few grams). Additionally, the commercial materials were tested for humidity absorption under the same RH conditions to examine whether their weight increase was due to water or VOC absorption. For 55% RH, polyurethane absorbed significantly less humidity than activated carbon and silica gel. So, it can be concluded that polyurethane has good selectivity for ammonia and acetic acid over water molecules. In the case of 75% RH, polyurethane again exhibited significant selectivity for formaldehyde absorption over humidity, contrary to activated carbon and silica gel. Finally, for 99% RH, none of the commercial absorbents had selectivity for formic acid absorption over humidity. Thus, it can be also deduced that the detected weight increase of the examined commercial materials after exposure to formic acid was entirely because of humidity absorption. As a consequence, this benchmarking study showed that the graphene/TMDs-based aerogels exhibit superior absorption properties and gas selectivity compared to the available commercial products.

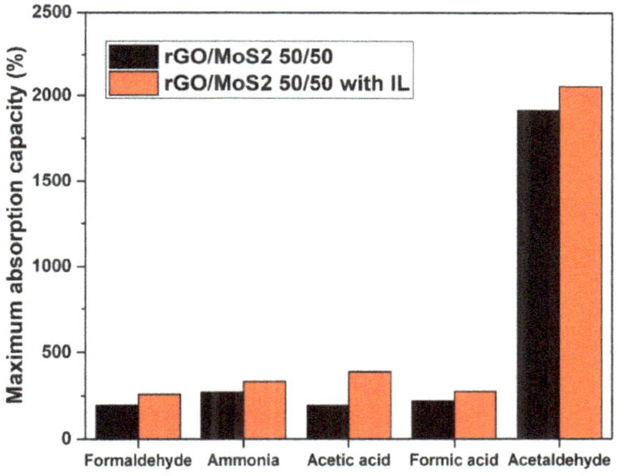

Figure 7. Effect of the IL on the maximum absorption capacities of rGO/MoS_2 *50/50* aerogels.

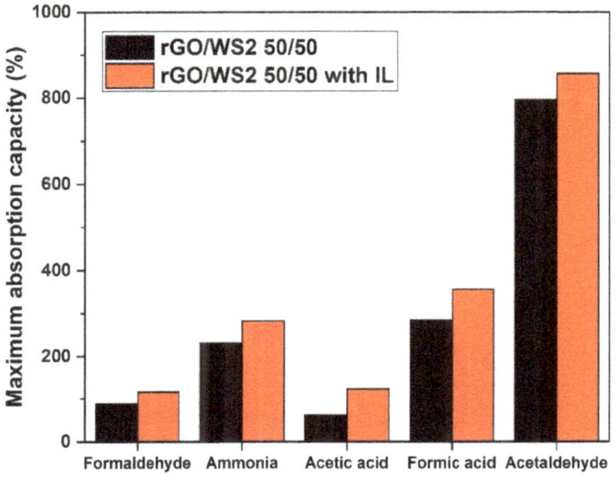

Figure 8. Effect of the IL on the maximum absorption capacities of rGO/WS_2 *50/50* aerogels.

The as-prepared 'hybrid' aerogels (without the addition of the IL) were put into the absorption system to investigate their interactions with VOCs by FTIR analysis, as described in the experimental procedure and in a previous work [7], followed by desorption treatments to remove the physiosorbed pollutants and evaluate their ability to capture VOCs with a stable interaction. Representative results of the absorption tests after VOC treatments are reported in Figure 9A for rGO/WS_2 *50/50* and in Figure 9B for rGO/MoS_2 *50/50*. For all the rGO/TMD aerogels (untreated = UTT and treated = TT), there are three important bands of the oxygen-containing epoxide functional group at ~1154 cm^{-1} (C-O-C stretching peak), ~1000 cm^{-1} (C-O-H deformation peak), and ~800 cm^{-1} (C-OH stretching peak). The band at 1567 cm^{-1} is attributed to the C=C skeletal vibration of rGO/TMD [84]. Additionally, an extra peak emerged at ~470 cm^{-1}, which is characteristic of the Mo-S stretching mode of vibration, confirming the successful incorporation of MoS_2 in the rGO/TMD (identified with a red star in Figure 9B) [85].

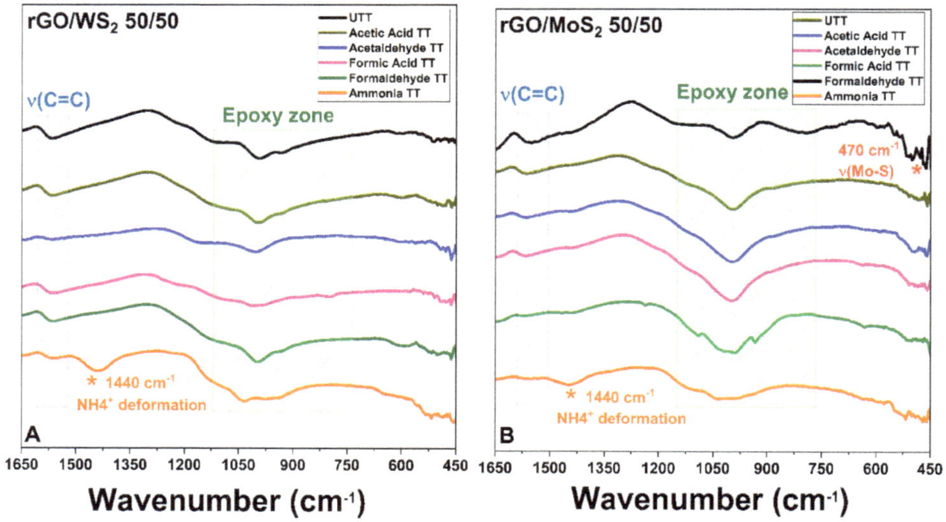

Figure 9. FTIR spectra of rGO/TMD ((**A**) for rGO/WS_2 *50/50* and (**B**) for rGO/MoS_2 *50/50*) after absorption tests with acetic acid (gold line), acetaldehyde (blue line), formic acid (pink line), formaldehyde (olive line), and ammonia (orange line). The spectra of aerogels as prepared (UTT), after exposure to the pollutant for 168 h (TT), and after the desorption test are compared.

The doping of rGO aerogels with TMD nanosheets (MoS_2 or WS_2) introduced TMD-originated acidic centers and resulted in rGO with a higher content of functional groups due to a lower degree of reduction. As a consequence, the surfaces of these materials became more sensitive and selective to the attachment of ammonia molecules, which was shown in the absorption-capacity experiments (Figure 5 for rGO/MoS_2 aerogels and Figure 6 for rGO/WS_2 aerogels). The FTIR spectra acquired after the exposure of rGO/TMD aerogels to ammonia show the ability to capture this VOC with a stable interaction. The most visible band at 1440 cm^{-1} can be attributed to the characteristic NH deformation vibration of NH_4^+ (identified with an orange star for rGO/TMDs TT) [86]. The formation of NH_4^+ was due to ammonia hydrolysis in contact with atmospheric or adsorbed water molecules. For all the tested VOCs, the FTIR spectra of the rGO/TMD aerogels exposed to a VOC show changes compared with the spectra of their untreated equivalents. This confirms the occurrence of chemical changes in the surfaces of these aerogels after treatments with VOCs.

The antifungal properties of rGO and rGO-based hybrid aerogels have been investigated by evaluating their capacity to inhibit the proliferation of *Aspergillus niger*. This fungus was selected as a model microorganism, representative of those commonly detected in museums [87]. In particular, the effect of TMDs (MoS_2 and WS_2) on the antifungal properties of rGO aerogels (rGO/TMDs *50/50*), without or with the IL, was studied. This was evaluated by the ability of these aerogels to inhibit fungal growth in the surrounding environment. The results reported in the following Table 1 show that the addition of 10 wt.% of IL improved the antifungal property of the rGO aerogel, significantly reducing the average percentage of the Petri dish covered by the fungus, which was demonstrated in our previous work [7]. By embodying a TMD (MoS_2 or WS_2) in the aerogels with the IL additive, increased surface areas of the Petri dishes were covered by the fungus, although an inhibition effect was maintained. The results presented in Figure 10 show that all the aerogels were able to inhibit the fungal growth in the culture medium. Representative pictures of the Petri dishes after 24 h of incubation are shown in Figure 11, visualizing the inhibition effect of the different formulations. The obtained results clearly indicate the positive effect of IL. Although this effect was partially lost by the incorporation of a TMD (likely the result of a strong interaction between the TMD and the IL), it is worth noting that all rGO-based aerogels considered in this study were effective in preserving, to varying degrees, their surfaces and surroundings from fungal attack. The latter properties were mainly improved through the addition of IL.

Table 1. Percentage of Petri dish surface covered by the fungus.

	Average %	St.D.
rGO	52.27	34.20
rGO + IL 10%	26.56	22.48
rGO/WS_2 + IL 10%	79.28	5.29
rGO/MoS_2 + IL 10%	72.93	6.50

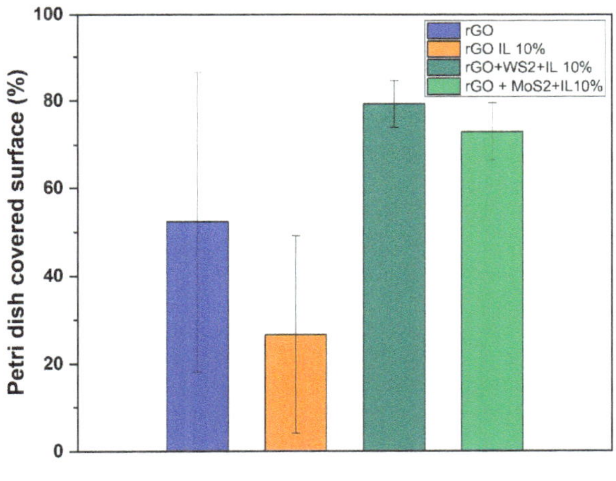

Figure 10. Graph of the percentage of the Petri dishes colonized by the fungus.

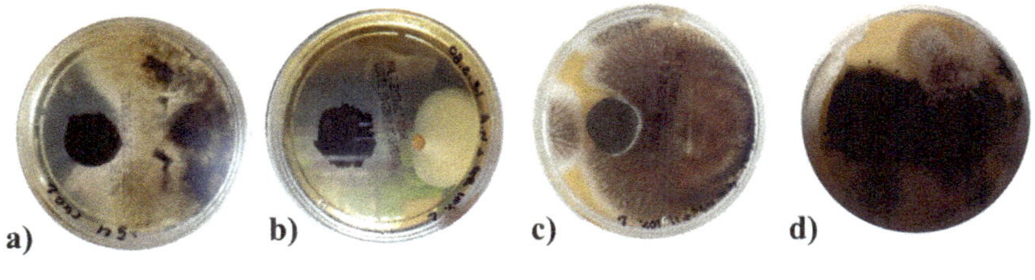

Figure 11. Representative pictures of the Petri dishes colonized by the fungus *Aspergillus niger* in the presence of (**a**) rGO, (**b**) rGO + IL 10%, (**c**) rGO/WS_2 + IL 10%, and (**d**) rGO/MoS_2 + IL 10% aerogels.

4. Conclusions

In conclusion, macroporous rGO/TMD (MoS_2 or WS_2) hybrid aerogels, without or with the IL C_{16}MImCl, were successfully prepared through freeze-drying as novel absorbers of VOCs with antifungal properties. These materials can be created with varying content of the examined TMDs in the rGO backbone, specifically *90/10*, *70/30*, and *50/50% w/w* (TMD/starting GO). For the rGO/TMDs *50/50* samples, SEM confirmed the macroporosity with clearly shown TMD crystals, while with the assistance of XPS, XRD, and Raman spectroscopy, the characteristic peaks of the TMDs were detected, except the already shown ones of the rGO. All hybrids showed high absorption rates with ammonia, formic acid, acetic acid, formaldehyde, and acetaldehyde, which was absorbed the most, exceeding the corresponding performances of other benchmark products, like activated carbon, silica gel, and polyurethane. As these superior performances were maintained in three cycles of absorption–desorption tests, these hybrids can be effectively reused after a simple drying process. The addition of a TMD enhanced the absorption of ammonia, providing materials for the selective absorption of ammonia with the rGO/TMDs *50/50* exhibiting the highest performance. When the rGO/MoS_2 structure is compared with the rGO/WS_2 one, the absorption of ammonia for the first structure is clearly higher than the latter one. FTIR spectroscopy was also used for the detection of the absorption of all the examined pollutants, while especially for ammonia, a stable interaction was deduced. Especially after the ammonia absorption, the most visible band at 1440 cm^{-1} was attributed to the characteristic NH deformation vibration of NH_4^+. The formation of NH_4^+ was due to ammonia hydrolysis in contact with atmospheric or adsorbed water molecules. This finding confirms the occurrence of chemical changes in the surfaces of these aerogels after the treatment with this VOC, rendering a spectroscopic proof of their absorbing ability. C_{16}MImCl as an additive potentialized the absorption ability for all the examined VOCs, without altering the previously-reported selectivity, and the anti-fungal activity against the fungus *Aspergillus niger*, which is a model microorganism and representative of those commonly detected in museums, galleries, private collections, et al. This anti-fungal intrinsic property was evaluated by the ability of these materials to inhibit fungal growth in their surrounding environment. From the results, it is shown that the incorporation of TMDs in the aerogels with IL additive increased the surface areas of the Petri dishes that were covered by the fungus, but a significant inhibition effect was maintained. All examined formulations were able to inhibit fungal growth in the culture medium, indicating, thus, the positive effect of IL. All the aforementioned findings turn these hybrid aerogels into promising tools for the preventive conservation of cultural heritage.

Supplementary Materials: The following supporting information can be downloaded at: https://www.mdpi.com/article/10.3390/ma17133174/s1, Figure S1: Schematic representation of the experimental process of rGO aerogels production. Figure S2: (A,B) Experimental set-up for the measurement of VOCs absorption from the prepared graphene aerogels. (C) Regeneration of aerogels using an electric hairdryer. Figure S3: XPS Survey Scan of sample. Figure S4: XPS Deconvoluted C1s of sample. Figure S5: VOCs absorption results for neat rGO aerogel sample. Figure S6: (A) Maximum absorption capacity of commercial absorption materials (activated carbon, silica gel and polyurethane) for the examined VOCs. (B–E) Kinetics of absorption of commercial absorption materials for the examined VOCs; Table S1: Quantification, Table S2: Maximum absorption capacity of commercial absorption materials (activated carbon, silica gel and polyurethane) for the examined VOCs, Table S3: Maximum absorption capacity of commercial absorption materials (activated carbon, silica gel and polyurethane) for the examined RH conditions. References [46,47,88–96] are cited in the Supplementary Materials.

Author Contributions: G.G., M.K., E.M., V.M.M., G.D.C., C.A., E.L.N., C.M.L.S. and H.S.S. designed and performed the experiments. G.G., E.M. and G.D.C. interpreted the data. G.P. and N.K. synthesized graphene oxide. G.G., E.M., G.D.C., J.A.C. and H.S.S. drafted the manuscript. G.G., G.D.C., E.M., H.S.S. and C.G. conceived the work, participated in its design and coordination, supervised all experimental procedures, and revised the manuscript. All authors have read and agreed to the published version of the manuscript.

Funding: This research is part of the GREENART, "GREen ENdeavor in Art ResToration", project that received funding from the European Union's Horizon Europe research and innovation program under grant agreement No [101060941].

Institutional Review Board Statement: Not applicable.

Informed Consent Statement: Not applicable.

Data Availability Statement: The original contributions presented in the study are included in the article/Supplementary Materials, further inquiries can be directed to the corresponding authors.

Acknowledgments: Monica Albini is acknowledged for the support in the antifungal tests. Also, the authors acknowledge and thank the Laboratory of Electron Microscopy and Microanalysis (L.E.M.M.) located at the Department of Biology at the University of Patras for providing us with access to their freeze-drying instrument.

Conflicts of Interest: The authors declare no conflict of interest.

ferences

Uhde, E.; Salthammer, T. Impact of reaction products from building materials and furnishings on indoor air quality—A review of recent advances in indoor chemistry. *Atmos. Environ.* **2007**, *41*, 3111–3128. [CrossRef]

Salthammer, T. Emissions of Volatile Organic Compounds from Products and Materials in Indoor Environments. *Handb. Environ. Chem.* **2004**, *4*, 37–71.

Raphael, T.; Burke, M. A set of conservation guidelines for exhibitions. *Objects Spec. Group Postprints* **2000**, *7*, 5–20.

Druzik, J.R.; Hatchfield, P.B. Pollutants in the Museum Environment: Practical Strategies for Problem Solving in Design, Exhibition and Storage. *J. Am. Inst. Conserv.* **2003**, *42*, 485–488. [CrossRef]

Chiantore, O.; Poli, T. Indoor air quality in museum display cases: Volatile emissions, materials contributions, impacts. *Atmosphere* **2021**, *12*, 364. [CrossRef]

Kotsidi, M.; Gorgolis, G.; Carbone, M.G.P.; Paterakis, G.; Anagnostopoulos, G.; Trakakis, G.; Manikas, A.C.; Pavlou, C.; Koutroumanis, N.; Galiotis, C. Graphene nanoplatelets and other 2D-materials as protective means against the fading of coloured inks, dyes and paints. *Nanoscale* **2023**, *15*, 5414–5428. [CrossRef]

Gorgolis, G.; Messina, E.; Kotsidi, M.; Staccioli, M.P.; Nhuch, E.L.; Di Carlo, G.; Schrekker, H.S.; Paterakis, G.; Koutroumanis, N.; Galiotis, C. Antifungal Graphene-based Absorbers as Advanced Materials for Preventive Conservation of Cultural Objects. *ChemNanoMat* **2022**, *8*, 202200265. [CrossRef]

Androulidakis, C.; Kotsidi, M.; Gorgolis, G.; Pavlou, C.; Sygellou, L.; Paterakis, G.; Koutroumanis, N.; Galiotis, C. Multi-functional 2D hybrid aerogels for gas absorption applications. *Sci. Rep.* **2021**, *11*, 13548. [CrossRef] [PubMed]

Gorgolis, G.; Ziemann, S.; Kotsidi, M.; Paterakis, G.; Koutroumanis, N.; Tsakonas, C.; Anders, M.; Galiotis, C. Novel Graphene-Based Materials as a Tool for Improving Long-Term Storage of Cultural Heritage. *Materials* **2023**, *16*, 3528. [CrossRef]

Kotsidi, M.; Gorgolis, G.; Carbone, M.G.P.; Anagnostopoulos, G.; Paterakis, G.; Poggi, G.; Manikas, A.; Trakakis, G.; Baglioni, P.; Galiotis, C. Preventing colour fading in artworks with graphene veils. *Nat. Nanotechnol.* **2021**, *16*, 1004–1010. [CrossRef]

Odlyha, M.; Mottner, P.; Thickett, D.; Ryhl-svendsen, M.; Schmidbauer, N.; Scharff, M. Measurement of Organic and Inorganic Pollutants in Microclimate Frames for paintings. *Museum* **2010**, *7*, 59–70.

Baglioni, P.; Carretti, E.; Chelazzi, D. Nanomaterials in art conservation. *Nat. Nanotechnol.* **2015**, *10*, 287–290. [CrossRef] [PubMed]

Cappitelli, F.; Cattò, C.; Villa, F. The control of cultural heritage microbial deterioration. *Microorganisms* **2020**, *8*, 1542. [CrossRef] [PubMed]

Sterflinger, K.; Pinzari, F. The revenge of time: Fungal deterioration of cultural heritage with particular reference to books paper, and parchment. *Environ. Microb.* **2012**, *14*, 559–566. [CrossRef] [PubMed]

Crook, B.; Burton, N.C. Indoor moulds, Sick Building Syndrome and building related illness. *Fungal Biol.* **2010**, *24*, 106–113. [CrossRef]

Ahmed, H.E.; Kolisis, F.N. A Study on Using of Protease for Removal of Animal Glue Adhesive in Textile Conservation. *Appl. Polym. Sci.* **2012**, *124*, 3565–3576. [CrossRef]

Pekhtasheva, E.; Neverov, A.; Kubica, S.; Zaikov, G. Biodegradation and Biodeterioration of some Natural Polymers. *Chem. Chem. Technol.* **2012**, *6*, 263–280. [CrossRef]

Branysova, T.; Demnerova, K.; Durovic, M.; Stiborova, H. Microbial biodeterioration of cultural heritage and identification of the active agents over the last two decades. *J. Cult. Herit.* **2022**, *55*, 245–260. [CrossRef]

Pietrzak, K.; Puchalski, M.; Otlewska, A.; Wrzosek, H.; Guiamet, P.; Piotrowska, M.; Gutarowska, B. Microbial diversity of pre-Columbian archaeological textiles and the effect of silver nanoparticles misting disinfection. *J. Cult. Herit.* **2017**, *23*, 138–147. [CrossRef]

Bastholm, C.J.; Madsen, A.M.; Andersen, B.; Frisvad, J.C.; Richter, J. The mysterious mould outbreak—A comprehensive fungal colonisation in a climate-controlled museum repository challenges the environmental guidelines for heritage collections. *J. Cult. Herit.* **2022**, *55*, 78–87. [CrossRef]

21. Yu, S.; Wang, X.; Ai, Y.; Tan, X.; Hayat, T.; Hu, W.; Wang, X. Experimental and theoretical studies on competitive adsorptio aromatic compounds on reduced graphene oxides. *J. Mater. Chem. A* **2016**, *4*, 5654–5662. [CrossRef]
22. Zhao, J.; Wang, Z.; Zhao, Q.; Xing, B. Adsorption of phenanthrene on multilayer graphene as affected by surfactant and exfolia *Environ. Sci. Technol.* **2014**, *48*, 331–339. [CrossRef] [PubMed]
23. Chen, X.; Chen, B. Macroscopic and spectroscopic investigations of the adsorption of nitroaromatic compounds on grapl oxide, reduced graphene oxide, and graphene nanosheets. *Environ. Sci. Technol.* **2015**, *49*, 6181–6189. [CrossRef] [PubMed]
24. Wang, J.; Chen, Z.; Chen, B. Adsorption of polycyclic aromatic hydrocarbons by graphene and graphene oxide nanosh *Environ. Sci. Technol.* **2014**, *48*, 4817–4825. [CrossRef] [PubMed]
25. Wang, X.; Qin, Y.; Zhu, L.; Tang, H. Nitrogen-doped reduced graphene oxide as a bifunctional material for removing bisphe Synergistic effect between adsorption and catalysis. *Environ. Sci. Technol.* **2015**, *49*, 6855–6864. [CrossRef]
26. Zhang, X.; Teng, S.Y.; Loy, A.C.M.; How, B.S.; Leong, W.D.; Tao, X. Transition metal dichalcogenides for the applicatio pollution reduction: A review. *Nanomaterials* **2020**, *10*, 1012. [CrossRef] [PubMed]
27. Nasri, A.; Pétrissans, M.; Fierro, V.; Celzard, A. Gas sensing based on organic composite materials: Review of sensor ty progresses and challenges. *Mater. Sci. Semicond. Process.* **2021**, *128*, 105744. [CrossRef]
28. Liao, W.; Zhao, H.B.; Liu, Z.; Xu, S.; Wang, Y.Z. On controlling aerogel microstructure by freeze casting. *Compos. Part B Eng.* **2** *173*, 107036. [CrossRef]
29. Worsley, M.A.; Shin, S.J.; Merrill, M.D.; Lenhardt, J.; Nelson, A.J.; Woo, L.Y.; Gash, A.E.; Baumann, T.F.; Orme, C.A. Ultra density, monolithic WS_2, MoS_2, and MoS_2/graphene aerogels. *ACS Nano* **2015**, *9*, 4698–4705. [CrossRef]
30. Zhu, W.; Lin, Y.; Kang, W.; Quan, H.; Zhang, Y.; Chang, M.; Wang, K.; Zhang, M.; Zhang, W.; Li, Z.; et al. An aerogel adsorl with bio-inspired interfacial adhesion between graphene and MoS2 sheets for water treatment. *Appl. Surf. Sci.* **2020**, *512*, 14 [CrossRef]
31. Zhang, Q.; Cai, S.; Zhang, W.; Lan, Y.; Zhang, X. Density, viscosity, conductivity, refractive index and interaction study of bi mixtures of the ionic liquid 1–ethyl–3–methylimidazolium acetate with methyldiethanolamine. *J. Mol. Liq.* **2017**, *233*, 471– [CrossRef]
32. Clarke, C.J.; Bui-Le, L.; Hallett, J.P.; Licence, P. Thermally-stable imidazolium dicationic ionic liquids with pyridine functi groups. *ACS Sustain. Chem. Eng.* **2020**, *8*, 8762–8772. [CrossRef]
33. Wu, H.B.; Zhang, B.; Liu, S.H.; Chen, C.C. Flammability estimation of 1-hexyl-3-methylimidazolium bis(trifluoromethylsulfonyl)in *J. Loss Prev. Process Ind.* **2020**, *66*, 104196. [CrossRef]
34. Ahrenberg, M.; Beck, M.; Neise, C.; Keßler, O.; Kragl, U.; Verevkin, S.P.; Schick, C. Vapor pressure of ionic liquids at temperatures from AC-chip-calorimetry. *Phys. Chem. Chem. Phys.* **2016**, *18*, 21381–21390. [CrossRef] [PubMed]
35. Mezzetta, A.; Łuczak, J.; Woch, J.; Chiappe, C.; Nowicki, J.; Guazzelli, L. Surface active fatty acid ILs: Influence of the hydroph tail and/or the imidazolium hydroxyl functionalization on aggregates formation. *J. Mol. Liq.* **2019**, *289*, 11115. [CrossRef]
36. Kumar, V.; Malhotra, S.V. Study on the potential anti-cancer activity of phosphonium and ammonium-based ionic liquids. *Bi Med. Chem. Lett.* **2009**, *19*, 4643–4646. [CrossRef]
37. Feder-Kubis, J.; Tomczuk, K. The effect of the cationic structures of chiral ionic liquids on their antimicrobial activities. *Tetrahe* **2013**, *69*, 4190–4198. [CrossRef]
38. Pernak, J.; Feder-Kubis, J. Synthesis and properties of chiral ammonium-based ionic liquids. *Chem. A Eur.* **2005**, *11*, 4441–4 [CrossRef]
39. Turner, M.B.; Spear, S.K.; Holbrey, J.D.; Rogers, R.D. Production of bioactive cellulose films reconstituted from ionic liqu *Biomacromolecules* **2004**, *5*, 1379–1384. [CrossRef]
40. Kelley, S.P.; Narita, A.; Holbrey, J.D.; Green, K.D.; Reichert, W.M.; Rogers, R.D. Understanding the effects of ionicity in s solvates, co-crystals, ionic co-crystals, and ionic liquids, rather than nomenclature, is critical to understanding their beha *Cryst. Growth Des.* **2013**, *13*, 965–975. [CrossRef]
41. Holbrey, J.D.; Reichert, W.M.; Tkatchenko, I.; Bouajila, E.; Walter, O.; Tommasid, I.; Rogers, R.D. 1,3-Dimethylimidazoliu carboxylate: The unexpected synthesis of an ionic liquid precursor and carbene-CO_2 adduct. *Chem. Commun.* **2003**, *347*, 28 [CrossRef] [PubMed]
42. Jiang, B.; Tantai, X.; Zhang, L.; Hao, L.; Sun, Y.; Denga, L.; Shic, Z. Synthesis of chlorostannate(II) ionic liquids and their n application in the preparation of high-quality l-lactide. *RSC Adv.* **2015**, *5*, 50747–50755. [CrossRef]
43. Węgrzyn, M.; Rudnik, E.; Kamocka-Bronisz, R.; Kukfisz, B. Mechanical and Thermal Properties of Biocomposites Based Polyethylene from Renewable Resources Modified with Ionic Liquids. *J. Polym. Environ.* **2021**, *29*, 1808–1816. [CrossRef]
44. Raucci, M.G.; Fasolino, I.; Pastore, S.G.; Soriente, A.; Capeletti, L.B.; Dessuy, M.B.; Giannini, C.; Schrekker, H.S.; Ambrosi Antimicrobial Imidazolium Ionic Liquids for the Development of Minimal Invasive Calcium Phosphate-Based Bionanocompos *ACS Appl. Mater. Interfaces* **2018**, *10*, 42766–42776. [CrossRef] [PubMed]
45. Egorova, K.S.; Gordeev, E.G.; Ananikov, V.P. Biological Activity of Ionic Liquids and Their Application in Pharmaceutics Medicine. *Chem. Rev.* **2017**, *117*, 7132–7189. [CrossRef] [PubMed]
46. Hummers, W.S.; Offeman, R.E. Preparation of Graphitic Oxide. *J. Am. Chem. Soc.* **1958**, *80*, 1339. [CrossRef]
47. Kovtyukhova, N.I.; Ollivier, P.J.; Martin, B.R.; Mallouk, T.E.; Buzaneva, E.V.; Gorchinskiy, A.D. Layer-by-layer assembl ultrathin composite films from micron-sized graphite oxide sheets and polycations. *Chem. Mater.* **1999**, *11*, 771–778. [CrossR

Libel, G.P.; Facchi, S.P.; de Almeida, D.A.; Madruga, L.C.; Kipper, M.J.; Schrekker, H.S.; Martins, A.F.; Radovanovic, E. Cross-linked poly(vinyl alcohol)/citric acid electrospun fibers containing imidazolium ionic liquid with enhanced antiadhesive and antimicrobial properties. *Mater. Chem. Phys.* **2024**, *316*, 129087. [CrossRef]
Bergamo, V.Z.; Donato, R.K.; Nemitz, M.C.; Acasigua, G.A.X.; Selukar, B.S.; Lopes, W.; Lana, D.F.D.; Teixeira, M.L.; Teixeira, H.F.; Schrekker, H.S.; et al. Assessing an imidazolium salt's performance as antifungal agent on a mouthwash formulation. *J. Appl. Microbiol.* **2016**, *121*, 1558–1567. [CrossRef]
Hong, J.Y.; Sohn, E.H.; Park, S.; Park, H.S. Highly-efficient and recyclable oil absorbing performance of functionalized graphene aerogel. *Chem. Eng. J.* **2015**, *269*, 229–235. [CrossRef]
Shu, D.; Feng, F.; Han, H.; Ma, Z. Prominent adsorption performance of amino-functionalized ultra-light graphene aerogel for methyl orange and amaranth. *Chem. Eng. J.* **2017**, *324*, 1–9. [CrossRef]
Li, J.; Jia, H.; Ding, Y.; Luo, H.; Abbas, S.; Liu, Z.; Hu, L.; Tang, C. NaOH-embedded three-dimensional porous boron nitride for efficient formaldehyde removal. *Nanotechnology* **2015**, *26*, 475704. [CrossRef] [PubMed]
Li, X.; Li, X.; Li, Z.; Wang, J.; Zhang, J. WS_2 nanoflakes based selective ammonia sensors at room temperature. *Sens. Actuators B Chem.* **2017**, *240*, 273–277. [CrossRef]
Zhou, Y.; Wang, Y.; Liu, T.; Xu, G.; Chen, G.; Li, H.; Liu, L.; Zhuo, Q.; Zhang, J.; Yan, C. Superhydrophobic hBN-Regulated Sponges with Excellent Absorbency Fabricated Using a Green and Facile Method. *Sci. Rep.* **2016**, *7*, 45065. [CrossRef] [PubMed]
Thommes, M.; Kaneko, K.; Neimark, A.V.; Olivier, J.P.; Rodriguez-Reinoso, F.; Rouquerol, J.; Sin, K.S.W. Physisorption of gases, with special reference to the evaluation of surface area and pore size distribution (IUPAC Technical Report). *Pure Appl. Chem.* **2015**, *87*, 1051–1069. [CrossRef]
Gorgolis, G.; Galiotis, C. Graphene aerogels: A review. *2D Mater.* **2017**, *4*, 7883. [CrossRef]
Ma, X.; Gao, F.; Liu, G.; Xie, Y.; Tu, X.; Li, Y.; Dai, R.; Qu, F.; Wang, W.; Lu, L. Sensitive determination of nitrite by using an electrode modified with hierarchical three-dimensional tungsten disulfide and reduced graphene oxide aerogel. *Microchim. Acta* **2019**, *186*, 291. [CrossRef] [PubMed]
Marsh, K.L.; Souliman, M.; Kaner, R.B. Co-solvent exfoliation and suspension of hexagonal boron nitride. *Chem. Commun.* **2015**, *51*, 187–190. [CrossRef]
Ha, H.; Shanmuganathan, K.; Ellison, C.J. Mechanically stable thermally crosslinked poly(acrylic acid)/reduced graphene oxide aerogels. *ACS Appl. Mater. Interfaces* **2015**, *7*, 6220–6229. [CrossRef]
Gorgolis, G.; Kotsidi, M.; Paterakis, G.; Koutroumanis, N.; Tsakonas, C.; Galiotis, C. Graphene aerogels as efficient adsorbers of water pollutants and their effect of drying methods. *Sci. Rep.* **2024**, *14*, 8029. [CrossRef]
Ferrari, A.C. Raman spectroscopy of graphene and graphite: Disorder, electron-phonon coupling, doping and nonadiabatic effects. *Solid State Commun.* **2007**, *143*, 47–57. [CrossRef]
Si, W.; Wu, X.; Zhou, J.; Guo, F.; Zhuo, S.; Cui, H.; Xing, W. Reduced graphene oxide aerogel with high-rate supercapacitive performance in aqueous electrolytes. *Nanoscale Res. Lett.* **2013**, *8*, 247. [CrossRef] [PubMed]
Berkdemir, A.; Gutiérrez, H.R.; Botello-Méndez, A.R.; Perea-López, N.; Elías, A.L.; Chia, C.I.; Wang, B.; Crespi, V.H.; López-Urías, F.; Charlier, J.C.; et al. Identification of individual and few layers of WS_2 using Raman Spectroscopy. *Sci. Rep.* **2013**, *3*, 1755. [CrossRef]
Paterakis, G.; Vaughan, E.; Gawade, D.R.; Murray, R.; Gorgolis, G.; Matsalis, S.; Anagnostopoulos, G.; Buckley, J.L.; O'Flynn, B.; Quinn, A.J.; et al. Highly Sensitive and Ultra-Responsive Humidity Sensors Based on Graphene Oxide Active Layers and High Surface Area Laser-Induced Graphene Electrodes. *Nanomaterials* **2022**, *12*, 2684. [CrossRef] [PubMed]
Zhou, C.; Yang, W.; Zhu, H. Mechanism of charge transfer and its impacts on Fermi-level pinning for gas molecules adsorbed on monolayer WS_2. *J. Chem. Phys.* **2015**, *142*, 274704. [CrossRef]
Qin, Z.; Zeng, D.; Zhang, J.; Wu, C.; Wen, Y.; Shan, B.; Xie, C. Effect of layer number on recovery rate of WS_2 nanosheets for ammonia detection at room temperature. *Appl. Surf. Sci.* **2017**, *414*, 244–250. [CrossRef]
Kim, Y.H.; Park, J.S.; Choi, Y.R.; Park, S.Y.; Lee, S.Y.; Sohn, W.; Shim, Y.S.; Lee, J.H.; Park, C.R.; Choi, Y.S.; et al. Chemically fluorinated graphene oxide for room temperature ammonia detection at ppb levels. *J. Mater. Chem. A* **2017**, *5*, 19116–19125. [CrossRef]
Varghese, S.S.; Varghese, S.H.; Swaminathan, S.; Singh, K.K.; Mittal, V. Two-dimensional materials for sensing: Graphene and beyond. *Electronics* **2015**, *4*, 651–687. [CrossRef]
Zhou, Y.; Liu, G.; Zhu, X.; Guo, Y. Ultrasensitive NO_2 gas sensing based on rGO/MoS_2 nanocomposite film at low temperature. *Sens. Actuators B Chem.* **2017**, *251*, 280–290. [CrossRef]
Yan, W.; Worsley, M.A.; Pham, T.; Zettl, A.; Carraro, C.; Maboudian, R. Effects of ambient humidity and temperature on the NO_2 sensing characteristics of WS_2 /graphene aerogel. *Appl. Surf. Sci.* **2018**, *450*, 372–379. [CrossRef]
Donarelli, M.; Ottaviano, L. 2d materials for gas sensing applications: A review on graphene oxide, MoS_2, WS_2 and phosphorene. *Sensors* **2018**, *18*, 3638. [CrossRef] [PubMed]
Cho, B.; Yoon, J.; Lim, S.K.; Ki, A.R.; Kim, D.H.; Park, S.G.; Kwon, J.D.; Lee, Y.J.; Lee, K.H.; Lee, B.H.; et al. Chemical Sensing of 2D Graphene/MoS_2 Heterostructure device. *ACS Appl. Mater. Interfaces* **2015**, *7*, 16775–16780. [CrossRef] [PubMed]
Late, D.J.; Huang, Y.K.; Liu, B.; Acharya, J.; Shirodkar, S.N.; Luo, J.; Yan, A.; Charles, D.; Waghmare, U.V.; Dravid, V.P.; et al. Sensing behavior of atomically thin-layered MoS_2 transistors. *ACS Nano* **2013**, *7*, 4879–4891. [CrossRef] [PubMed]

74. Teimoori, S.; Hassani, A.H.; Panahi, M.; Mansouri, N. A Review, Methods for removal and adsorption of volatile org compounds from environmental matrixes. *Anal. Methods Environ. Chem. J.* **2020**, *3*, 34–58. [CrossRef]
75. Li, L.; Li, Z.; Zhu, X.; He, M.; Ma, C.; Wang, Q.; Yu, H.; Wang, J. Theoretical calculation and experimental investigation on i liquid [C16mim]Cl affecting wettability of low-rank coal. *Adv. Compos. Hybrid Mater.* **2022**, *5*, 1241–1252. [CrossRef]
76. Uragami, T.; Matsuoka, Y.; Miyata, T. Removal of dilute benzene in water through ionic liquid/poly(Vinyl chloride) membr by pervaporation. *J. Membr. Sci. Res.* **2016**, *2*, 20–25.
77. Zhang, W.; Luo, J.; Sun, T.; Yu, F.; Li, C. The Absorption Performance of Ionic Liquids–PEG200 Complex Absorbent for V *Energies* **2021**, *14*, 3592. [CrossRef]
78. Indra, S.; Subramanian, R.; Daschakraborty, S. Interaction of volatile organic compounds acetone and toluene with r temperature ionic liquid at the bulk and the liquid-vacuum interface. *J. Mol. Liq.* **2021**, *331*, 115608. [CrossRef]
79. Li, C.H.; Gao, K.X.; Meng, Y.N.; Wu, X.K.; Zhang, F.; Wang, Z.X. Solution Thermodynamics of Imidazolium-Based Ionic Liq and Volatile Organic Compounds: Benzene and Acetone. *J. Chem. Eng. Data* **2015**, *60*, 1600–1607. [CrossRef]
80. Salthammer, T. Very volatile organic compounds: An understudied class of indoor air pollutants. *Indoor Air* **2016**, *26*, 25 [CrossRef]
81. Lu, D.; Liu, H. Fabrication of novel graphene aerogel by the assistance of l-tyrosine and excellent adsorption for organic solv *Res. Chem. Intermed.* **2020**, *46*, 253–266. [CrossRef]
82. Huang, Y.; Ho, S.S.H.; Lu, Y.; Niu, R.; Xu, L.; Cao, J.; Lee, S. Removal of indoor volatile organic compounds via photocatal oxidation: A short review and prospect. *Molecules* **2016**, *21*, 56. [CrossRef] [PubMed]
83. Kumar, V.; Lee, Y.S.; Shin, J.W.; Kim, K.H.; Kukkar, D.; Tsang, Y.F. Potential applications of graphene-based nanomaterial adsorbent for removal of volatile organic compounds. *Environ. Int.* **2020**, *135*, 105356. [CrossRef] [PubMed]
84. Liu, C.; Liu, H.; Xu, A.; Tang, K.; Huang, Y.; Lu, C. In situ reduced and assembled three-dimensional graphene aerogel for effic dye removal. *J. Alloys Compd.* **2017**, *714*, 522–529. [CrossRef]
85. Kaushik, J.; Sharma, C.; Lamba, N.K.; Sharma, P.; Das, G.S.; Tripathi, K.M.; Joshi, R.K.; Sonkar, S.K. 3D Porous MoS_2-Decor Reduced Graphene Oxide Aerogel as a Heterogeneous Catalyst for Reductive Transformation Reactions. *Langmuir* **2023** 12865–12877. [CrossRef]
86. Mattson, E.C.; Pande, K.; Unger, M.; Cui, S.; Lu, G.; Gajdardziska-Josifovska, M.; Weinert, M.; Chen, J.; Hirschmugl, C.J. Explo adsorption and reactivity of NH_3 on reduced graphene oxide. *J. Phys. Chem. C* **2013**, *117*, 10698–10707. [CrossRef]
87. Romero, S.M.; Giudicessi, S.L.; Vitale, R.G. Is the fungus Aspergillus a threat to cultural heritage. *Cult. Herit.* **2021**, *51*, 107– [CrossRef]
88. Pham, H.D.; Pham, V.H.; Cuong, T.V.; Nguyen-Phan, T.-D.; Chung, J.S.; Shin, E.W.; Kim, S. Synthesis of the chemically conve graphene xerogel with superior electrical conductivity. *Chem. Commun.* **2011**, *47*, 9672–9674. [CrossRef]
89. Tang, W.; Peng, L.; Yuan, C.; Wang, J.; Mo, S.; Zhao, C.; Yu, Y.; Min, Y.; Epstein, A.J. Facile synthesis of 3D reduced graphene o and its polyaniline composite for super capacitor application. *Synth. Met.* **2015**, *202*, 140–146. [CrossRef]
90. Chabot, V.; Higgins, D.; Yu, A.; Xiao, X.; Chen, Z.; Zhang, J. A review of graphene and graphene oxide sponge: Material synth and applications to energy and the environment. *Energy Environ. Sci.* **2014**, *7*, 1564–1596. [CrossRef]
91. Pham, T.; Goldstein, A.P.; Lewicki, J.P.; Kucheyev, S.O.; Wang, C.; Russell, T.P.; Worsley, M.A.; Woo, L.; Mickelson, W.; Zett Nanoscale structure and superhydrophobicity of sp^2-bonded boron nitride aerogels. *Nanoscale* **2015**, *7*, 10449–10458. [Cross [PubMed]
92. Pipolo, S.; Salanne, M.; Ferlat, G.; Klotz, S.; Saitta, M.; Pietrucci, F. Navigating at Will on the Water Phase Diagram. *Phys. Rev.* **2017**, *119*, 245701. [CrossRef] [PubMed]
93. Thøgersen, A.; Syre, M.; Olaisen, B.R.; Diplas, S. Studies of the oxidation states of phosphorus gettered silicon substrates u X-ray photoelectron spectroscopy and transmission electron microscopy. *J. Appl. Phys.* **2013**, *113*, 044307. [CrossRef]
94. Siuzdak, K.; Szkoda, M.; Sawczak, M.; Lisowska-Oleksiak, A.; Karczewski, J.; Ryl, J. Enhanced photoelectrochemical photocatalytic performance of iodine-doped titania nanotube arrays. *RSC Adv.* **2015**, *5*, 50379–50391. [CrossRef]
95. Deerattrakul, V.; Yigit, N.; Rupprechter, G.; Kongkachuichay, P. The roles of nitrogen species on graphene aerogel suppo Cu-Zn as efficient catalysts for CO_2 hydrogenation to methanol. *Appl. Catal. A Gen.* **2019**, *580*, 46–52. [CrossRef]
96. Wu, J.; Wang, W.; Wang, Z. Porphin-Based Carbon Dots for "Turn Off–On" Phosphate Sensing and Cell Imaging. *Nanomate* **2020**, *10*, 326. [CrossRef]

Disclaimer/Publisher's Note: The statements, opinions and data contained in all publications are solely those of the indivic author(s) and contributor(s) and not of MDPI and/or the editor(s). MDPI and/or the editor(s) disclaim responsibility for any injur people or property resulting from any ideas, methods, instructions or products referred to in the content.

materials

MDPI

icle

nalysis of the Surface of Historic Fabric from the uschwitz-Birkenau State Museum after Treatment with thanol Mist Used to Eliminate Microorganisms Harmful Human Health

na Wawrzyk [1,2,*], Janina Poskrobko [3], Krystyna Guzińska [4], Dorota Kaźmierczak [4], Aleksandra Papis [2], l Jastrzębiowska [2], Natalia Uroda [2], Maria Szymankiewicz [5], Dagmara Zeljaś [6], Iga Wawrzyk-Bochenek [7] l Sławomir Wilczyński [7]

1 Silesian Park of Medical Technology Kardio-Med Silesia in Zabrze, M. Curie Skłodowskiej 10C Str., 41-800 Zabrze, Poland
2 Auschwitz-Birkenau State Museum, Więźniów Oświęcimia 20, 32-603 Oświęcim, Poland
3 The Department of Textile Conservation, The Metropolitan Museum of Art (The Met), 1000 Fifth Avenue, New York, NY 10028, USA; janina.poskrobko@metmuseum.org
4 Lukasiewicz Research Network-Lodz Institute of Technology, M. Sklodowskiej-Curie 19/27, 90-570 Lodz, Poland
5 Department of Microbiology, Prof. F. Łukaszczyk Oncology Centre, 85-796 Bydgoszcz, Poland
6 Department of Drilling and Geoengineering, Faculty of Drilling, Oil and Gas, AGH University of Krakow, Al. A. Mickiewicza 30, 30-059 Krakow, Poland; dagzel@agh.edu.pl
7 Department of Basic Biomedical Science, Faculty of Pharmaceutical Sciences in Sosnowiec, Medical University of Silesia, Kasztanowa 3, 41-205 Sosnowiec, Poland
* Correspondence: anna.wawrzyk@gazeta.pl

Abstract: Aim: the aim of the work was to present the changes occurring on the model and historical cotton surface of cotton resulting from disinfection with 90% ethanol mist. Materials and methods: Samples of historical materials consisted of fabric elements from suitcases stored in A-BSM. A mist of 90% ethanol was applied for 15 s at a distance of 16 cm from the surface. The spectra of cotton samples before and after ethanol application were recorded using Fourier transform infrared spectroscopy (FTIR-ATR). Analyses of the surface layers were performed using X-ray photoelectron spectroscopy (XPS). Results: the decontamination performed did not show any significant differences in the chemical composition and surface structure of cotton before and after the use of 90% ethanol mist. Conclusions: Ethanol mist, which eliminates microorganisms from the historical surface, does not cause significant changes to the surface of historical objects.

Keywords: ethanol mist; disinfection; microorganisms; FTIR; XPS

1. Introduction

Preserving cultural heritage for years is very important for a country like Poland. The state of preservation of historical objects depends on the protection works, among others: disinfection that protects against biodeterioration.

KL Auschwitz was the largest concentration camp and extermination center established by Nazi Germany during World War II on the outskirts of Oświęcim in occupied Poland. Auschwitz was originally intended to serve as concentration camp and a place of slow death for Polish political prisoners and other Poles. In later years, however, it gradually became the main center of mass extermination of European Jews and the largest concentration camp for prisoners of other nationalities from almost all of Europe. KL Auschwitz is currently the most recognizable symbol of terror, genocide, and the Shoah.

A total of at least 1,300,000 people were deported to Auschwitz; 1,000,000 of them died [1].

After the Second World War, following the liberation, the Auschwitz-Birkenau State Museum in Oświęcim (A-BSM) was established at the site of the former concentration camp, focusing on preserving the remnants of the Auschwitz-Birkenau Konzentrationslager. During the museum's formation, everything found within the area of the former camp structure was secured to the best extent possible, laying the foundation for the museum's collection. Personal items discovered after liberation often serve as the sole testament to the presence of individuals deported to KL Auschwitz. A significant part of these objects, known as looted property, comprises items containing fabric in their composition. These include, among others: clothing, shoes, prosthetics, belts, and suitcases. The presence of fabric in individual objects varies. Objects can be entirely made of fabric, while others incorporate fabric as one of several materials used in their creation (Figure 1).

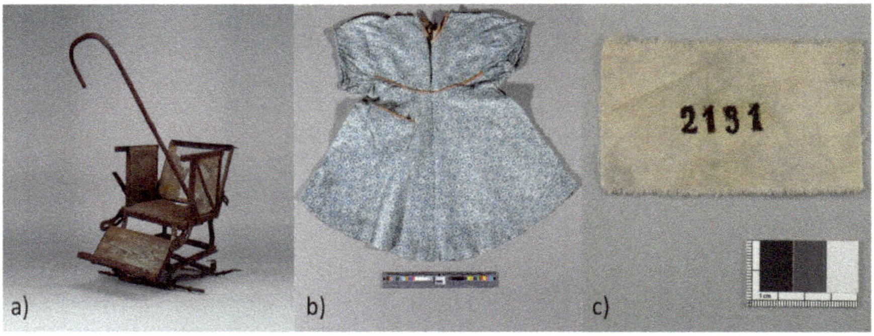

Figure 1. Cotton objects from the collections of the Auschwitz-Birkenau State Museum in Oświęcim. A child's carriage with elements made of cotton fabric (**a**), and objects entirely made of cotton: a child's dress (**b**), and a prisoner's camp number patch (**c**).

A significant part of the collection is in poor condition, primarily due to the prolonged use of items and the way they were treated after confiscation in the camp. They were often piled in large heaps in front of warehouse buildings, directly on the ground. Additional damage occurred during the searching process after returning them to their owners, leading to numerous instances of mechanical destruction.

The main conservation efforts at A-BSM regarding textile objects focus on appropriate prevention, ensuring optimal storage and exhibition conditions, including microbiological cleanliness. In order to minimize the potential spread of microorganisms, suitable decontamination techniques are sought. A-BSM is testing various disinfection techniques for historical objects using different methods and biocidal substances. The primary goal of these investigations is to eliminate microorganisms harmful to human health and those acting destructively on surfaces isolated from historical objects stored in the Collection Department. This is crucial for the preservation of cultural heritage, as confirmed by numerous studies on the biodegradation of cultural heritage [2–5].

The species composition of microorganisms varies depending on the material they inhabit. The microbiome composition is largely determined by the type of substance covering the historical material. Therefore, in the field of museology, constant efforts are made to find new disinfection methods. In each case, the biocidal effectiveness of a method depends on the properties of the disinfected surface material. Often, disinfection methods are adapted from the medical field, where they were applied to abiotic surfaces [6].

Microbiological studies conducted in museums worldwide and at A-BSM have revealed the presence of various types of bacteria on the surfaces of textiles. One frequently encountered type is the spore-forming *Bacillus* sp. with cellulolytic potential [1,7,8].

These rods can be harmful to historical objects, and also exhibit pathogenic potential. *B. cereus* is a common cause of acute food poisoning and post-traumatic eye infections, while

B. subtilis has been isolated from individuals with bacteremia, endocarditis, pneumonia, and sepsis [9,10].

On the surface of historical textiles at A-BSM, 12 species of fungi were detected, including *Aspergillus flavus* and *Aspergillus niger* [11].

A. flavus is the main etiological agent of allergic bronchopulmonary aspergillosis and lung infections. It has also been linked to infections of the ear canal, pulmonary aspergillosis, and other localized aspergillosis [12,13].

The elimination of *A. flavus* from the museum environment is essential because, under favorable conditions, this fungus produces aflatoxins in large quantities, which are hepatotoxic and carcinogenic [14].

A. niger, when detected on objects, is responsible for infections of the middle and external ear, as well as nail fungus. It causes invasive pulmonary aspergillosis, and certain strains can lead to subcutaneous infections [15,16].

Fungi isolated from the museum environment can, under favorable conditions, produce toxic aflatoxins and ochratoxins, like other mycotoxins, posing health risks [17].

A. niger is highly pervasive and invasive, posing a threat not only to humans but also to historical objects. Even one colony of this fungus detected on an object at A-BSM qualifies the object for comprehensive disinfection.

Given that potentially harmful microorganisms are isolated in museum environments, A-BSM tested the biocidal effectiveness of vaporized hydrogen peroxide (VHP), ethylene oxide (EtO), and diode laser. Each of these techniques is dedicated to different types of objects. VHP and EtO are used for comprehensive disinfection of large object surfaces through fogging [11].

A diode laser was tested for the elimination of microorganisms on the surfaces of historical textiles appearing as very small spots [18].

Research conducted currently at A-BSM focuses on adapting the ethanol mist disinfection method to surfaces of approximately 1 m or slightly larger, which is the subject of this study. To eliminate fungi and bacteria with potential pathogenic and cellulolytic capabilities, 90% ethanol mist has been applied. Wawrzyk et al., presented the results of microbiological studies, confirming the biocidal effectiveness of the method. The reduction of most microorganisms was over 99% [19].

To confirm that ethanol mist does not have a harmful effect on the fibers of historical fabric, the authors conducted scanning electron microscopy (SEM) tests, which did not show any significant, visible changes on the surface after the use of ethanol mist [19].

In the case of employing innovative techniques or using a biocidal agent different from the traditional one, the composition and structure of the disinfected material should not undergo adverse changes. The authors also presented results from Scanning Electron Microscopy (SEM), which did not reveal significant, visible changes on the surface after the application of ethanol mist [19].

For a more detailed analysis of the impact of ethanol mist on the chemical properties of historical materials' surfaces, Fourier-Transform Infrared Spectroscopy (FTIR) analysis was conducted before and after the decontamination process. FTIR is successfully used for analyzing textile surfaces, as confirmed by researchers [20].

It has also been applied to analyze historical paper [21].

In A-BSM, this research technique has been previously used for the analysis of textiles, leather, and materials based on cellulose nitrate [18].

Researchers also recommend this technique for archaeological textile studies because it is non-invasive and non-destructive [22].

To further assess the potential impact of ethanol mist on the near-surface layer at PMA-B, X-ray photoelectron spectroscopy (XPS) studies were conducted, allowing for the capture of changes up to a depth of 2–3 nm. XPS is successfully utilized for studying both biotic and abiotic surfaces [23].

Topalovic applied XPS to determine chemical changes occurring on the surface of cotton due to bleaching, explaining the correlation between the increase in the capillary constant and the removal of non-cellulosic impurities characterized by the C1 component in the C1s XPS spectrum [24].

XPS has also been used for the analysis and conservation studies of artistic emulsion acrylic paints [25].

In A-BSM and other museums, it is extremely important to eliminate from the environment microorganisms that are potentially harmful to human health and those that may contribute to the biodegradation of objects.

Therefore, the aim of the work is to analyze, using highly specialized methods, the potential impact of 90% ethanol mist on the chemical changes that may occur on the surfaces of highly degraded historic materials.

2. Materials and Methods

2.1. Research Objects

The surfaces of both the model and historical materials underwent Fourier-Transform Infrared Spectroscopy (FTIR Thermo Fisher Scientific, 168 Third Avenue Waltham, MA, USA, 02451) and X-ray photoelectron spectroscopy (XPS, 100 Red Schoolhouse Road, Bldg. A-8 Chestnut Ridge, NY 10977, USA) analyses. The model material used was cotton (SDC Enterprises Limited, Thongsbridge, UK) with a grammage of 100 g/m^2. Before the decontamination effectiveness test, the samples were sterilized (121 °C, 20 min).

The historical material sample was cut from the inner side of a suitcase lid from the collections of A-BSM, dating back to the first half of the 20th century. Fiber analysis revealed that it was dusty cotton covered with secondary conservation agents.

The dimensions of the model fabric samples were adjusted to the dimensions of the historic fabric and selected so that the total surface was 100 cm ^{2}. Two cotton samples with dimensions of 50 × 100 mm and two cotton samples with dimensions of 40 × 12.5 mm on which ethanol was applied were used as controls. The test was performed in two repetitions.

2.2. Application of 90% Ethanol in Mist Form on Cotton

Both short-term and long-term effects of 90% ethanol (Chempur, Piekary Śląskie, Poland) were investigated. For this purpose, ethanol mist was applied to the samples using Paasche VL 0819 and VE 0707 airbrushes at a pressure of 0.2 MPa and a PA HEAD VLH-5 nozzle (with a diameter of 1.05 mm). The application was carried out in a chamber with laminar air flow of microbiological safety class II, with a double HEPA filtration system ensuring the flow of sterile air. During application, the sample was placed vertically and secured with a metal handle. As part of the optimization of the method, the pressure parameters, nozzle and application time were adjusted to ensure the minimum moisture of the samples. After applying ethanol, the sample was weighed to determine the mass of ethanol applied.

During the assessment of the short-term effect of ethanol, the mass of the applied solution depended on the alcohol concentration and ranged from 0.2 to 1.0 g per 100 cm^2. The application time was set at 4–16 s/100 cm^2. Subsequently, the cotton was dried in a sterile chamber. To examine the long-term effect of ethanol, after the short-term treatment, the contact time of microorganisms with ethanol was extended. Following the application, the cotton was wrapped in foil, and after 22 h, it was also dried in a sterile chamber.

2.3. Chemical Analysis of the Surfaces

2.3.1. FTIR Analysis of the Chemical Composition of Cotton Surface before and after Decontamination with Ethanol

To investigate whether ethanol induced adverse changes on the surface of the disinfected fabric, Fourier-Transform Infrared Spectroscopy (FTIR) analysis was employed. The FTIR analysis was conducted using a Nicolet 8700 FTIR spectrometer (Thermo Scientific, Waltham, MA, USA) with an Attenuated Total Reflection (ATR) diamond crystal and a liquid nitrogen-cooled Mercury-Cadmium-Telluride (MCT-A) detector. Spectra were collected before and after the application of ethanol mist from a layer with a thickness of 2–3 μm in the range of 4000–650 cm^{-1} and with a resolution of 4 cm^{-1}. ATR spectra underwent ATR correction, scaled normalization, and baseline correction. The obtained spectra were equivalent to transmittance spectra. The software OMNIC 3.2 (Thermo Scientific, USA) was utilized for the analysis.

2.3.2. XPS Analysis of the Chemical Composition of Cotton Surface before and after Decontamination with Ethanol

For the analysis of the near-surface layer of materials before and after decontamination, X-ray photoelectron spectroscopy (XPS) was employed. The analysis utilized a multi-chamber Ultra High Vaccum (UHV) analytical system (Prevac, Poland). Photoelectrons were excited by X-rays with Al Kα characteristic line at an energy of 1486.7 eV, generated by a VG Scienta SAX 100 lamp with an aluminum anode along with a VG Scienta XM 780 monochromator. The X-ray lamp operated at U = 12 kV and Ie = 30 mA. Photoelectrons were recorded using a hemispherical analyzer Scienta R4000. The pressure in the analysis chamber during measurements was below 1.0×10^{-8} mbar. The fundamental parameters for the survey spectrum were as follows: sweeping mode, pass energy: 200 eV, measured range of photoelectron binding energy: 0–1350 eV, step size: 0.5 eV, and dwell time in a single step: 200 ms. To compensate for the electric charge formed during measurement, samples were bombarded with a low-energy electron beam. The recorded spectra were processed using Casa XPS Version 2.3.16 PR16 software. All spectra were calibrated by establishing the position of the C1s carbon line at an energy of 284.5 eV.

2.3.3. Quality Assurance in Research

Quantitative microbiological methods were validated by repetition test 20 times. The accuracy values were <0.25 log under repeatability conditions and <0.33 log under intralaboratory reproducibility conditions. The workload under repeatability conditions was also determined for all physicochemical elements of quantitative methods, and amounted to 1% for FTIR and 7% for XPS. Moreover, to ensure quality in ethanol biocidal effectiveness testing, each combination of microbial strain and ethanol mist parameter variant was performed using three samples, and each sample was tested twice. FTIR and XPS spectra were collected for one area of each material at a number of replicates appropriate for each method.

3. Results

3.1. Fourier-Transform Infrared Spectroscopy (FTIR) Analysis of the Surface of Disinfected Model and Historical Cotton

The first stage of research, focusing on the analysis of the impact of 90% ethanol mist decontamination on the historic cotton surface, involved conducting measurements using the FTIR-ATR technique. The application of this method aimed to determine the influence of the applied biocidal solution on the chemical structure of the decontaminated surface. Three samples of model cotton and three samples of historical cotton were used for the study. The model cotton, with a known and precisely defined chemical structure, served as a reference material in relation to the historical cotton, which was the target material for the discussed decontamination method. Samples of historical materials are available in small quantities because they are cultural heritage, which is the greatest limitation in research conducted on historical objects.

Sample I of the model cotton and historical cotton served as control samples not subjected to the decontamination process. Sample II of the model cotton and historical cotton was prepared by applying 90% ethanol mist to the material surface, and measurements were taken immediately after the biocidal agent dried. The last samples, i.e., model cotton III and historical cotton III, were prepared similarly to samples II, but the measurements were taken after 22 h of decontamination. During this time, these samples were stored in a closed container.

3.1.1. Fourier-Transform Infrared Spectroscopy (FTIR) Analysis of the Surface of Disinfected Model Cotton

The model material is used for preliminary research so as not to destroy the monuments. It has properties similar to disinfected historical materials, but does not have a layer of dust or impregnation on the surface.

As a result of conducting FTIR measurements using the ATR attachment, spectra were obtained for the model material samples, which are presented in Figure 2 below. Figure 2 below shows the collective spectrum for three model cotton samples.

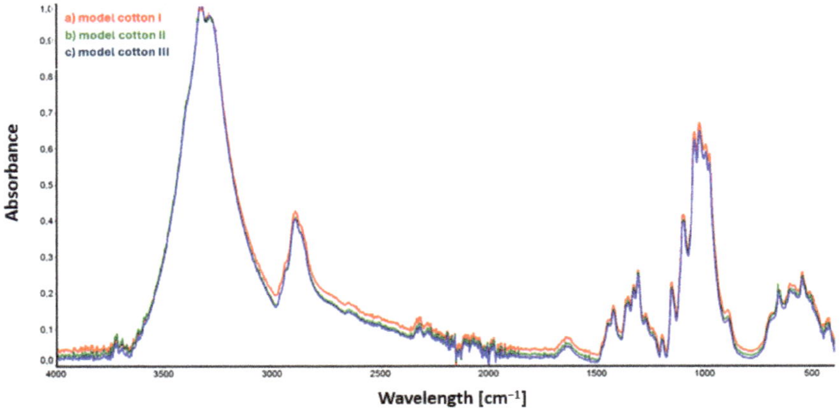

Figure 2. ATR collective spectra of samples: (a) model cotton I (sample not subjected to decontamination); (b) model cotton II (sample subjected to 90% ethanol mist decontamination); (c) model cotton III (sample subjected to 90% ethanol mist decontamination and stored in foil for an additional 22 h).

3.1.2. Fourier-Transform Infrared Spectroscopy (FTIR) Analysis of the Surface of Disinfected Historical Cotton

The spectra obtained from the analysis of historical material are presented in Figure 3.

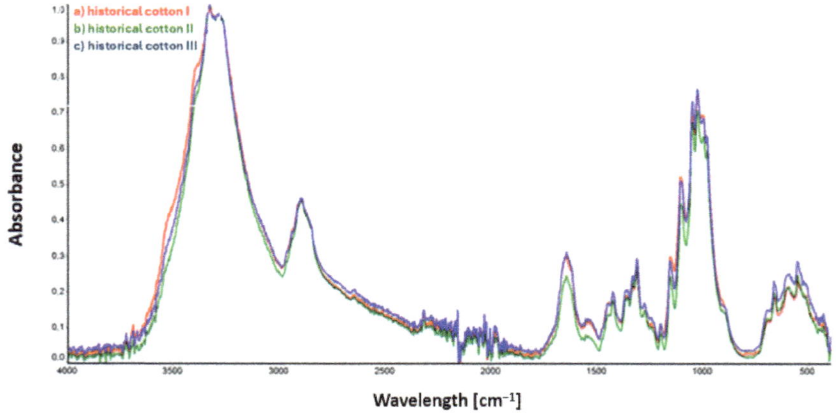

Figure 3. ATR collective spectra of samples: (a) historical cotton I (sample not subjected to decontamination); (b) historical cotton II (sample subjected to 90% ethanol mist decontamination); (c) historical cotton III (sample subjected to 90% ethanol mist decontamination and stored in foil for an additional 22 h).

Confirmation of the relationship between the peak intensities for the three historical cotton samples shown in the spectra (Figure 3) is presented in the composite spectrum containing graphs for the three historical cotton samples mentioned.

Analysis of the spectra from samples of model cotton, presented in Figure 2, did not reveal noticeable and significant differences in the intensity and position of spectral bands in the IR spectrum.

In the case of historical cotton samples (Figure 3), the only visible change in the spectrum concerns the intensity of bands in the range of 1650–1730 cm^{-1}, corresponding to stretching vibrations of C=O groups. Reduced intensity of the band around 1650 cm^{-1} is observed for historical cotton sample II, and may result from incomplete evaporation of the applied 90% ethanol solution on the surface of the examined material. The peaks of the other two samples, historical cotton I and historical cotton III, located at the same wavenumber value, overlap with each other.

3.2. X-ray Photoelectron Spectroscopy (XPS) Analysis of the Surface of Decontaminated Model and Historical Cotton

3.2.1. X-ray Photoelectron Spectroscopy (XPS) Analysis of the Surface of Decontaminated Model Cotton

To fully illustrate the impact of decontamination with a 90% ethanol mist on cotton surfaces, a complementary study was conducted using XPS. It allowed the determination of the elemental composition of the external layers of the investigated model samples. The obtained spectra are presented in Figure 4.

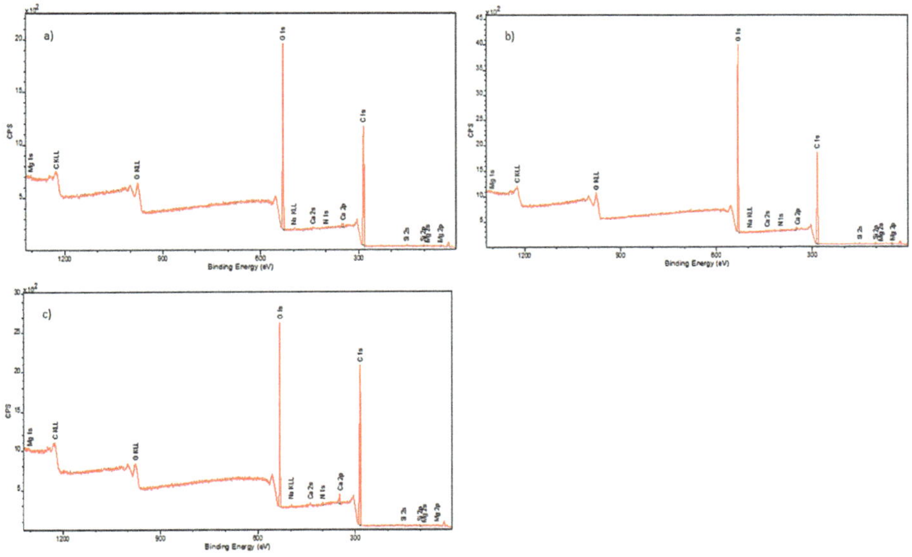

Figure 4. XPS spectra obtained for samples: (**a**) model cotton I (sample not subjected to decontamination); (**b**) model cotton II (sample subjected to decontamination with 90% ethanol mist); (**c**) model cotton III (sample subjected to decontamination with 90% ethanol mist and additionally stored in foil for 22 h).

3.2.2. X-ray Photoelectron Spectroscopy (XPS) Analysis of the Surface of Decontaminated Historical Cotton

Elemental composition of historical cotton was also determined. The obtained spectra are presented in Figure 5.

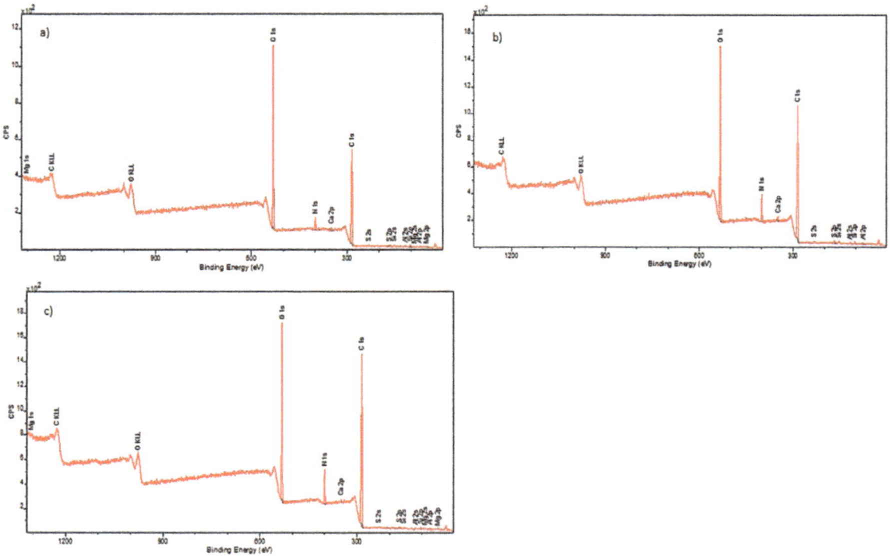

Figure 5. XPS spectra for samples: (**a**) historical cotton I (not subjected to disinfection); (**b**) historical cotton II (sample subjected to disinfection with 90% ethanol in the form of mist); (**c**) historical cotton III (sample subjected to disinfection with 90% ethanol in the form of mist and additionally stored in foil for 22 h).

Analysis of the spectral lines visible in the spectra allowed for the detection of elements present in the examined samples. The outer surface of the model cotton consisted of the following elements: carbon, nitrogen, oxygen, magnesium, silicon, and calcium. The historical cotton samples additionally contained aluminum and sulfur on their surface.

Detailed information regarding the percentage distribution of individual elements on the model cotton, obtained in the course of the conducted study, is presented in Figure 6.

Figure 7 illustrates the percentage distribution of individual elements on historical cotton obtained during the XPS study.

It was observed that, in the case of model cotton samples, the vast majority of the elemental composition, expressed in mass percentages, consists of carbon and oxygen. This is due to the structure of cotton, which is composed of over 90% cellulose. XPS studies conducted in a narrow binding energy range showed the presence of the following chemical bonds characteristic of carbon: C-C, C-O-C, C=O, O-C=O. Calcium present in the samples occurs in the form of carbonates, while silicon is present in the form of silicon dioxide. No significant differences were observed between the model cotton samples not subjected to decontamination and those treated with 90% ethanol in the form of mist.

Historical cotton samples also show the highest content of carbon and oxygen in their composition. Additionally, an increase in nitrogen content was observed compared to model cotton samples, which likely originates from the chemical compounds used for dyeing materials, such as azo dyes, nitroso dyes, or nitric dyes [26].

In the case of historical samples, aluminum and sulfur were additionally detected, which most likely constitute a secondary layer of the examined historical surface. The analysis of historical cotton material samples also did not reveal significant changes in the elemental composition after decontamination with a 90% ethanol mist.

The increase in carbon content in both types of material and oxygen content in the case of model material is most likely due to the incomplete evaporation of ethanol applied to the examined surfaces. Ethanol has carbon and oxygen atoms in its structure, which can be detected during surface studies.

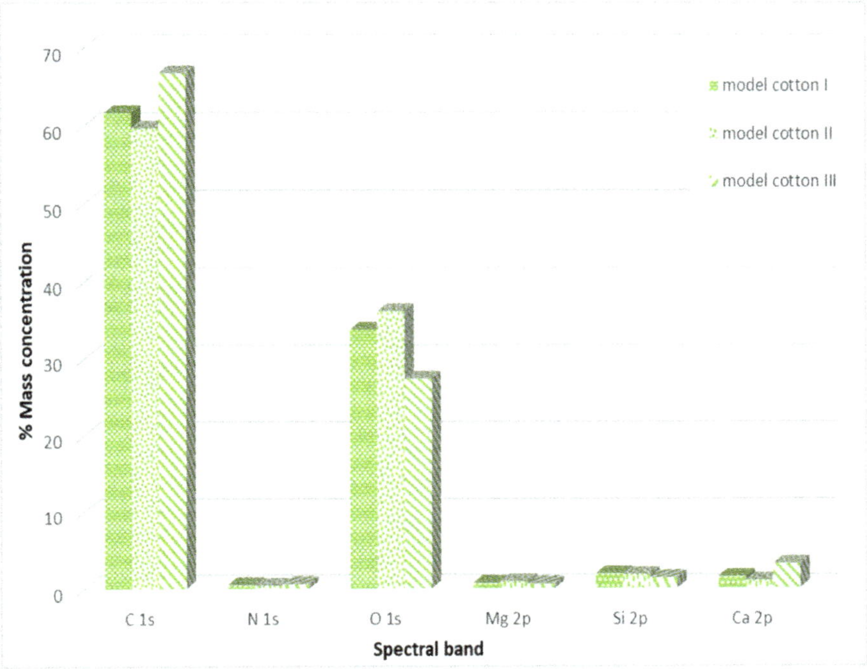

Figure 6. Percentage distribution of individual elements in the outer layer of the examined model cotton samples.

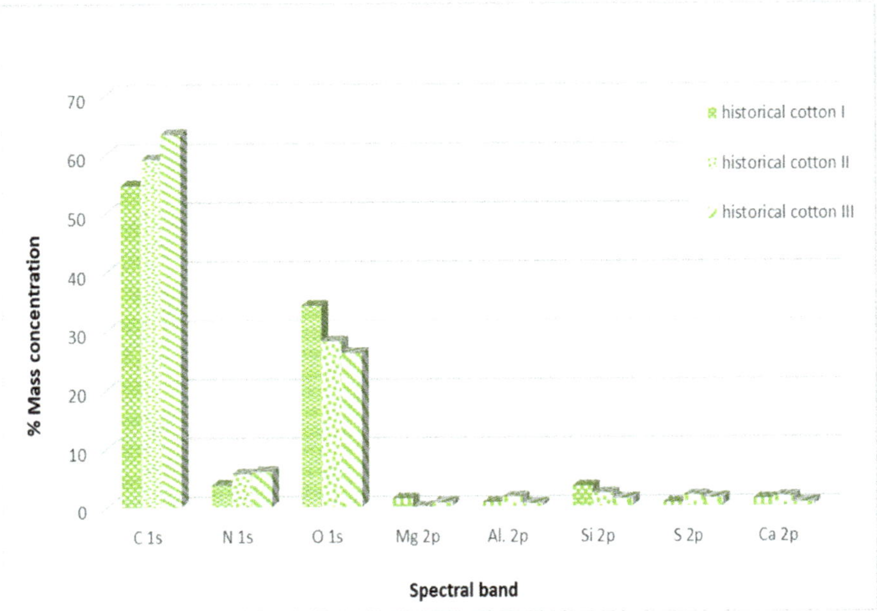

Figure 7. Percentage distribution of individual elements in the outer layer of the examined historical cotton samples.

The results were obtained after applying ethanol mist from the same distance, but the limitation of the method will be its repeatability in real conditions, when the airbrush is used by conservators freely. You cannot then maintain a constant distance.

4. Discussion

In A-BSM and other museums, microorganisms harmful to human health and those that can adversely affect the surfaces of historical objects are isolated. In museology, in addition to the biocidal aspect, minimizing the impact of the applied decontamination on the structure and morphology of disinfected surfaces is crucial. Various decontamination methods have been tested in museums, including gamma radiation, X-ray radiation, low-temperature plasma, volatile compounds, and essential oils, as well as silver nanoparticles [27–32].

Researchers have achieved different degrees of microbial reduction. In A-BSM, the biocidal efficacy of vaporized hydrogen peroxide (VHP), ethylene oxide (EtO), diode laser, and the focus of this study, ethanol in the form of mist, was tested [11].

VHP, previously used in medical spaces, resulted in a reduction (R) of the majority of tested microbial strains by a minimum of R = 3 log and all tested mixed cultures above R = 98% when applied at a concentration of 300 ppm for 20 min on porous textile material [6].

Historical cardboard subjected to disinfection using VHP (300 ppm, 20 min) showed a reduction in the number of isolated microorganisms by 1.6–7.0 log on new cardboard, corresponding to a reduction of 97.26–100.00% [33].

A-BSM successfully achieves disinfection with a diode laser, where in the medical field, a reduction of bacteria and fungi from 60% to 100% was achieved on various materials [34,35].

On 25-year-old corroded metal, the reduction of microorganisms using a diode laser was 88.85–100% [36].

In A-BSM, historical collagen material achieved a reduction of 78–92%, and on cellulose, it achieved 90–100% reduction [18].

Among the tested methods in museology, fogging or vapor methods were included. Disinfection with Cinnamomum zeylanicum essential oil-based alcohol mist demonstrated a reduction effect of 5–7 logs on cotton and linen textiles [37].

Using a benzalkonium chloride solution at a concentration of 0.01 mg/m^3 in vapor form showed no reduction in bacteria on wool and cotton [38].

Researchers have also explored the impact of ethanol on artifacts, primarily in liquid form. No harmful effects of 70% ethanol were observed on the tested paper, either in the short or long term [39].

Karbowska tested the biocidal effectiveness of ethanol vapors applied for 18 h on old paper, and showed an R > 4.00 log reduction of the tested fungi. In the case of *Cladosporium cladosporioides*, only a 3 h exposure to ethanol vapor was sufficient. Other fungi, including *Penicillium spinulosum*, and the most resistant strains of *Trichoderma viride* and *Chaetomidium subfimeti*, were completely eliminated after 18 h [40].

In A-BSM, using the 90% ethanol mist decontamination technique, a reduction level was achieved at concentrations of 80% and 90%, ranging from 93.27% to 99.91% for fungi and from 94.96% to 100% for bacteria, with 74.24% for *B. subtillis* [19].

Comparing the results of the biocidal effectiveness of ethanol mist to previously tested decontamination techniques in A-BSM, such as diode laser and VHP, it can be observed that the antimicrobial effect is slightly lower. Disinfection with a diode laser with an exposure power of 0.3W in continuous CW mode for 2 min in two repetitions, a 90.20–100% reduction in the number of microorganisms was achieved. However, the use of vaporized hydrogen peroxide at a dose of 300 ppm for 20 min reduced the number of microorganisms by 70–100%. The use of various decontamination techniques gives similar results in the case of fungi, but spore-forming bacterial species, especially those of the Bacillus genus, are the most difficult to combat. The tested ethanol in the form of mist also has the lowest effectiveness against these microorganisms.

However, the conducted studies are promising because, under the applied parameters of a distance of 16 cm from the object, misting for 15 s, at 2 bar pressure, and 90% concentration of applied ethanol, no changes in the morphology and chemical structure of the surface were observed.

A crucial aspect when implementing new decontamination methods on historical artifacts is the absence of an impact on surface properties and no change in the color of the treated objects. To assess color changes, the CIELab scale is used. A-BSM cannot use any decontamination technique that changes the color of the object's surface. Both the spectrophotometric and visual methods confirmed that the tested ethanol in the form of mist did not change the color of the disinfected surfaces of the historical fabric. Researchers confirmed using this method that a bath in absolute ethanol (99.80%) for 24 h did not alter the color of white and yellow silks, and red silk velvet. After disinfection with ethanol mist, there were no changes significant for the condition of the objects. Conservators did not observe any changes with the naked eye.

To evaluate the impact of decontamination techniques on surface morphology and fiber changes, SEM electron microscopy is employed [41].

In previous studies in A-BSM, SEM was utilized, demonstrating no changes in fiber morphology after the application of ethanol mist on historical fabric [19].

FTIR is a commonly used technique in various research fields, including museum studies. Its utility has been confirmed in the examination of cotton and other fabrics, particularly in determining changes on surfaces caused by decontamination techniques. The method has been applied to investigate the impact of ethanol and isopropanol solutions on silk artifacts. FTIR results show that even after immersing silk in these solutions for 180 min, no significant chemical or physical changes are observed in the silk fibers [42].

Analysis of FTIR results indicates that VHP decontamination of cotton fabric practically does not induce changes in the structure of cotton cellulose, thus not affecting the material's susceptibility to biodeterioration [11].

Results from FTIR analysis of cotton fabrics after decontamination with Cinnamomum essential oil vapors and low-temperature plasma show that these processes cause only minimal changes in the molecular structure of cellulose [30,37].

FTIR has also been used to study surface changes resulting from the cleaning of cotton materials. Analyses of obtained spectra allowed the for estimation of, among other things, the amount of waxes remaining on the examined fabric layer after the cleaning process [43].

For cotton fibers, the region between 1750 and 1600 cm^{-1} is most suitable for assessing cellulose degradation through oxidation, as confirmed in the current study [44].

Kavkler et al. utilized FTIR to determine the degree of biodegradation in historical textiles based on proteinaceous components stored in museums in Slovenia. More intense biodegradation processes caused by microorganisms and other degrading factors were observed in the inner part of the fibers compared to their superficial part [45].

XPS spectroscopy enabled the analysis of the surfaces of bleached cotton fibers. This study compared and identified surface chemical changes in a sample of used fabric and a model fabric previously cleaned of easily removable contaminants [24].

XPS has also been successfully used to assess the impact of VHP on cotton fabric in the medical field [6].

In the conservation of cultural heritage, including at A-BSM, XPS has been repeatedly employed to analyze the surfaces of objects. This technique provides information about changes occurring in the near-surface layer, typically 2–3 nm deep. This is crucial, as it allows for the assessment of changes not only in the base material, but also in the layers of conservation preparations covering the objects.

In the current study, FTIR measurements did not register any impact of decontamination with a 90% ethanol mist on changes in the structure of the decontaminated cotton samples, both model and historical. It is highly probable that the ethanol solution used in the study evaporated completely or to a significant extent from the surfaces of the tested materials. Cotton primarily consists of cellulose, which does not react with ethanol.

Research conducted using XPS, complementary to the FTIR measurements, confirms that the decontamination of cotton samples with a 90% ethanol mist has no impact on the chemical structure of the outer layer of cotton. Any potential changes in the percentage content of individual elements likely result from the presence of ethanol on the surface of the samples, which did not completely evaporate.

The use of these techniques in analyzing the chemical composition of cotton surfaces suggests that they are valuable methods for the research described in this article, focusing on the impact of decontaminating cotton materials with a 90% ethanol mist. The obtained results did not show any drawbacks to the application of this form of decontamination.

The tests performed showed that ethanol effectively eliminates most of the microorganisms inhabiting historical fabrics, but worse results are achieved if the sample contains spore-forming bacteria of the Bacillus genus. The next stage of research will be related to the inclusion of antibiotics, which may improve effectiveness.

5. Conclusions

The decontamination method using 90% ethanol mist is biocidal, and does not negatively affect the surface structure of both model and historical cotton. Therefore, it can be used for decontamination of cotton elements in historic buildings in the A-BSM area. In further research, the method should be tested on other textile materials, for example linen, viscose, silk, and wool.

The next stage will be testing of ethanol in the form of mist with the addition of antibiotics, which may prove to be more effective against bacteria of the Bacillus genus.

Ethanol fog is safe for people who disinfect moving objects because it can be carried out in a fume hood or in a safe work chamber. A-BSM is conducting research on employee safety that will allow for the disinfection of large wall surfaces in stationary facilities. The research simultaneously focuses on the safety of facilities and the protection of people performing conservation work.

Ethanol mist is an easy-to-use, cheap and, importantly, non-destructive disinfection method that can be used on cotton objects.

Author Contributions: Conceptualization, A.W., J.P., A.P., N.J. and A.P.; methodology, A.W., K.G., D.K., N.U., M.S. and J.P.; software, D.Z., S.W. and I.W.-B., validation, and A.W.; formal analysis, J.P.; investigation, A.W., D.K. and K.G.; resources, A.P.; data curation, A.W. and N.U.; writing—original draft preparation, A.W. and I.W.-B.; writing—review and editing, N.J. and A.W.; visualization, A.W. and I.W.-B.; supervision, S.W.; project administration, A.W. and A.P.; funding acquisition, A.P. and N.J. All authors have read and agreed to the published version of the manuscript.

Funding: This research received no external funding.

Institutional Review Board Statement: Not applicable.

Informed Consent Statement: Not applicable.

Data Availability Statement: Data are contained within the article.

Acknowledgments: The authors of the article would like to thank the Deputy Director of the Auschwitz-Birkenau State Museum in Oświęcim, Rafał Pióro, for making it possible to carry out the research, providing access to historical materials for research and substantive support.

Conflicts of Interest: The authors declare no conflicts of interest.

References

1. Cywiński, P.A.; Lachendro, J.; Setkiewicz, P. *Auschwitz from A to Z*; Auschwitz-Birkenau State Museum: Oświęcim, Poland, 2013; p
2. Szostak-Kotowa, J. Biodeterioration of textiles. *Int. Biodeterior. Biodegrad.* **2004**, *53*, 165–170. [CrossRef]
3. Abdel-Kareem, O. Evaluating the combined efficacy of polymers with fungicides for protection of museum textiles against fur deterioration in Egypt. *Pol. J. Microbiol.* **2010**, *59*, 271–280. [CrossRef] [PubMed]
4. Arshad, K.; Mujahid, M. *Biodegradation of Textile Materials*; University of Borås/Swedish School of Textiles: Boras, Sweden, 2
5. Gutarowska, B.; Pietrzak, K.; Machnowski, W.; Milczarek, J.M. Historical textiles–a review of microbial deterioration analysis disinfection methods. *Text. Res. J.* **2016**, *87*, 2388–2406. [CrossRef]

Wawrzyk, A.; Rahnama, M.; Rybitwa, D.; Wieczorek, K.; Michalczewski, G.; Łobacz, M. Decontamination of microbiologically contaminated abiotic porous surfaces in an oral surgery clinic using vaporised hydrogen peroxide (VHP). *J. Environ. Health Sci. Eng.* **2020**, *18*, 639–653. [CrossRef]
Mazzoli, R.; Giuffrida, M.G.; Pessione, E. Back to the past: "find the guilty bug-microorganisms involved in the biodeterioration of archeological and historical artifacts". *Appl. Microbiol. Biotechnol.* **2018**, *102*, 6393–6407. [CrossRef] [PubMed]
Safarov, B.; Indrie, L.; Costea, M.; Turza, A.; Avazov, K.; Baias, S.; ILIEȘ, D.C.; Zdrinca, M.; Pantea, E.; ILIEȘ, G.; et al. Non-invasive analytical methods applied in the study of cultural heritage artefacts. *Ind. Text.* **2023**, *74*, 321–331. [CrossRef]
Tewari, A.; Abdullah, S. *Bacillus cereus* food poisoning: International and Indian perspective. *J. Food Sci. Technol.* **2015**, *52*, 2500–2511. [CrossRef]
Kamar, R.; Gohar, M.; Jéhanno, I.; Réjasse, A.; Kallassy, M.; Lereclus, D.; Sanchis, V.; Ramarao, N. Pathogenic potential of *Bacillus cereus* strains as revealed by phenotypic analysis. *J. Clin. Microbiol.* **2013**, *51*, 320–323. [CrossRef] [PubMed]
Wawrzyk, A.; Gutarowska, B.; Rybitwa, D.; Pietrzak, K.; Machnowski, W.; Wrzosek, H.; Papis, A.; Walawska, A.; Otlewska, A.; Szulc, J.; et al. Vapourised hydrogen peroxide (VHP) and ethylene oxide (EtO) methods for disinfecting historical. *Int. Biodeterior. Biodegrad.* **2018**, *133*, 42–51. [CrossRef]
Rudramurthy, S.M.; Paul, R.A.; Chakrabarti, A.; Mouton, J.W.; Meis, J.F. Invasive Aspergillosis by *Aspergillus flavus*: Epidemiology, Diagnosis, Antifungal Resistance, and Management. *J. Fungi* **2019**, *5*, 55. [CrossRef]
Amaike, S.; Keller, N.P. Aspergillus flavus. *Annu. Rev. Phytopathol.* **2011**, *49*, 107–133. [CrossRef]
Klich, M.A. Aspergillus flavus: The major producer of aflatoxin. *Mol. Plant Pathol.* **2007**, *8*, 713–722. [CrossRef] [PubMed]
Paulussen, C.; Hallsworth, J.E.; Álvarez-Pérez, S.; Nierman, W.C.; Hamill, P.G.; Blain, D.; Rediers, H.; Lievens, B. Ecology of aspergillosis: Insights into the pathogenic potency of Aspergillus fumigatus and some other Aspergillus species. *Microb. Biotechnol.* **2017**, *10*, 296–322. [CrossRef]
Workum, J.D.; de Jong, S.W.; Gresnigt, M.S.; Becker, K.L.; Pickkers, P.; van de Veerdonk, F.L.; Heijdra, Y.F.; Kolwijck, E. Microbiological and immunological characteristics of a lethal pulmonary *Aspergillus niger* infection in a non-neutropenic patient. *Med. Mycol. Case Rep.* **2018**, *21*, 4–7. [CrossRef] [PubMed]
Frisvad, J.C.; Møller, L.L.; Larsen, T.O.; Kumar, R.; Arnau, J. Safety of the fungal workhorses of industrial biotechnology: Update on the mycotoxin and secondary metabolite potential of *Aspergillus niger*, *Aspergillus oryzae*, and *Trichoderma reesei*. *Appl. Microbiol. Biotechnol.* **2018**, *102*, 9481–9515. [CrossRef]
Rybitwa, D.; Wawrzyk, A.; Wilczyński, S.; Łobacz, M. Irradiation with medical diode laser as a new method of spot-elimination of microorganisms to preserve historical cellulosic objects and human health. *Int. Biodeterior. Biodegrad.* **2020**, *154*, 105055. [CrossRef]
Wawrzyk, A.; Dymel, M.; Guzińska, K.; Cywiński, P.; Papis, A.; Konka, A.; Wawrzyk-Bochenek, I.; Wilczyński, S. Optimization of the Process of Eliminating Microorganisms Harmful to Human Health and Threatening Objects Isolated from Historical Materials from the Auschwitz-Birkenau State Museum in Poland (A-BSM) Collection with the Use of Ethanol in the Form of Mist. *Materials* **2023**, *16*, 2700. [CrossRef]
Peets, P.; Kaupmees, K.; Vahur, S.; Leito, I. Reflectance FT-IR spectroscopy as a viable option for textile fiber identification. *Herit Sci.* **2019**, *7*, 93. [CrossRef]
Yan, Y.; Wen, C.; Jin, M.; Duan, L.; Zhang, R.; Luo, C.; Xiao, J.; Ye, Z.; Gao, B.; Liu, P.; et al. FTIR spectroscopy in cultural heritage studies: Non-destructive analysis of Chinese Handmade Papers. *Chem. Res. Chin. Univ.* **2019**, *35*, 586–591. [CrossRef]
Margariti, C. The application of FTIR microspectroscopy in a non-invasive and non-destructive way to the study and conservation of mineralised excavated textiles. *Herit. Sci.* **2019**, *7*, 63. [CrossRef]
Rybitwa, D.; Wawrzyk, A.; Rahnama, M. Application of a medical diode laser (810 nm) for disinfecting small microbiologically contaminated spots on degraded collagenous materials for improved biosafety in objects of exceptional historical value from the Auschwitz-Birkenau State Museum and Protection of Human Health. *Front. Microbiol.* **2020**, *11*, 596852. [CrossRef] [PubMed]
Topalovic, T.; Nierstrasz, V.A.; Bautista, L.; Jocic, D.; Navarro, A.; Warmoeskerken, M. XPS and contact angle study of cotton surface oxidation by catalytic bleaching. *Colloids Surf. A Physicochem. Eng. Asp.* **2007**, *296*, 76–85. [CrossRef]
Willneff, E.A.; Ormsby, B.A.; Stevens, J.S.; Jaye, C.; Fischer, D.A.; Schroeder, S. Conservation of artists' acrylic emulsion paints: XPS, NEXAFS and ATR-FTIR studies of wet cleaning methods. *Surf. Interface Anal.* **2014**, *46*, 776–780. [CrossRef] [PubMed]
Benkhaya, S.; M'rabet, S.; Elharfi, A. A review on classifications, recent synthesis and applications of textile dyes. *Inorg. Chem. Commun.* **2020**, *115*, 107891. [CrossRef]
Cortella, L.; Tran, Q.K.; Gluszewski, W.; Moise, I.; Ponta, C. *Nuclear Techniques for Preservation of Cultural Heritage Artifacts*; Report No. RER 8015; International Atomic Energy Agency (AT): Vienna, Austria, 2011; 44p.
Drabkova, K.; Durovic, M.; Kucerova, I. Influence of gamma radiation on properties of paper and textile fibres during disinfection. *Radiat. Phys. Chem.* **2018**, *152*, 75–80. [CrossRef]
Vadrucci, M.; Bellis, G.; Mazzuca, C.; Mercuri, F.; Borgognoni, F.; Schifano, E.; Uccelletti, D.; Cicero, C. Effects of the ionizing radiation disinfection treatment on historical leather. *Front. Mater.* **2020**, *7*, 21. [CrossRef]
Szulc, J.; Urbaniak-Domagala, W.; Machnowski, W.; Wrzosek, H.; Lacka, K.; Gutarowska, B. Low temperature plasma for textiles disinfection. *Int. Biodeter. Biodegr.* **2017**, *131*, 97–106. [CrossRef]
Palla, F.; Bruno, M.; Mercurio, F.; Tantillo, A.; Rotolo, V. Essential oils as natural biocides in conservation of cultural heritage. *Molecules* **2020**, *25*, 730. [CrossRef]

32. Gutarowska, B.; Pietrzak, K.; Machnowski, W.; Danielewicz, D.; Szynkowska, M.; Konca, P.; Surma-Slusarska, B. Applicatio silver nanoparticles for disinfection of materials to protect historical objects. *Curr. Nanosci.* **2014**, *10*, 277–286. [CrossRef]
33. Wawrzyk, A.; Rybitwa, D.; Rahnama, M.; Wilczynski, S. Microorganisms colonising historical cardboard objects from Auschwitz-Birkenau State Museum in Oswięcim, Poland and their disinfection with vaporised hydrogen peroxide (VHP). *Biodeter. Biodegr.* **2020**, *152*, 104997. [CrossRef]
34. Wawrzyk, A.; Rahnama, M.; Sofińska-Chmiel, W.; Wilczyński, S.; Łobacz, M. The Use of the Diode Laser against the Microbi on Composites Closing the Screw Access Hall (Sah) in the Reconstruction of Dental Implants: Ex Vivo Studies. *Int. J. Environ. Public Health* **2022**, *19*, 7494. [CrossRef] [PubMed]
35. Wawrzyk, A.; Łobacz, M.; Adamczuk, A.; Sofińska-Chmiel, W.; Rahnama, M. The Use of a Diode Laser for Removal of Microorgani from the Surfaces of Zirconia and Porcelain Applied to Superstructure Dental Implants. *Microorganisms* **2021**, *9*, 2359. [Cross [PubMed]
36. Wawrzyk, A.; Rahnama, M.; Sofińska-Chmiel, W.; Wilczyński, S.; Gutarowska, B.; Konka, A.; Zeljas, D.; Łobacz, M. Analysis of Microbiome on the Surface of Corroded Titanium Dental Implants in Patients with Periimplantitis and Diode Laser Irradiatio an Aid in the Implant Prosthetic Treatment: An Ex Vivo Study. *Materials* **2022**, *15*, 5890. [CrossRef] [PubMed]
37. Matusiak, K.; Machnowski, W.; Wrzosek, H.; Polak, J.; Rajkowska, K.; Smigielski, K.; Kunicka-Styczynska, A.; Gutarowsk Application of *Cinnamomum zeylanicum* essential oil in vapour phase for heritage textiles disinfection. *Int. Biodeter. Biodegr.* **2** *131*, 88–96. [CrossRef] [PubMed]
38. Pironti, C.; Motta, O.; Proto, A. Development of a new vapour phase methodology for textiles disinfection. *Clean. Eng. Tech* **2021**, *4*, 100170. [CrossRef]
39. Sequeira, S.; Phillips, A.; Cabrita, E.; Macedo, M. Ethanol as an antifungal treatment for paper: Short-term and long-term eff *Stud. Conserv.* **2017**, *62*, 33–42. [CrossRef]
40. Karbowska-Berent, J.; Gorniak, B.; Czajkowska-Wagner, L.; Rafalska, K.; Jarmiłko, J.; Kozielec, T. The initial disinfectio paper-based historic items e Observations on some simple suggested methods. *Int. Biodeterior. Biodegrad.* **2018**, *131*, 60 [CrossRef]
41. Nabil, A.; Basma, M.; El-Aziz, E.; Tarek, M.; Elmaaty, A.; Ramadanc, S. Multifunctional cellulose-containing fabrics using modi finishing formulations. *RSC Adv.* **2017**, *7*, 33219–33230. [CrossRef]
42. Costantini, R.; Balliana, E.; Dalla Torre, D.; Aricò, F.; Zendri, E. Evaluating the Impacts of Alcohol-Based Solutions on Chemical, Mechanical and Wettability Changes before and after Artificial Ageing. *Heritage* **2022**, *5*, 3588–3604. [CrossRef]
43. Chung, C.; Lee, M.; Choe, E.K. Characterization of cotton fabric scouring by FT-IR ATR spectroscopy. *Carbohydr. Polym.* **2004** 417–420. [CrossRef]
44. Abdel-Kareem, O. The long-term effect of selected conservation materials used in the treatment of museum artefacts on s properties of textiles. *Polym. Degrad. Stab.* **2005**, *87*, 121–130. [CrossRef]
45. Kavkler, K.; Gunde-Cimerman, N.; Zalar, P. FTIR spectroscopy of biodegraded historical textiles. *Polym. Degrad. Stab.* **2011** 574–580. [CrossRef]

Disclaimer/Publisher's Note: The statements, opinions and data contained in all publications are solely those of the indivic author(s) and contributor(s) and not of MDPI and/or the editor(s). MDPI and/or the editor(s) disclaim responsibility for any injur people or property resulting from any ideas, methods, instructions or products referred to in the content.

MDPI AG
Grosspeteranlage 5
4052 Basel
Switzerland
Tel.: +41 61 683 77 34

Materials Editorial Office
E-mail: materials@mdpi.com
www.mdpi.com/journal/materials

Disclaimer/Publisher's Note: The title and front matter of this reprint are at the discretion of the Guest Editors. The publisher is not responsible for their content or any associated concerns. The statements, opinions and data contained in all individual articles are solely those of the individual Editors and contributors and not of MDPI. MDPI disclaims responsibility for any injury to people or property resulting from any ideas, methods, instructions or products referred to in the content.

www.ingramcontent.com/pod-product-compliance
Ingram Content Group UK Ltd.
Pitfield, Milton Keynes, MK11 3LW, UK
UKHW060224061026
15625UKWH00008B/483

9 783725 862368